OXYGEN TRANSPORT
TO TISSUE XXVII

ADVANCES IN EXPERIMENTAL MEDICINE AND BIOLOGY

A Continuation Order Plan is available for this series. A continuation order will bring delivery of each new volume immediately upon publication. Volumes are billed only upon actual shipment. For further information please contact the publisher.

Giuseppe Cicco
Duane F. Bruley
Marco Ferrari
David K. Harrison
(Eds.)

Oxygen Transport
to Tissue XXVII

 Springer

Giuseppe Cicco
C.E.M.O.T.– Centro Interdipantimentale di
Ricerca in Emoreologia, Microcircolazione,
Transporto di Ossigeno e Tecnologie Ottiche
    non Invasive,
Università di Bari
Bari, Italy, 70124

Duane F. Bruley
College of Engineering
University of Maryland Baltimore County
Baltimore, MD 21250
USA

Marco Ferrari
Dip. Di Scienze e Tecnologie Biomediche
Università de L'Aquila
L'Aquila, Italy, 67100

David K. Harrison
Regional Medical Physics Department
University Hospital of North Durham
Durham, United Kingdom, DH1 5TW

Proceedings of the 32nd annual meeting of the International Society on Transport to Tissue held in Bari, Italy, at the Sheraton Nicolaus Hotel - by C.E.M.O.T.- Medical Faculty, University of Bari, Italy from August 21st to 26th 2004.

Library of Congress Control Number: 2005935283

ISBN-10: 0-387-29543-7   e-ISBN: 0-387-29540-2   Printed on acid-free paper
ISBN-13: 978-0387-29543-5

Printed in the United States of America.   (MP/ JP)

9 8 7 6 5 4 3 2 1

springer.com

This book is dedicated to all the ISOTT scientists, teachers and researchers, and to my wife Rosella for her precious and continuous help before and during the congress.

Giuseppe Cicco
February 2005

INTERNATIONAL SOCIETY ON OXYGEN TRANSPORT TO TISSUE 2004-2005

The International Society on Oxygen Transport to Tissue is an interdisciplinary society comprised of approximately 260 members worldwide. Its purpose is to further the understanding of all aspects of the process involved in the transport of oxygen from the air to its ultimate consumption in the cells of the various organs of the body.

The annual meetings bring together scientists, engineers, clinicians and mathematicians in a unique international forum for the exchange of information and knowledge, the updating of participants on latest developments and techniques, and the discussion of controversial issues within the field of oxygen transport to tissue.

Examples of areas in which members have made highly significant contributions include electrode techniques, spectrophotometric methods, mathematical modeling of oxygen transport, and the understanding of local regulation of oxygen supply to tissue and fluorocarbons/blood substitutes.

Founded in 1973, the society has been the leading platform for the presentation of many of the technological and conceptual developments within the field, both at meetings themselves and in the proceedings of the society. These are currently published by Kluwer Academic/Plenum Publishers in its "Advances in Experimental Medicine and Biology", which is listed with an impact factor in the Science Citation Index.

Officers

Giuseppe Cicco	President
Paul Okunieff	Past President
David Maguire	President Elect
Oliver Thews	Secretary
Peter E. Keipert	Treasurer
Duane F. Bruley	Chairperson, Knisely Award Committee

Executive Committee

Clare Elwell
Arthur Fournell
Avraham Mayevsky
Harold M. Swartz
Peter Vaupel
Roland N. Pittman
Kyung A. Kang
Valentina Quaresima
Eiji Takahashi

Local Committee

Prof. Marco Ferrari, Università L'Aquila (Italy)
Prof. Pietro Mario Lugarà, Università Bari (Italy)
Prof. Vincenzo Memeo, Università Bari (Italy)
Prof. Luigi Nitti, Università Bari (Italy)
Prof. Giuseppe Palasciano, Università Bari (Italy)
Prof. Valentina Quaresima, Università L'Aquila (Italy)

Executive Committee Elected
New Members – ISOTT 2004

1) Per Liss	(Sweden)
2) Jerry Glickson	(USA)
3) Chris Cooper	(UK)

PREFACE

The International Society on Oxygen Transport to Tissue (ISOTT) held its 32nd Annual Meeting in Bari (Apulia), Italy from August 21st to 26th 2004. This was the first time that Italy has organized the ISOTT annual meeting. The logo of the Bari meeting includes an image of the Castel del Monte (a famous and historically important Swabian Castle, a UNESCO World Heritage site, near Bari) built by Emperor Frederick II of Hohenstaufen in the XIII century. The venue of the meeting was the Sheraton Nicolaus Hotel, the meeting was hosted by the University of Bari, and organized by the research center - C.E.M.O.T. – of the University of Bari.

At this Meeting we had 93 registered participants (including speakers, accompanying persons and students).

The Scientific Program included 86 presentations by participants from North America, Europe, Asia, and Australia. We congratulate Dr. Fredrick Palm (Sweden), Dr. Richard Olsson (Sweden) and Dr. Derek Brown (Switzerland) for being selected as award winners for, respectively, the Melvin H. Knisely Award, the Dietrich W. Lübbers Award and the Britton Chance Award, based on the scientific excellence of their submitted papers and scientific activity. This year was the first time that the "Duane Bruley Student Travel Award" was awarded. During the Gala Dinner at the Cocevola Restaurant (Andria), ten postgraduate students who had presented a paper at the meeting were selected as winners of this Award.

As President of the 32nd ISOTT Meeting, also on behalf of the editors, I would like to thank all the prestigious participants for their great scientific and social contributions. The editors, on behalf of ISOTT, also gratefully acknowledge the help of all those who helped to make possible the 32nd ISOTT Annual Meeting held from 21-26 August 2004 in Bari, Italy.

For the Editors

Giuseppe Cicco

February 2005

CONTENTS

NOVEL NIRS APPLICATIONS (2)

TISSUE ENGINEERING : ISOTT'S ROLE

ORGAN TRANSPLANTATION AND O₂

CENTRAL NERVOUS SYSTEM AND O₂

ONCOLOGY AND O₂

AWARDS

KNISELY AWARD

The <u>Melvin H. Knisely Award</u> was first presented by ISOTT at the 1983 annual banquet to acknowledge a young investigator (35 years of age or less) for outstanding achievements in research related to oxygen transport to tissue. This award acknowledges the pleasure that Dr. Knisely derived from assisting and encouraging young scientists and engineers to contribute to the study of the transport of anabolites and metabolites in the microcirculation. His many accomplishments in the field have inspired developing investigators to follow in his footsteps. The continuation of this award aims to encourage young scientists and engineers to join ISOTT and aspire to generate high quality research in the area of oxygen transport to tissue.

TERMS OF THE AWARD

A candidate must not pass his/her 36th birthday before the date of presentation of the award. A candidate must provide evidence of sustained activity in research related to oxygen transport to tissue. A candidate must be a member of ISOTT or eligible for membership. A candidate must be present at the annual meeting and make an oral or poster presentation of his/her work. The work is to be published in the annual ISOTT proceedings. A candidate who is unsuccessful in any given year can be re-nominated in a subsequent year if he or she continues to meet the necessary criteria.

MELVIN H. KNISELY AWARDEES

1983	Antal G. Hudetz (Hungary)
1984	Andras Eke (Hungary)
1985	Nathan A. Bush (USA)
1986	Karlfried Groebe (Germany)
1987	Isumi Shubuya (Japan)
1988	Kyung A. Kang (Korea/USA)
1989	Sanja Batra (Canada)
1990	Stephen J. Crinale (Australia)
1991	Paul Okunieff (USA)

1992	Hans Degens (The Netherlands)
1993	David A. Banaron (USA)
1994	Koen van Rossem (Belgium)
1995	Clare E. Elwell (UK)
1996	Sergei A. Vinogradov (USA)
1997	Chris Cooper (UK)
1998	Martin Wolf (Switzerland)
1999	Huiping Wu (USA)
2000	Valentina Quaresima (Italy)
2001	Fahmeed Hyder (Bangladesh)
2002	Geofrey De Visscher (Belgium)
2003	Mohamed Nadeem Khan (USA)
2004	Frederik Palm (Sweden)

LÜBBERS AWARD

The Dietrich W. Lübbers Award was established in honour of Professor Lübbers' long-standing commitment, interest, and contribution to the problems of oxygen transport to tissue and to the Society.

TERMS OF THE AWARD

The Lübbers Award is made to a young investigator 30 years of age or less (with the nomination and sponsorship of an ISOTT member) and will consist of travel support to the meeting at which the award is made. The selection will be based on the scientific excellence of the individual's first authored manuscript on the topic of oxygen transport as judged by the members of the organizing committee of the annual meeting. The recipient of the award will be made known prior to the annual meeting in order to enable timely distribution of travel funds.

DIETRICH W. LÜBBERS AWARDEES :

1994	Michael Dubina (Russia)
1995	Philip E. James (UK/USA)
1996	Resit Demit (Germany)
1997	Juan Carlos Chavez (Perù)
1998	Nathan A. Davis (UK)
1999	Paola Pichiule (USA)
2000	Ian Balcer (USA)
2001	Theresa M. Bush (USA)
2002	Link K. Krah (USA)
2003	James J. Lee (USA)
2004	Richard Olsson (Sweden)

BRITTON CHANCE AWARD

The <u>Britton Chance Award</u> was established in Rochester 2003 in honour of Professor Britton Chance's long-standing commitment, interest, and contribution to the problems of oxygen transport to tissue and to the Society. This year is the first time this award has been made.

TERMS OF THE AWARD

The Chance Award is made to a young investigator 30 years of age or less (with the nomination and sponsorship of an ISOTT member) and will consist of travel support to the meeting at which the award is made. The selection will be based on the scientific excellence of the individual's first authored manuscript on the topic of oxygen transport as judged by members of the organizing committee of the annual meeting. The recipient of the award will be made known prior to the annual meeting in order to enable timely distribution of travel funds.

BRITTON CHANCE AWARD :

   2004 Derek Brown (Switzerland)

DUANE BRULEY STUDENT TRAVEL AWARD

The Duane Bruley Student Travel Award is made to postgraduate students who are registered at the Meeting and present a paper. This award is made at the discretion of the ISOTT President who can also consider the particular situation (for travel costs, etc) of these students. The ISOTT President can select no more than 10 students and a total of US$ 4000 can be allocated for these Awards. This Award was finalized during the ISOTT Meeting held in Rochester 2003 and was allocated for the first time in Bari 2004.

DUANE BRULEY AWARDEES

   1)   Helga Blocks (Belgium)
   2)   Jennifer Caddick (UK)
   3)   Charlotte Ives (UK)
   4)   Nicholas Lintell (Australia)
   5)   Leonardo Mottola (Italy)
   6)   Samin Rezania (USA)
   7)   Ilias Tachtsidis (UK)
   8)   Liang Tang (USA)
   9)   Ijichi Sonoko (Japan)
   10) Antonio Franco (Italy)

WELCOME BY PROF. LUIGI NITTI DIRECTOR OF C.E.M.O.T., UNIVERSITY OF BARI PRESENTED AT THE OPENING CEREMONY 2004 TO ALL ISOTT PARTICIPANTS

"On behalf of the University of Bari and on behalf of the Faculty of Medicine and Surgery, I am very pleased to welcome the participants of the 32nd Annual Meeting of the International Society on Oxygen Transport to Tissue 2004.

First of all I wish to thank the President of the Conference, Dr. Giuseppe Cicco, for his great efforts and personal commitment in making this scientific event possible.

In addition, I wish to thank the distinguished scientists, coming from all over the world, for their attendance at the Conference.

I am very glad that the ISOTT Annual Meeting takes place in Bari, because in our University a Research Center is operating on topics concerning Hemorheology, Microcirculation, Oxygen Transport and non-invasive optical technologies.

About 25 researchers in the fields of medicine and physics are involved in the activities of the Center with relevant scientific results. I thank Dr. Giuseppe Cicco, who has been for several years the Scientific Coordinator of this Center, for his sound competence and great engagement.

I hope that the time you will spend in Bari will be full of scientific achievements and interesting discussions. I also hope that you will enjoy and appreciate our natural beauties and artistic monuments".

Prof. Luigi Nitti
C.E.M.O.T. Director
University of Bari

BARI August 21-26, 2004 Annual ISOTT Meeting
Attenders
in front of the "Basilica di St. Nicola," Bari, Italy.

The ISOTT editors thank the following scientific reviewers very much for all their work:

Duane Bruley, University of Maryland Baltimore County, USA
Chris Cooper, University of Essex, UK
Andras Eke, Semmelweis University of Medicine, Hungary
Clare Elwell, University College London, UK
Marco Ferrari, University of L'Aquila, Italy
David Harrison, University Hospital of North Durham, UK
Kyung Kang, University of Louisville, USA
Peter Keipert, Sangart Inc., USA
Joseph LaManna, Case Western Reserve University, USA
Edwin Nemoto, University of Pittsburgh, USA
Valentina Quaresima, University of L'Aquila, Italy
Oliver Thews, University of Mainz, Germany
David Wilson, University of Pennsylvania, USA
Martin Wolf, University Hospital Zurich, Switzerland

In particular, I would like to thank my co-editors and friends Prof. David Harrison, Prof. Duane Bruley and Prof. Marco Ferrari.

I would also like to thank the language reviewer, Mrs. Laraine Visser-Isles, for her valuable work in copy-editing this book and Mrs. Eileen Harrison for formatting the final manuscripts for publication.

Sponsors

The ISOTT 2004 team thanks our sponsors:

•A. Menarini Industrie Farmaceutiche Riunite
•Errekappa Euroterapici
•Hamamatsu Photonics Italia
•Janssen Cilag Italia
•Natural Bradel Italia
•Perimed Italia
•Pharma Italia

MICROCIRCULATION HEMORHEOLOGY AND TISSUE OXYGENATION IN CLINICAL AND EXPERIMENTAL PRACTICE: CEMOT ACTIVITY IN THE LAST TWO YEARS

Giuseppe Cicco[*]

1. INTRODUCTION

The study of microcirculation has always been an important target for many scientists, because it is there that the main functions linked with tissue life are realized. It is well known that the main functions of the microcirculation are the transport and exchange of metabolic substances, oxygen, carbon dioxide and nutrients between the blood and tissues. Many alterations in microcirculation induce changes in physiology and consequently induce pathologies.

During proliferation of capillaries (angiogenesis) there is often thickening of the capillary basement membrane and arterial wall. In other situations (such as system vasculitis) it is possible to have rarefaction of capillaries (decrease in capillary loop number), in hypertension it is also possible to find an increase in vascular tortuosity, connected with microaneurisms. Endothelial swelling and capillary plugging are often present in diabetes and metabolic disease (proteinosis). In inflammatory diseases, leukocyte adhesion is present in many previous pathologies, often it is possible to see hyperviscosity, hemoconcentration and sludges with a decrease in Red Blood cell (RBC) deformability, and a decrease in tissue oxygenation. In these conditions vasomotion abnormalities are often present.

---

[*] **C.E.M.O.T.** – Interdepartmental Centre for Research on Hemorheology, Microcirculation, Oxygen Transport to tissue and non invasive optical technologies University of Bari, Policlinico, P.za G. Cesare, 11 70124 Bari, Italy.Telefax 0039 080 5478156 email gcicco.emo@tiscali.it .

2. CEMOT AIM

The aim of CEMOT is to encourage collaboration between different scientists in order to optimize the study of tissue oxygenation, microcirculation and hemorheology in different pathologies strongly related to interested specialities (Fig 1).

CEMOT operators are connected and continuously exchange their scientific and clinical experience with colleagues operating in the Departments of General Surgery and Liver Transplantation, Hematology, Nephrology, Physics, Internal Medicine, Ophthalmology, Obstetrics, and Neurology (Fig.2).

Figure. 1: The CEMOT logo

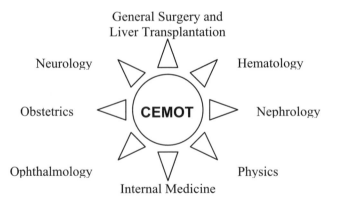

Figure 2. The Structure of CEMOT

3. CEMOT: A SHORT HISTORY

CEMOT began as a simple laboratory of microcirculation in hypertension. Dr. Giuseppe Cicco and only two co-workers were employed in this laboratory. Later, in 1993, the laboratory became the 'Laboratories of hemorheology and microcirculation' still under the direction of Dr. Giuseppe Cicco.

In 1996 the first LORCA (Laser assisted Optical Rotational Red Cell Analyzer) from the Netherlands arrived in these laboratories. Many results were produced on this instrument between 1996 and 1998. Much of this work was presented at the Second

International Conference on Hemorheology and Tissue Oxygenation in Hypertension and Vascular Diseases held at the Hotel Mercure Villa Romanazzi Carducci, in Bari, organized by Dr. Giuseppe Cicco, under the Patronage of the University of Bari on 1998, October 29-31.

In 1999 the Interdepartmental Center for Hemorheology, Microcirculation, Oxygen Transport and non-Invasive Optical Technologies (CEMOT) was created. At the beginning of 2001, Dr. Giuseppe Cicco was chosen as the CEMOT Scientific Coordinator. In 2003 the second LORCA, and now the only one in function, arrived in Bari.

4. HEMORHEOLOGY

The first experiments in hemorheology were performed in 1675. Antoni van Leeuwenhoek was the first to study the shape of RBCs with his newly developed microscope. From these early experiments of Antoni van Leeuwenhoek,[1] hemorheology underwent continuous development to arrive ultimately at the realization by Herbert Meiselman (in Los Angeles, USA) (Fig. 4) and Max R. Hardeman (in Amsterdam) (Fig. 3) of the Laser assisted Optical Rotational Red Cell Analyzer (LORCA). LORCA is a computerized instrument served by dedicated software able to detect RBC deformability (Elongation Index=EI) evaluated at different Shear Stress, from 0.3 to 30 Pascals (Pa). Nowadays, we routinely consider only the standard shear stress at 30 Pa. LORCA is also able to evaluate the aggregation half-time (T½), expressed as a ratio of values expressed in seconds[2-4] (Fig.5).

Figure 3: Max Hardeman Ph. D
Amsterdam, NL

Figure 4: Herbert Meiselman Ph. D
Los Angeles, USA

My first experience with hemorheological measurements was in Siena (Italy) in 1989 in the Institute of Internal Medicine with the Director, Prof. Sandro Forconi, a great scientist and Professor in Internal Medicine and in Hemorheology. There I learned about the Erythrocyte Morphologic Index (EMI) according to Zipursky (Toronto)[5-7] and applied in Italy by Forconi (Siena). Using an optical microscope with an objective lens under immersion with glycerol, it is possible to study a fresh venous blood sample (50 µl) in EDTA (Ethyl – Dimetyl – Trichloro – Acetic Acid) with a phosphate buffer (0.115 M , pH 7.3 – 7.4) fixed with glutaraldehyde at a concentration of 0.3%.

Figure 5. LORCA

The number of bowls and discocytes per 100 cells are detected under the microscope (Fig. 6 A,B). After this, the EMI, that is the ratio between the bowl-shaped erythrocytes and discocytes (n.v. >1), is calculated as:

$$EMI = \frac{Number\ of\ bowls}{Number\ of\ discocytes}$$

Bowls 55%, discocytes 44%
Echinocytes and Knizocytes 1%

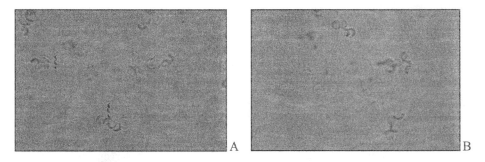

Figure 6A,B. Microscope images useful for EMI as Bowls (#) and Discocytes (§)

During a close collaboration with the Department of Internal Medicine, we studied the RBC alterations and hemorheology in alcoholism. In our center (CEMOT), we examined 19 alcoholic subjects (14 males and 5 females aged 40±6 years) and 18 controls (11 males and 7 females aged 36±4 years). Alcohol withdrawal was initiated 3 months before the trial.

No significant alterations were found in the Elongation Index in alcoholics when compared with controls whilst we found a decrease in RBC aggregability T½ when compared with controls.

5. COMPUTERIZED VIDEOCAPILLAROSCOPY

Computerized videocapillaroscopy is another very useful technique in the study of microvasculature "in vivo". The traditional capillaroscope was simply a microscope that statically evaluated the nail bed capillaries of the 4th finger of the left hand. Today it is possible to explore many other parts of the human body (eyes, head, hands, arms, legs, etc.) using the computerized videocapillaroscope[8-9] (Fig.7).

Figure 7. Videocapillaroscope

Although many other sites can be examined in research or specialized clinical examinations, evaluation of the nail bed capillaries of the 4th finger of the left hand has been accepted as a standard method for routine clinical analysis.

Routine videocapillaroscopy is widely used in clinical practice including angiology (POAD – Peripheral Occlusive Arterial Disease), phlebology, rheumatology (systemic progressive sclerosis and systemic lupus erythematosus), cardiology (hypertension), vascular surgery, dermatology, surgery (organ transplantation), oncology, endocrinology (diabetes, hypothyroidism), and psychiatry (schizophrenia, epilepsy, oligophrenia)(Fig. 8).

Capillaroscopy could offer many other important contributions to scientific research. For example, capillaroscopy could be useful for mapping the regional architecture of microvessels in physico-mechanical, hemodynamic, neuroendocrine, postural and other stimulations. It could also be used detect the response of microvessels to pharmacological stimuli as well as endogenous and exogenous stimulations. Capillaroscopy is able to monitor the efficacy of drugs and to study the microcirculation – tissue relationships at different ages of patients. Finally, capillaroscopy could be used to evaluate regional dynamic hemorheology.

6. LASER DOPPLER AND TRANSCUTANEOUS OXYMETER

We are also able to evaluate the transcutaneous oxygen partial pressure and CO_2 using the Perimed instrument Periflux System 5000 (Fig. 9), which is able to detect concomitantly $TcPO_2$ and $TcPCO_2$ using a Combi Sensor (Fig. 11), while blood flow is measured using laser Doppler and expressed in perfusion units.

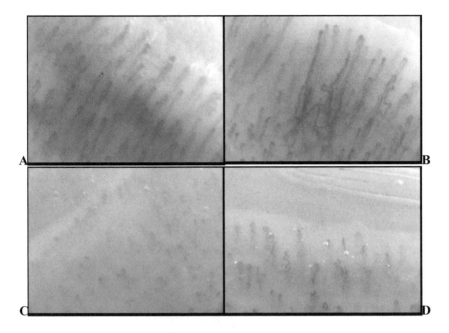

Figure 8. Capillaries as observed by videocapillaroscopy : A) normals, B) hypertensives C) systemic vasculites (SSP), D) hypertensives smokers

Figure 9. Periflux   **Figure 10:** Oxyraf

7. OXYRAF DEVELOPMENT

A prototype oxymeter (Oxyraf) was developed during a collaboration between the Physics Department (Prof. PM Lugarà, Prof. R. Tommasi and Prof. L. Nitti), myself and RAF Electronics (Bari, Italy) in 1998. I have presented many papers arising from work carried out in CEMOT, Bari using this prototype in previous ISOTT conferences (Hanover 1999, Nijmegen 2000, Manchester 2002)[9-13]. An important characteristic of this instrument is its ability to detect the percentage O_2 saturation anywhere on the human body, not only on the index finger or on the ear as applies with standard pulsed oximetry.

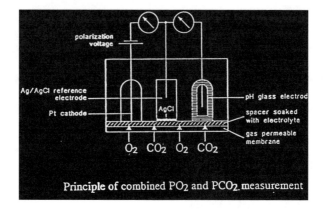

Figure 11. Combi sensor

The functional principle of the Oxyraf (Fig. 10) is measurement of the intensity of the light rays using two different wavelengths backscattered after crossing living tissues and interacting with Hb and HbO_2.

The Oxyraf uses two light emitting diodes (LEDs) similar to the Pulse Oxymeter (PO). One LED operates at 650 nm (Hb absorption) and the other one operates at 830 nm, in the infra-red (i. r.) and near the isosbestic point at which Hb and HbO_2 show the same absorbance. The percentage HbO_2 saturation can be expressed as a linear function of the ratio of absorption coefficients at the two wavelengths of these LEDs. We have thus developed a new prototype based on fiber coupled laser diodes to illuminate patient tissue with low power near infrared radiations always at two different wavelengths. Backscattered radiation is collected using four optical fibers and the resulting signal is analyzed to evaluate tissue oxygen saturation.

8. INTRAERYTHROCYTIC CYTOSOLIC CALCIUM (ICC)

We evaluate ICC in erythrocytes washed in saline phosphate buffer (PBS) and incubated with FURA 2/AM (Perkin Elmer) at 37° C over a 30-minute period. Fluorescence is detected using a spectrofluorimeter (Perkin Elmer LS30) at the wavelength (λ) 335 nm for FURA 2 – Ca^{++} complex and 385 nm for FURA 2 free Ca^{++}. Calcium content is expressed as the ratio of fluorescence values at 335 and 385 nm[14-19].

9. ACETYLCHOLINESTERASE (AChE)

Increase in AChE enzyme activity can be translated into reduced RBC membrane deformability, making the passage of red blood cells through the vessels more difficult[20-21].

10. PERIFLUX NIRO

An experimental study was carried out with the 1st Surgical Unit and Liver Graft Center of Bari University to investigate changes in liver microcirculation following warm

ischemia. In liver surgery, total or partial vascular exclusion is necessary to prevent blood loss.

We also know that warm ischemia induces an increase in cytolysis and mitochondrial alterations.

We studied 8 pigs, mean weight of 30 kg under general anaesthesia, using isolated liver preparations. The Periflux System 5000 (Perimed, Fig. 9) was used to evaluate blood flow (needle probe) at normal perfusion (20'), during clamping (60'), and after perfusion (60'). Blood flow decreased to less than 30% of the basal value during these experiments.

Reperfusion flow was 30% lower than basal flow after 30'. It increased slowly but always remained lower than basal values. This indicates that there was always some reperfusion injury damage present (even if small).

A study was performed using the NIRO 300 Hamamatsu instrument during liver transplantation. We demonstrated mild hypoxia in the brain of patients submitted to a liver transplantation, that improves suddenly after the connection of the liver vessels, improving the inhaled oxygen pressure.

11. CONCLUSIONS

It is obvious from these results that CEMOT has been a driving force in the development and application of haemorheology in research and clinical practice.

It is my intention that these studies will be developed further. It is also my hope that the various collaborations between the Institutes and the Departments now involved in the CEMOT Research and Clinical Activity can continue and be further developed. The underlying aim is always the resolution of human diseases and the care of patients.

12. REFERENCES

1. Mokken Ch. F.,1996, Historical aspects of Hemorheology in : "Mokken C.F. Clinical and Experimental studies on Hemorheology" Elinkwijk bv. Ed. - Utrecht Holland.
2. Hardeman M.R., Goedhart P.T., Dobbe J.G.G. and Lettinga K.R., 1994, Laser Assisted Optical Rotational Red Cell Analyzer (LORCA) I: A new instrument for measurement of various structural hemorheological parameters. Clin. Hemorheol. 14 : 605-618.
3. Hardeman M.R., Goedhart P.T. and Shut N.H., 1994, Laser Assisted Optical Rotational Red Cell Analyzer (LORCA) II : Red Blood Cell Deformability, elongation index versus cell transit analyzer, Clin. Hemorheol. 14 : 619-630.
4. Cicco G., 1999, LORCA in clinical practice : Hemorheological kinetics and tissue oxygenation. In: Oxygen transport to tissue XXI ed. by A. Eke and D. Delpy Adv. Exp., 471,73:631-637.
5. Zipursky A., Brown E., Polko J. and Brown E.J. 1983, The erythrocyte differential count in newborn infants. Am. J. Pediatr. Hemat. Oncol. 5(1): 45-51.
6. Forconi S., 1989, Can red cell deformability be measured in clinical medicine ? Clin. Emorheol. 9:27.
7. Turchetti V., De Matteis C., Leoncini F., Tribalzini L., Guerrini M. and Forconi S., 1997, Variations of erythrocyte morphology in different pathologies Clin Hemorheol. and Microcirculation 17 : 209-215.
8. Cicco G., 2003, Videocapillaroscopia computerizzata ed ossigenazione tissutale nelle vasculopatie in "Ann. Italiani di Medicina Interna" Cepi – AIM Eds. Group 364:155.
9. Cicco G., Placanica G., Memeo V., Lugarà P.M., Nitti L., Migliau G., 2003, Microcirculation assessment in vasculapathies : capillaroscopy and peripheral tissue oxygenation. Adv. Exp. Med. And Biology 540,38 : 271-276.
10. Lugarà P.M., 1999, Current Approaches to non invasive optical oximetry. Clin. Hemorheol. and Microcircul. 21 : 307-310.
11. Cicco G., 1999, Non Invasive optical oximetry in humans : Preliminary data. Clin. Hemorheol. and Microcircul. 21,311-314.
12. Cicco G., 1999, A new Instrument to assess tissue oxygenation in 27th ISOTT Annual Meeting Hanover UK Abstract Book": 10.

13. Cicco G., 2000, Tissue oxygenation assessment and hemorheology in hypertensives in"28th ISOTT Annual Meeting Nijmegen – Holland Abstract Book" : 7.
14. Cicco G., Carbonara MC., Hardeman M.R., 2000, Hemorheology, Cytosolic calcium and carnitinaemia in subjects with hypertension and POAD. J. des Mal. Vasculaires 154:20-22.
15. Cicco G., Carbonara MC., Hardeman M.R., 2001 Cytosolic Calcium and hemorheological patterns during arterial hypertension, Clin. Hemorheology and Microcirculation 24, 25-31
16. Cicco G., 1999, Relationship between cytosolic calcium and red blood cell deformability during arterial hypertension, High blood pressure, Ed. Kurtis, Milano, 8, 1:13
17. Cicco G., 1999, La deformabilità eritrocitaria studiata con il LORCA e il Ca^{++} intraeritrocitario in corso di ipertensione arteriosa. Giorn. Ital. Microcirc., XIX Congr. Naz. SISM, Grado, Atti del Congresso : 36
18. Cicco G. 2000, Alterazioni emoreologiche e calcio citosolico eritrocitario nell'ipertensione arteriosa, Giornale Italiano di Nefrologia, Ed. Wicatig, Milano, 17, 164, C9:46
19. Cicco G., 2000, Hemorheological Alterations and erytrocyte cytosolic calcium in arterial hypertension, nephrology, dialysis and Transplantation, 15, 9: A67.
20. Saldanha C., Quintão T., Garcia C., 1992, Avaliação des esterases sanguincas na dença de Alzheimer, Acta Med. Portuguesa 5: 591 – 3
21. Martins e Silva J., Proeuça C., Braz Nogueira J., Garjão Clara J., Manso C., 1980, Erythrocyte acetylcholinesterase in essential hypertension, 49: 127

2.

INTRACELLULAR pH IN GASTRIC AND RECTAL TISSUE POST CARDIAC ARREST

Elaine M. Fisher[1], Richard P. Steiner[2] and Joseph C. LaManna[3*]

1. INTRODUCTION

Hypoperfusion to the gut during cardiac arrest is an important clinical problem. The inability to control pH during metabolic stress, e.g. ischemia, leads to the disruption or halting of processes vital to balancing cellular metabolism. Alterations in cellular pH have been linked to changes in intramucosal permeability, which may result in the leakage of inflammatory mediators or bacteria, or both, into the systemic circulation and contribute to the development of organ failure, shock, or death.

A strong relationship exists between energy metabolism and cellular acid-base balance. Acidosis during ischemia is likely due to glycolytic accumulation of CO_2, lactic acid, and H^+. Tissue P_{CO2} is a recognized clinical marker of perfusion failure, resulting from a variety of conditions (hemorrhagic shock, sepsis, trauma) that arise when tissue O2 requirements can no longer be met and anaerobic metabolism is initiated.[1,2] The link between gastric mucosal hypercarbia and intracellular pH (pHi) has not been established. The relationship between gut tissue P_{CO2} and pHi is clinically important since cellular/tissue pHi monitoring technologies are not available at the bedside.

To our knowledge, gut pHi under hypoxic conditons has not previously been reported. This study was undertaken, then, to: 1) define baseline pHi of stomach and rectum, two clinically useful monitoring sites; 2) evaluate changes in pHi at these sites during ischemia; 3) identify differences in pHi in the gut layers ; 4) compare pHi to mucosal P_{CO2}; and finally 5) estimate the buffering of pHi in the stomach and rectum and compare these values with arterial bicarbonate (HCO_3^-).

[1] The University of Akron, College of Nursing, Akron, OH. ,
  E-mail:efisher@uakron.edu

[2] The University of Akron, Department of Statistics, Akron, OH

[3] Case Western Reserve University, Department of Neurology, School of
  Medicine, Cleveland OH

2.   METHODS

2.1 Surgical Preparation

The surgical protocols for anesthesia, cannulation of vessels, and insertion of P_{CO2} electrodes in rats has been previously described.[3] While the abdominal cavity was open for instrumentation of the stomach, a small catheter (IP catheter) was inserted into the peritoneal cavity and the cavity was closed with sutures.

2.2 Experimental Protocol

Post-instrumentation, rats were injected via the IP catheter with 3 mL of 2% Neutral Red Dye (Sigma Chemical). The rat was repositioned several times during the first 5 minutes post-IP infusion to distribute dye to the gut tissue. All animals were stabilized for 30 minutes in order for mean arterial pressure to return to baseline. Arrest was induced by an intra-atrial injection of norcuron (.1-.2 mg/kg) followed by a potassium chloride bolus (0.5M/L; 0.12 mL/100 gm of body weight). Control animals remained under anesthesia while the stomach and rectum were harvested. Tissues were immersed in liquid nitrogen, and stored at -80°C. The rats were then euthanized by cutting the abdominal aorta. Organs from rats in the arrest group were similarly harvested and frozen.

2.3 pHi Analysis

The pHi is a measure of the acid-base status of cells within a specified area. The pHi was determined by a reflectance histophotometric imaging technique using the absorption dye neutral red. Tissue was mounted on a cryotome chuck and maintained at -25°C. An unstained tissue blank was mounted next to each tissue block prior to photographing the sample monochromatically at 550 nm and 450 nm. The pHi was calculated as: $pHi = -1.3(OD\ 550/OD\ 450) +10.5$.[4] Transmittance was determined by dividing pixel values from the 550 and 450 images by the 100% transmittance level obtained from the blank. Using Image Pro Plus (4.0 Media Cybernetics) images were processed by first converting them to floating point images, determining a ratio, and comparing values to a calibration curve. Grayscale tissue images were used to identify the area of interest (AOI) and these AOIs were identified within the pixel images. Mean AOI pHi values were determined after eliminating values outside of the defined dye range, 6-8 units. Each tissue was sectioned 3-4 times and a separate determination was made of the pHi for the whole tissue (all layers within the area) and for layers (mucosal, submucosal, muscularis mucosa).

2.4 Data Analysis

Statistical analysis was performed using JMP IN computer software (Version 4, SAS Institute). For each rat 3 to 4 slices of the same tissue were analyzed. The average pHi (mean $\pm$ SEM) for whole tissue and for layers, was determined for each slice by taking the average of pixel values within the AOI that fell within the pH range of 6-8 units. The AOI was drawn to include the largest possible area of each tissue sample. Because the same spot on each tissue was not sampled (e.g., the fundus of the stomach was not sampled consistently), tissue and group were treated as random effects in the model, thereby increasing the number of observations included in the analysis (whole tissue, n=99; layers, n=283). To evaluate changes in the pHi by site (rectal vs. gastric) and group, ANOVA was used. To compare pHi by layer over the experimental protocol a nested ANOVA design was conducted (tissue type: rectal, gastric; by group: control, experimental; by site: mucosa, submucosa, muscularis mucosa). Pearson product moment correlation coefficients were calculated for pHi and P_{CO2} by site and group. Tissue buffer was calculated by modifying the Henderson Hasselbalch Equation whereby the pHi obtained in this study was substituted for pH and mucosal P_{CO2} was substituted for arterial P_{CO2} to produce a buffer value. Correlations were also used to determine the relationships between rectal, gastric, and arterial bicarbonate.

3. RESULTS

The anlaysis included data from 16 male Wistar rats (5 control; 11 15" Arrest). The mean arterial pressure (MAP) at baseline ranged from 85-119 torr. Baseline systemic variables for the control and 15" arrest group were similar and are summarized as group data in Table 1.

Table 1. Mean (+SD) of systemic variables at baseline (n = 16)

Variable	Mean	+	STD	Variable	Mean	+	STD
pH, units	7.40	+	0.06	K^+, mmol/L	4.21	+	0.5
pCO_2, torr	37.21	+	7.2	Na^+, mmol/L	137.42	+	1.2
pO_2, torr	141.38	+	29.8	Ca^{++}, mmol/L	4.68	+	0.2
HCO_3^-, mmol/L	22.44	+	3.5	Cl^-, mmol/L	102.33	+	2.4
O_2 Sat, %	98.87	+	1.7	Lactate, mmol/L	2.33	+	1.1
tHb, g/dL	14.76	+	4.4	Glucose, mmol/L	9.97	+	2.3

P_{CO2}, partial pressure of carbon dioxide; P_{O2}, partial pressure of oxygen; HCO_3^- bicarbonate; O_2 Sat, O_2 saturation; tHb, total hemoglobin ; K^+, potassium; Na^+, sodium; Ca^{++}, corrected calcium; Cl^-, chloride.

3.1 pHi in the Control and 15" Arrest Group for Whole Tissue

The pHi of the rectum and stomach under control and 15" arrest condition are shown for whole tissue in Figure 1. Stomach pHi was more alkaline than the rectum at baseline and acidified to a greater degree during 15" arrest. However, a similar change in pHi occurred between the control and 15" arrest group at the rectal and gastric site (NS, p = .10). The average control and 15" arrest pHi for rectal and gastric tissue were 7.27+.10 and 7.36+08; and 7.38 + .1 and 7.15 + .08 units.

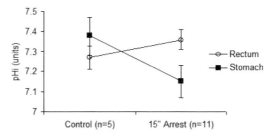

Figure 1. Rectal and stomach pHi at control and 15" arrest (n=16).

3.2 Relationship Between Layers of the Gut and Site

The mean rectal and gastric pHi of the mucosa, submucosa, and muscularis mucosa for the control and 15" arrest group are presented in Figure 2. Changes in pHi between the control and 15" arrest group were not the same in the stomach and the rectum ($p <$ .001). Gastric layers were more alkalotic under control conditions with a rapid decline noted for the arrest group. Conversely, the rectal pHi changed only slightly and remained acidotic for both the control and arrest group. The change in pHi was similar across the layers (mucosa, submucosa, muscularis mucosa) in the control as well as the 15" arrest group (p = .45). Homogeneous changes were noted between the pHi of the layers in the stomach and rectum (p = .69).

3.3 Relationship of pHi to Tissue P_{CO2}

Under control conditions, the rectal and gastric P_{CO2} were highly correlated (r = 0.95; n=5), yet poorly correlated at 15" of arrest (r = -0.39; n = 11). Similarly, rectal and gastric pHi were moderately correlated for the control group but not for the arrest group (r = .54; r = -0.05). A comparison of rectal pHi to rectal P_{CO2} and gastric pHi to gastric P_{CO2} yielded poor negative correlations (r=-.14; p=.60; r=-.33; p=.21, respectively).

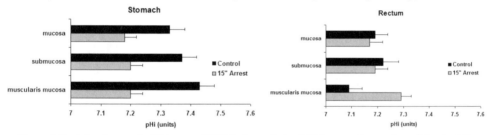

Figure 2. Rectal and stomach pHi (mean ± SEM) by layer during 15" Arrest and control condition (n=16).

3.4 Calculating the Buffering Component

In the control group, the average calculated buffer value (mean ± SEM) was similar at both sites (rectal 48.1 ± 15.4; gastric 48.9 ± 10.5) but more than double the value of arterial HCO_3^- (22.3 ± 1.1). While the amount of calculated buffer at the rectal site nearly doubled at 15" arrest, gastric buffer increased less dramatically (84.8 + 20.3 vs. 57.14 + 8.02) and no statistical difference from control was noted. Rectal buffer was moderately correlated with both gastric buffer (r = .66) and arterial HCO_3^- (r = .52) under control conditions, but an inverse correlation was found between the calculated buffer value at the two sites at 15" arrest (r = -0.40).

3.5 Dye Considerations

Tissue samples were pink in color on visual inspection. The percent of pixels within the physiologic range of 6-8 units was slightly greater in the stomach than the rectum (53.7% vs. 45.9%). Under arrest conditions the amount of in-range pixels was reduced in the rectum while increased in the stomach, both in the direction of a lower pHi. The percent of pixels in the range was similar for all layers (p = .20), with the greatest number of pixels in range within the mucosal lining for both sites.

4. DISCUSSION

The pHi for gastric and rectal tissue exceeded the values reported for muscle (7.00 ± 0.01)[5] and brain (hippocampus, 7.01 ± 0.15; cerebral cortex, 7.04 ± 0.02).[6] Changes in pHi in relation to acid secretion in the gastric mucosa have been reported in several studies.[7,8] However, in these reports, using the dye bromcresol purple in which the pK was unreported, allowed for only a qualitative report of changes in pHi. We found no values for gastric or rectal pHi in the literature.

It may be that fifteen minutes of cardiac arrest was insufficient time to produce significant changes in pHi in gut tissue. Whether the pHi analysis was done using whole tissue or by layer, we found no difference in pHi by site (rectal, gastric) or condition (control, 15" arrest). However, the downward trend for pHi noted between all layers of the stomach and between the mucosal and submucosal layer of the rectum at 15" of ischemia suggests an extended period of time would have yielded significant reductions in pHi from baseline.

Heterogeneity of microcirculatory blood flow to different regions of the gut and the different layers has been proposed by other scientists based on varying oxygen consumption needs during hypoxic and ischemic challenges. Reductions in O_2 and flow may reduce pHi as well as produce unequal changes in pHi between the layers. During the hypodynamic phase of sepsis, Hiltebrand's group reported reductions from baseline in microcirculatory blood flow to the muscularis of the stomach (55%) and colon (70%), yet the colon mucosa blood flow remained constant. In contrast, during the hyperdynamic phase of shock, microcirculatory flow exceeded baseline in both the gastric and colonic mucosa while flow to the muscularis mucosa returned to baseline in the stomach. In the colon however, there was reduced flow to the muscularis mucosa suggesting autoregulation favored the mucosa.[9] If we link changes in pHi to blood flow, these findings are not wholly consistent with our results. The pHi was reduced in the muscularis in the stomach during arrest, yet increased in the rectum at 15" arrest.

The layer that exhibited the highest gastric pHi was the muscularis mucosa. Although rats were fasted 14-18 hours, stomachs contained food/stool/bedding. Because gastric acid was likely to be secreted in response to stomach contents, the findings may be explained by acid secretion, where HCO_3^- enters the diffusion barrier (from lamina propria, muscularis mucosa, and part of the submucosa) at a rate equal to the rate of hydrogen ion secretion through the mucosa.[10] The raised pHi at baseline may be linked to alkalinization within the muscularis mucosa in response to H+ secretion. Under anoxic conditions the reduction in pHi preceded the cessation of H+ secretion. We noted a greater reduction in pHi from baseline for the muscularis mucosa when compared to the gastric mucosal and submucosal pHi. For the rectum, the submucosal and mucosal layers had a greater pHi value for the control group. Unlike the stomach, the pHi of the muscularis mucosa in the rectum increased during arrest.

High levels of CO_2 were generated during flow stagnation induced by cardiac arrest. The high correlation in the control group between the two sites suggests comparable generation of CO_2 occurs. However, during arrest the site differences may not be attributable to differences in CO_2 generation alone. A difference in tissue buffering capacity or impaired removal of CO_2 cannot be ruled out.

In our study the gastric site had the greater reduction in pHi hence,less buffer during arrest. The buffer values reflect an estimate of intracellular buffering, extracellular buffering, or a combination thereof. Both intracellular (pHi) and extracellular components (tissue PCO_2) were entered into the Henderson Hasselbalch equation to obtain a buffer value. Changes in PCO_2 and pHi may be buffered to a different degree depending on the tissue type.

5.   CONCLUSION

We directly measured pHi using the pH sensitive dye, neutral red. We defined pHi for rectal and gastric tissue in whole tissue and by layer under control and arrest conditions. Fifteen minutes of arrest was not sufficient time to alter the pHi at the rectal or gastric site. On initial inspection, the stomach may be more sensitive to ischemic changes than the rectum. Understanding the mechanism by which PCO_2 generation is used to track clinical changes is vital to the early detection of tissue dysoxia in order to effectively treat and manage critically ill patients.

6.   ACKNOWLEDGEMENTS

This work was supported by a Faculty Research Grant, The University of Akron, Akron, OH; Delta Omega Chapter, Sigma Theta Tau, Akron, OH. We thank Maxwell Neil for his assistance with pHi analysis and Charlene Calabrese for her technical expertise and editing of this manuscript.

7.   REFERENCES

1. Gutierrez G, Brown SD. Gastrointestinal tonometry: a monitor of regional dysoxia. New Horiz 1996; 4:413-419.
2. Maynard N, Bihari D, Beale R, Smithies M, Baldock G, Mason R, McColl I. Assessment of splanchnic oxygenation by gastric tonometry in patients with acute circulatory failure. JAMA 1993; 270:1203-1210.
3. Fisher EM, LaManna JC. Gut dysoxia: Comparison of sites to detect regional gut dysoxia. Advances in Experimental Medicine and Biology In press.
4. Hoxworth JM, Xu K, Zhou Y, Lust WD, LaManna JC. Cerebral metabolic profile, selective neuron loss, and survival of acute and chronic hyperglycemic rats following cardiac arrest and resuscitation. Brain Res 1999; 821:467-479.
5. Zhang RG, Kelsen SG, LaManna JC. Measurement of intracellular pH in hamster diaphragm by absorption spectrophotometry. J Appl Physiol 1990; 68:1101-1106.
6. Crumrine RC, LaManna JC, Lust WD. Regional changes in intracellular pH determined by neutral red histophotometry and high energy metabolites during cardiac arrest and following resuscitation in the rat. Metab Brain Dis 1991; 6:145-155.
7. Hogben CA. Gastric secretion of hydrochloric acid. Introduction: the natural history of the isolated bullfrog gastric mucosa. Fed Proc 1965; 24:1353-1359.
8. Hersey SJ, High WL. Effect of unstirred layers on oxygenation of frog gastric mucosa. Am J Physiol 1972; 223:903-909.
9. Hiltebrand LB, Krejci V, tenHoevel ME, Banic A, Sigurdsson GH. Redistribution of microcirculatory blood flow within the intestinal wall during sepsis and general anesthesia. Anesthesiology 2003; 98:658-669.
10. Sanders SS, Hayne VB, Jr., Rehm WS. Normal H+ rates in frog stomach in absence of exogenous CO2 and a note on pH stat method. Am J Physiol 1973; 225:1311-1321.

3

O₂ UPTAKE KINETICS IN SKELETAL MUSCLE: WHY IS IT SO SLOW? AND WHAT DOES IT MEAN?

Bruno Grassi •

1. INTRODUCTION

Oxidative phosphorylation represents the most important mechanism of energy supply in mammals. With respect to the other mechanisms of ATP resynthesis [phosphocreatine (PCr) hydrolysis and anaerobic glycolysis], oxidative metabolism has two main "limitations": a lower maximal power, represented by the maximal O_2 uptake (VO_2max); and a slowness in "getting into action", as exemplified by the well-known fact that, upon a step increase in metabolic demand, in the moderate-intensity domain, O_2 uptake (VO_2) needs 2-3 minutes to reach a steady-state. This behaviour is usually termed "VO_2 kinetics". The rate at which skeletal muscle oxidative metabolism adjusts to a new metabolic requirement is one of the factors that determines exercise tolerance: a faster adjustment of oxidative phosphorylation during increases in work rate reduces the need for substrate level phosphorylation, with less disturbance of cellular and organ homeostasis (lower degradation of PCr and glycogen stores, lower accumulation of H^+).[1,2] Whereas there is general agreement on the concept that the capacity to deliver O_2 to skeletal muscles is the main determinant of VO_2max, how does one interpret mechanistically a slower (or a faster) VO_2 kinetics? Would the capacity to deliver O_2 to muscle fibers be the factor to be "blamed" (or "congratulated") for, or the capacity by muscle fibers to utilize the O_2 they receive?

This aspect has been matter of controversy for many years, mainly between those in favor of the concept that the finite kinetics of VO_2 adjustment during transitions is attributable to an intrinsic slowness of intracellular oxidative matabolism to adjust to the new metabolic requirement ("metabolic oxidative inertia"),[1,3,4] and those who suggest that an important limiting factor resides in the finite kinetics of O_2 delivery to muscle fibers ("O_2 delivery limitation").[5,6] The aim of the present paper is to provide a summary of some recent studies, carried out by our group as well as by others, which have dealt with this topic.

• Dipartimento di Scienze e Tecnologie Biomediche, Università degli Studi di Milano, I-20090 Segrate (MI), Italy. E-mail: Bruno.Grassi@unimi.it

2. WHICH IS FASTER? O_2 DELIVERY OR O_2 UTILIZATION?

For some time, the approach to the problem has been to define whether the adjustment of O_2 delivery (estimated on the basis of heart rate [HR] or cardiac output [Q]) was indeed faster than that of O_2 utilization (inferred from the kinetics of pulmonary VO_2).[3,5] This approach, besides providing only indirect evidence in favor or against the hypotheses outlined above, was troubled by the fact that, in humans, the investigated variables were quite "distant" from the relevant ones, that is muscle blood flow and muscle VO_2. At least in part this problem was overcome by some studies which determined the kinetics of O_2 delivery and VO_2 in humans at the level of exercising limbs.[7-9] A common finding from all these studies is that, during the first seconds of exercise, increases in O_2 delivery exceed increases in VO_2, whereas for the ensuing part of the transition the results are less clear and led to different interpretations. In these studies, however, measurements were carried out *across* exercising limbs, so that transit delays from the sites of gas exchange to the measurements sites confounded the overall picture, as demonstrated by Bangsbo et al.,[9] who estimated such delays by dye injection into the arterial circulation.

More recently, some studies "got inside the muscle" by utilizing techniques such as intravascular phosphorescence quenching for the determination of microvascular PO_2 in rat muscle[10] or near-infrared spectroscopy (NIRS) for the determination of muscle oxygenation in humans.[11] A common denominator among these techniques is that the variables are the result of the balance (or lack of thereof) between O_2 delivery and VO_2 in the area of interest, being therefore conceptually similar to O_2 extraction, or to arterio-venous O_2 concentration difference [C(a-v)O_2]. An increased microvascular PO_2, or an increased oxygenation, following an increase in work, would indicate a faster adjustment of O_2 delivery *vs.* that of VO_2, thereby providing indirect evidence in favor of the "oxidative metabolic inertia" hypothesis. The studies suggest unchanged (or slightly decreased) O_2 extraction for several seconds after an increase in work rate, reflecting a tight coupling between O_2 delivery and VO_2. Thus, the rapid and pronounced increase in O_2 delivery at the transition allows VO_2 to increase even in the presence of an unchanged (or slightly decreased) O_2 extraction. Only after this initial delay an increased O_2 extraction would contribute, together with the ongoing O_2 delivery increase, to the further increase in VO_2. This O_2 extraction time-course is similar to that of C(a-v)O_2 data obtained across exercising legs in humans[7,12] and across the isolated *in situ* dog muscle.[13]

Whereas the absence of increase in O_2 extraction, during the first seconds of the transition, represents indirect evidence against a lack of O_2 during that period, the tight coupling between the increased O_2 delivery and the increased VO_2 does not allow to exclude, *per se*, that O_2 delivery is nonetheless limiting the VO_2 kinetics, and that an enhanced rate of O_2 delivery adjustment could lead to a faster VO_2 response. Moreover, the experiments mentioned above do not allow to make much inferences for the ensuing phases of the transition, beyond the initial 10 seconds. To demonstrate whether O_2 delivery does (or does not) represent a significant limiting factor for VO_2 kinetics, experiments showing that a faster than normal or an enhanced O_2 delivery is (or is not) associated with faster VO_2 kinetics were needed.

3. EXPERIMENTS IN *IN SITU* DOG MUSCLE

To do so, we went back to the classic isolated *in situ* dog gastrocnemius preparation. By utilizing this model, Piiper et al.[14] observed a faster O_2 delivery kinetics compared to that of VO_2, but these authors did not manipulate O_2 delivery to see the effects on VO_2 kinetics. Our experiments were carried out in the laboratories of Drs. Michael C. Hogan, Peter D. Wagner (University of California, San Diego) and L. Bruce Gladden (Auburn University). The main idea behind a first series of studies was to enhance O_2 delivery to muscle fibers, and to see if this enhancement determined (or not) faster muscle VO_2 kinetics, providing direct evidence in favor (or against) the "O_2 delivery hypothesis".

3.1 Convective O_2 Delivery

It was first tested whether convective O_2 delivery to muscle was a limiting factor for VO_2 kinetics: all delays in the adjustment of convective O_2 supply were eliminated by pump-perfusing the muscle, in order to keep muscle blood flow constantly elevated during transitions from rest to contractions corresponding to ~60%[15] or 100%[16] of the muscle peak VO_2. To prevent vasoconstriction a vasodilatory drug (adenosine) was infused intra-arterially. Elimination of all delays in convective O_2 delivery during the transition did not affect VO_2 kinetics at the lower contraction intensity, whereas the kinetics was slightly faster (vs. control) at the higher intensity, leading to a reduction in the calculated O_2 deficit, i.e. of the amount of energy deriving from non-oxidative energy sources during the transition. Thus, for transitions from rest to submaximal VO_2, VO_2 kinetics was not limited by convective O_2 delivery to muscle, whereas for transitions to peak VO_2 convective O_2 delivery played a relatively minor role as a limiting factor. The lower O_2 deficit during transitions to 100% of peak VO_2 was associated with significantly less muscle fatigue, directly confirming the important role of VO_2 kinetics in muscle fatigue and exercise tolerance.

3.2 Peripheral O_2 Diffusion

Further experiments were done to determine the role of peripheral O_2 diffusion as a potential limiting factor for muscle VO_2 kinetics. In the previously described intervention we eliminated all delays in convective O_2 delivery to muscle capillaries, that is to say, we managed to get a lot of O_2 (bound to Hb) in muscle capillaries from the very onset of contractions. But from muscle capillaries to mitochondria, O_2 still has a rather long path to cover, and resistances to overcome: the molecule has to unbind from Hb, cross the red blood cell membrane, the capillary wall, the interstitial space, the sarcolemma, the cytoplasm of the muscle fiber end enter mitochondria, where it is eventually consumed. According to Fick's law of diffusion, this process is driven, among other factors, by the difference in PO_2 between the capillaries ($PcapO_2$) and mitochondria ($PmitO_2$). How to enhance peripheral O_2 diffusion, that is to say, how to increase the "driving pressure" for O_2 from capillaries to mitochondria? We increased $PcapO_2$ by having the dogs breathe a hyperoxic gas mixture (100% O_2) and by the administration of a drug (RSR13, Allos Therapeutics) which, as an allosteric inhibitor of O_2 binding to Hb, causes a rightward shift of the Hb-O_2 dissociation curve, that is to say reduces the affinity of Hb for O_2. Both in the control and in the "treatment" condition muscle blood flow was kept constantly elevated through the transition and adenosine was infused. For the same O_2 delivery, VO_2 and $C(a-v)O_2$, a rightward shift of the Hb-O_2 dissociation would determine an increase of the measured venous PO_2 and, as a consequence, of the calculated (by numerical integration) mean $PcapO_2$. Both interventions (hyperoxic breathing, right-shifted Hb-O_2 dissociation curve) determined significant increases in mean $PcapO_2$. For both investigated transitions, that is from rest to ~60%[17] and ~100% (unpublished observations) of the muscle peak VO_2, however, enhancement of peripheral O_2 diffusion did not significantly affect muscle VO_2 kinetics.

3.3 Role of Exercise Intensity

The studies mentioned above provide compelling evidence in favor of the hypothesis that an oxidative metabolic *inertia* within the muscle represents the limiting factor for VO_2 kinetics during transitions to contractions of relatively low metabolic intensity. On the other hand, during transitions to contractions of relatively high metabolic intensity, convective O_2 delivery could play a relatively minor but significant role as a limiting factor. According to several groups, the "ventilatory threshold" (VT) could discriminate in humans between work intensities at which O_2 delivery is not (below VT) or is (above VT) one of the limiting factors for pulmonary VO_2 kinetics, even after taking into consideration that for exercise above VT the *scenario* is more complex due to the presence of the "slow component" of VO_2 kinetics.[18]

3.4 Possible Sites of Oxidative Metabolic Inertia

Within the oxidative *inertia* hypothesis, the rate of adjustment of oxidative phosphorylation during transitions would be mainly determined by the levels of cellular metabolic controllers and/or enzyme activation. There are several possible rate-limiting reactions within the complex oxidative pathways, and recent research pointed to acetyl group availability within mitochondria and to the activation of pyruvate dehydrogenase (PDH). PDH activation by pharmacological interventions determined indeed a "sparing" of PCr during exercise transition, suggesting a lower O_2 deficit and therefore, possibly, a faster adjustment of muscle VO_2 kinetics (see *e.g.* Timmons et al.[19]). We tested this hypothesis in the isolated dog gastrocnemius *in situ*. PDH activation was obtained by dichloroacetate (DCA). Transitions were from rest to 60-70% of the muscle peak VO_2. DCA infusion determined a significant activation of PDH, but it did not significantly affect "anaerobic" energy provision and VO_2 kinetics.[13] Thus, in this experimental model, PDH activation status did not seem to be responsible for the metabolic *inertia* of oxidative phosphorylation. Similar conclusions were drawn by another study on humans.[20] Rossiter et al.[21] confirmed that PDH activation by DCA does not determine, in humans, faster pulmonary VO_2 kinetics, nor a faster kinetics of PCr hydrolysis, even in the presence of a reduced O_2 deficit (less PCr hydrolysis and lower blood lactate accumulation). In the absence of changes in VO_2 and PCr kinetics, a lower O_2 deficit could be explained on the basis of the observed reduced amplitude of the VO_2 response. For the same power output, a reduced amplitude of the VO_2 response suggests an improved metabolic efficiency. An increased metabolic efficiency after DCA was also observed by our group[13] in the dog gastrocnemius. In the presence of less muscle fatigue (higher force production) after DCA, we indeed observed unchanged VO_2 and no significant differences for substrate level phosphorylation. "Closing the circle", then, the increased metabolic efficiency after DCA could explain, at least in part, the PCr "sparing" described by Timmons et al.,[19] with no need to hypothesize a faster VO_2 kinetics.

The metabolic inertia of skeletal muscle oxidative metabolism might be related, at least in part, to a regulatory role of nitric oxide (NO) on mitochondrial respiration. Amongst a myriad of functions, which comprehend vasodilation, NO competitively inhibits VO_2 in the electron transport chain at the cytochrome *c* oxidase level. Through its combined effects of vasodilation and VO_2 inhibition, NO may serve as part of a feedback mechanism aimed at reducing the reliance on O_2 extraction to meet tissue O_2 needs, thereby maintaining higher intramyocyte PO_2 levels during exercise.[22] Inhibition of NO synthase by the administration of the arginine analogue L-NAME to exercising horses determined indeed a slightly but significantly faster pulmonary VO_2 kinetics.[22] A slightly but significantly faster pulmonary VO_2 kinetics after L-NAME was also described by Jones et al.[23] in humans.

4. PHOSPHOCREATINE AND GLYCOLYTIC "ENERGETIC BUFFERING"

At exercise onset PCr hydrolysis and anaerobic glycolysis act to provide a temporal buffer for the initial ATP demand, thereby slowing the onset of oxidative phosphorylation by delaying key energetic controlling signal(s) between sites of ATP hydrolysis and mitochondria. A close coupling between the kinetics of pulmonary VO_2 and the kinetics of [PCr] decrease was observed in humans performing constant-load quadriceps exercise.[21] Thus, PCr hydrolysis may not be considered just as an ATP buffer, but it could represent (through changes in concentration of [Cr], or of other variables related to this) one of the main controllers of oxidative phosphorylation ("creatine shuttle" hypothesis, see e.g. Whipp & Mahler,[1] Cerretelli & di Prampero[2]). The intricate interplay between the various mechanisms of energy provision at exercise onset suggests that pharmacological interventions aimed at blocking PCr hydrolysis and/or glycolysis could speed the rate of adjustment of oxidative phosphorylation. This seems indeed to be the case, as demonstrated by studies conducted on isolated amphibian myocyites, after blocking PCr hydrolysis by iodoacetamide,[23] or in isolated rabbit hearts after blocking glycolysis by iodoacetic acid.[24]

5. CONCLUSIONS

The rate at which skeletal muscle oxidative phosphorylation adjusts to a new metabolic requirement ("VO_2 kinetics") is one of the determinants of exercise tolerance. The VO_2 kinetics does not seem to depend, in normal conditions, on O_2 availability, at least for moderate-intensity exercise. A relatively minor limiting factor for the VO_2 kinetics could be the inhibition of mitochondrial respiration by NO. PCr hydrolysis can be considered not only an energy buffer for ATP resynthesis early during the metabolic transition, but also one of the main controllers of oxidative phosphorylation.

6. ACKNOWLEDGEMENTS

Financial support by NATO LST.CLG 979220 Grant is recognized.

7. REFERENCES

1. B.J. Whipp, and M. Mahler, Dynamics of pulmonary gas exchange during exercise, in: *Pulmonary Gas Exchange*, Vol. II., edited by J.B. West (Academic Press, New York, 1989), pp. 33-96.
2. P. Cerretelli, and P.E.di Prampero, Gas exchange in exercise, in: *Handbook of Physiology*, Section 3, *The Respiratory System*, vol. IV, *Gas Exchange,* edited by L.E. Fahri and S.M. Tenney (American Physiological Society, Bethesda, 1987), pp. 297-339.
3. P. Cerretelli, D.W. Rennie, and D.R. Pendergast, Kinetics of metabolic transients during exercise, in: *Exercise Bioenergetics and Gas Exchange*, edited by P. Cerretelli and B.J. Whipp (Elsevier, Amsterdam, 1980), pp. 187-209.

4. B. Grassi, B., Regulation of oxygen consumption at exercise onset: is it really controversial? *Exerc. Sport Sci. Rev.* 29, 134-138 (2001).
5. R.L. Hughson, Exploring cardiorespiratory control mechanisms through gas exchange dynamics. *Med. Sci. Sports Exerc.* 22, 72-79 (1990).
6. R.L. Hughson, M.E. Tschakowsky, and M.E. Houston, Regulation of oxygen consumption at the onset of exercise. *Exerc. Sport Sci. Rev.* 29, 129-133 (2001).
7. B. Grassi, D.C. Poole, R.S. Richardson, D.R. Knight, B.K. Erickson, and P.D. Wagner, 1996, Muscle O_2 uptake kinetics in humans: implications for metabolic control. *J. Appl. Physiol.* 80, 988-998 (1996).
8. R.L. Hughson, J.K. Shoemaker, M.E. Tschakovsky, and J.M. Kowalchuck, Dependence of muscle VO_2 on blood flow dynamics at the onset of forearm exercise. *J. Appl. Physiol.* 81, 1619-1626 (1996).
9. J. Bangsbo, P. Krustrup, J. Gonzalez-Alonso, R. Boushel, and B. Saltin, Muscle oxygen kinetics at onset of intense dynamic exercise in humans. *Am. J. Physiol.* 279, R899-R906 (2000).
10. B.J. Behnke, C.A. Kindig, T.I. Musch, W.L. Sexton, and D.C. Poole, Effects of prior contractions on muscle microvascular oxygen pressure at onset of subsequent contractions. *J. Physiol.* 593, 927-934 (2002).
11. B. Grassi, S. Pogliaghi, S. Rampichini, V. Quaresima, M. Ferrari, C. Marconi, and P. Cerretelli, Muscle oxygenation and gas exchange kinetics during cycling exercise on- transition in humans. *J. Appl. Physiol.* 95, 149-158 (2003).
12. B. Grassi, Skeletal muscle VO_2 on-kinetics: set by O_2 delivery or by O_2 utilization? New insights into an old issue. *Med. Sci. Sports Exerc.* 32,108-116 (2000).
13. B. Grassi, M.C. Hogan, P.L. Greenhaff, J.J. Hamann, K.M. Kelley, W.G. Aschenbach, D. Constantin-Teodosiu, and L.B. Gladden, Oxygen uptake on-kinetics in dog gastrosnemius in situ following activation of pyruvate dehydrogenase by dichloroacetate. *J. Physiol.* 538:195-207 (2002).
14. J. Piiper, P.E. di Prampero, and P. Cerretelli, Oxygen debt and high-energy phosphates in gastrocnemius muscle of the dog. *Am. J. Physiol.* 215, 523-531 (1968).
15. B. Grassi, L.B. Gladden, M. Samaja, C.M. Stary, and M.C. Hogan, Faster adjustment of O_2 delivery does not affect VO_2 on-kinetics in isolated in situ canine muscle. *J. Appl. Physiol.* 85, 1394-1403 (1998).
16. B. Grassi, M.C. Hogan, K.M. Kelley, W.G. Aschenbach, J.J. Hamann, R.K. Evans, R.E. Pattillo, and L.B. Gladden, Role of convective O_2 delivery in determining VO_2 on-kinetics in canine muscle contracting at peak VO_2. *J. Appl. Physiol.* 89, 1293-1301 (2000).
17. B. Grassi, L.B. Gladden, C.M. Stary, P.D. Wagner, and M.C. Hogan, Peripheral O_2 diffusion does not affect VO_2 on-kinetics in isolated in situ canine muscle. *J. Appl. Physiol.* 85, 1404-1412 (1998).
18. G.A. Gaesser, and D.C. Poole, D.C., The slow component of oxygen uptake kinetics in humans, in: *Exercise and Sport Sciences Reviews*, Vol. 24., edited by J.O. Holloszy (Williams & Wilkins, Baltimore, 1996), pp. 35-71.
19. J.A. Timmons, T. Gustafsson, C.J. Sundberg, E. Jansson, and P.L. Greenhaff, Muscle acetyl group availability is a major determinant of oxygen deficit in humans during submaximal exercise. *Am. J. Physiol.* 274, E377-E380 (1998).
20. J. Bangsbo, M.J. Gibala, P. Krustrup, J. Gonzalez-Alonso, and B. Saltin, Enhanced pyruvate dehydrogenase activity does not affect muscle O_2 uptake at onset of intense exercise in humans. *Am. J. Physiol.* 282, R273-R280 (2002).
21. H.B. Rossiter, S.A. Ward, F.A. Howe, D.M. Wood, J.M. Kowalchuck, J.R. Griffiths, and B.J. Whipp, Effects of dichloroacetate on VO_2 and intramuscular ^{31}P metabolite kinetics during high-intensity exercise in humans. *J. Appl. Physiol.* 95, 1105-1115 (2003).
22. C.A. Kindig, P. McDonough, H.H. Erickson, and D.C. Poole, Effect of L-NAME on oxygen uptake kinetics during heavy-intensity exercise in the horse. *J. Appl. Physiol.* 91, 891-896 (2001).
23. A.M. Jones, D.P. Wilkerson, K. Koppo, S. Wilmshurst, and I.T. Campbell, Inhibition of nitric oxide synthase by L-NAME speeds phase II pulmonary VO_2 kinetics in the transition to moderate intensity exercise in man. *J. Physiol.* 552, 265-272 (2003).
24. H.B. Rossiter, S.A. Ward, J.M. Kowalchuck, F.A. Howe, J.R. Griffiths, and B.J. Whipp, Effects of prior exercise on oxygen uptake and phosphocreatine kinetics during high-intensity knee-extension exercise in humans. *J. Physiol.* 537, 291-303 (2001).
25. C.A. Kindig, R.A. Howlett, C.M. Stary, B. Walsh, and M.C. Hogan, Effects of acute creatine kinase inhibition on metabolism and tension development in isolated single myocytes. *J. Appl. Physiol.* 98: 541-549 (2004).
26. G.J. Harrison, M.H. van Wijhe, B. de Groot, F.J. Dijk, L.A. Gustafson, and J.H.G.M van Beek, Glycolytic buffering affects cardiac bioenergetic signaling and contractile reserve similar to creatine kinase. *Am. J. Physiol.* 285, H883-H890 (2003).

4.

BIOMECHANICAL DESIGN FACTOR FOR SOFT GEL CHROMATOGRAPHY COLUMNS TO SEPARATE HOMOLOGOUS, HIGH MOLECULAR WEIGHT, THERAPEUTIC PROTEINS

Wendy W. He [1] and Duane F. Bruley [2]

1. INTRODUCTION

Protein C (model molecule) is an anti-coagulant in blood. It also possesses anti-thrombotic and anti-inflammatory characteristics that are very effective in treating blood clotting phenomena found in many disease states. Recombinant production of Protein C and recovery from human plasma via immuno affinity chromatography are both very expensive technologies for the manufacture of a protein C product. Immobilized metal affinity chromatography (IMAC) is being examined as a less expensive process for Protein C purification from blood plasma. The purpose of this investigation is to examine how the mechanical characteristics of chromatographic resins influence the biomechanical design of chromatography columns. It is crucial for preparative chromatography that column scale-up takes into consideration the collapse potential of the gels because of the large cost and inconvenience of such an episode. Also, it is desirable to minimize process cycle-time considering resin mechanical properties to achieve economical preparative scale production of a desired protein. It is also true that a resin can have desirable mechanical characteristics but the surface chemistry for adsorption of the desired product might be inappropriate for the separation and purification of the specific product. Therefore, it is important to consider both the biomechanical and the biochemical characteristics of the resin systems to attain optimal process design and performance for reduced production cost.

This reseach has established theoretical and experimental guidlinelines and relationships for the understanding of column behavior and the scale-up of Immobilized Metal Affinity Chromatography (IMAC) [1] and Immuno Affinity Chromatography for the low cost production of Protein C.

Gel matrices for Protein C separation [2] have been analyzed here to determine the relationship between mechanical characteristics of the gel, flow conditions and column collapse potential have been analyzed here to determine the relationship between mechanical characteristics of the gel, flow conditions and column collapse potential.

---
[1] Wendy W. He, Assistant Professor, Lake Superior State University;
[2] Duane F. Bruley, Professor, University of Maryland Baltimore County/Synthesizer, Inc.

Using Biot's Consolidation Theory,[3] Darcy's Law and experimental data of flow rate vs. pressure drop, a column design relationship was determined. The design factor provides a first approximation for column scale-up without the occurrence of the catastrophic phenomena of column collapse.

2. MATERIALS AND METHODS

Due to availability, the primary gel tested was Sepharose 2B linked with Protein C monoclonal antibody, donated by the American Red Cross.

Experimental studies were performed to determine the correlation of the flow rate vs. pressure drop for a chromatography column (MT 20 column), packed with Sepharose 2B. The fluid flow rate and pressure drop (ΔP) were examined to determine how they influence column mechanical stability.

Specific MT 20 columns, 15 mm diameter and 130 mm length, were purchased from the Bio-Rad company (Hercules, California). The gel slurry was equilibrated with water until all ethanol was removed. The gel was then placed in a flask under vacuum for thirty minutes. Next, the degassed gel was injected in small quantities into the column, using a pipette, to a height of about 80 mm. The column was allowed to settle until there was no further reducton in gel height. This step was done overnight, before doing a flow rate vs. pressure drop experiment, to allow the gel matrix to achieve most of the expected bulk compression.

A packed column was connected to a triaxial pressure panel and cell. First, measurements were made at pressure increments starting from a minimum gauge pressure drop of zero psi. Using 5 psi increments, the gauge pressure drop reached a maximum (100psi) after which the pressure drop increments were reversed and data were recorded at 5 to 10 psi increments back to zero. The pressure drop across the column was read from a digital pressure gauge on the control panel. After each pressure increment the column was allowed to achieve steady state and then the time for 20ml of water to flow through the column matrix was recorded. The water flowed directly through a tube at the bottom of the chromatography column and into a burette at atmospheric pressure.

3. THEORETICAL MODEL

Biot's Consolidation Theory was reduced to one dimension and time to develop a set of systems equations for the microscopic theoretical analysis of the column. These equations describe the pore pressure distribution and the solid skeleton displacement in the porous media (gel).

Neglecting the initial effective stress and body forces and assuming a one-dimensional, time-dependent process along the axial (z) direction, the governing equations for the solid and fluid respectively are:

$$(\lambda + 2G)\frac{\partial^2 u}{\partial z^2} - \frac{\partial p}{\partial z} = 0 \tag{1}$$

$$\frac{\partial}{\partial z}[\frac{k}{\mu}(\frac{\partial p}{\partial z})] = (1 - \phi)C_s \frac{\partial p}{\partial t} + \frac{\partial^2 u}{\partial t \partial z} \tag{2}$$

A central finite difference and forward finite difference were utilized to obtain the finite difference equation computer solutions for the gel deformation and fluid pressure drop. Solutions have been obtained for both the process in steady state and the dynamic state.

The Biot's equations, which include Darcy's law, agreed with the flow vs pressure drop experimental data obtained for Sepharose 2B. Therefore, Darcy's law $\Delta p = \dfrac{Q\mu L}{Ak}$

[4] was used to derive the design factor representing the $(\dfrac{L}{D^2})$ ratio. Rearrangement of

terms gives $\dfrac{L}{D^2} \le \dfrac{\pi k_{peak} \Delta p_{peak}}{4Q\mu}$ which is suggested as an appropriate relationship to

determine a first approximation for the design of production scale chromatography columns using soft gels.

4. RESULTS AND DISCUSSION

The average values for five experimental tests of the target resin, Sepharose 2B, in a MT 20 column were calculated. A plot shows that flow rate has a maximum value when starting from a minimum pressure drop and continuing to a maximum pressure drop. The results show that there is a hysteresis phenomenon resulting from the gel characteristics (the flow rate is consistently lower when starting from the maximum pressure drop). This hysteresis loop indicates that the gel behavior changes from elastic to pseudo-plastic at high-pressure drops. This result suggests that, because of water reabsorption over time, the resin matrix has a visco-elastic behavior.

The first layer of the gel matrix was chosen to test the dynamic solution. The other layers showed a similar trend to the first layer. The steady-state time was chosen to simulate the pressure and displacement along the gel matrix column. It was shown that the pressure drop reaches a steady-state value of about 4 psi after approximately thirteen seconds for the first layer of the gel matrix. It was determined that displacement reaches 0.002 m after approximately twenty seconds which fits with the experimental data measured in steady-state

From the experiments on Sepharose 2B, the maximum flow rate was found at approximately 10 psi pressure drop across the column (Figure 1). Therefore, 10 psi is defined as the pseudo column collapse point and the column operating pressure drop should not exceed 10 psi.

The $\dfrac{L}{D^2}$ ratio is determined based upon pressure drop limitations to prevent column

collapse along with the appropriate gel volume capacity to meet production requirements in a reasonable cycle time. Considering the desired production rate of 62 kg protein C/year and the other design equation parameters determined from the experimental

results, it is found that $\dfrac{L}{D^2} \le 0.00014$ for Sepharose 2B with Q=1048ml/s,[5]

μ=0.01g/cms, $\Delta P = 4 psi$ and k=6.79*10[-13]m². Table 1 shows the column design using a trial and error method based on this ratio.

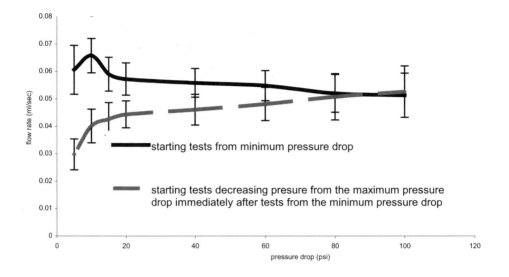

Figure 1: Flow rate vs. pressure drop measurement for Sepharose 2B in a MT 20 column

Table 1. IMAC column design for Sepharose 2B

Case	L (cm)	D(cm)	$\dfrac{L}{D^2}$	Number of Columns needed
1	4	170	0.00014	1
2	4	200	0.0001	1
3	1	100	0.0001	10

5. CONCLUSIONS

The results show that Biot's Consolidation Theory gives a reasonable estimate of gel matrix displacement and pressure drop along the column. About half the pressure drop is found in the last 20% of the column: this is consistent with the finding of Ostergren [6] for a dextran-based Sepharose gel. It also shows that the strain is larger towards the bottom of the column. This result agrees with Parker's [7] work that experimentally demonstrates that larger strain appears at the bottom of the column as a result of the flow shear when fluid flows through a porous media.

It is determined that a column 4cm thick and 170 cm in diameter with a maximum pressure drop of 4 psi would meet the required production rate of 62 kg Protein C/year with a reasonable cycle time for the process. Approximately 70 batches per year will meet the required Protein C production to serve approximately 25% of the Protein C deficient patient market in the United States.

The quantitative chromatography column design relationships determined by Darcy's law allows a first step theoretical approximation for the length to diameter squared ratio (L/D2) for operation of the column without matrix "collapse". This was verified by an actual American Red Cross trial and error design of one of the first Protein C production columns using immuno affinity chromatography.

6.   ACKNOWLEDGEMENT

This research was supported by The Whitaker Foundation through a "Special Opportunity Award in Biomedical Engineering", the National Science Foundation (Grant CTS-9904465) and the American Red Cross. The authors wish to thank them for all of their help and support.

7.   NOMENCLATURE

λ=Lame's constant
G=Shear Modulus
u = Displacement
p = Fluid pressure
z = Coordinate system along z direction
k = Permeability of the porous medium
μ = Dynamic viscosity of fluid

ϕ =Porosity
t = Time
Δp= Pressure drop along the gel matrix
Q =Flow rate
L= Length of the gel matrix
A= Cross-section area of the column
D= Column diameter

8.   REFERENCES

1. Winzerling JJ, Berna P and Porath J., 1992, How to use Immobilized Ion Affinity Chromatography Methods, *A Computation to Methods in Enzymology*, 4:4-13.
2. Bruley, D.F., and Drohan W.N., 1990, *Protein C and related anticoagulants.* Gulf Publishing Company.
3. Biot, M.A., 1955, Theory of Elasticity and Consolidation for a Porous Anisotropic Solid, *Journal Appl. Phys.,* 26:182-185.
4. Nakayama A., 1995, *PC-Aided Numerical Heat Transfer and Convection Flow 5.*
5. Davey, Ankur, 1999, M.S. *Dissertation* University of Maryland Baltimore County
6. Ostergren, KCE, 1998, Deformation of a Chromatographic Bed during steady-state Liquid Flow, *AICHE Journal, 44:20-25*
7. Parkar KH, Mehta RV, and Caro CG, 1987, Steady Flow in Porous, Elastically Deformable Materials, *Transaction of the ASME,* 54:794-796.

5.

RELATIONSHIPS BETWEEN MUSCLE SO$_2$, SKIN SO$_2$ AND PHYSIOLOGICAL VARIABLES

Charlotte L. Ives[*], David K. Harrison[*], Gerard Stansby[†]

1. INTRODUCTION

In 1995 we introduced the technique of lightguide spectrophotometry for the measurement of tissue oxygen saturation (SO$_2$) to predict healing viability of below-knee skin flaps in lower limb amputation for critical ischaemia[1]. This technique, using a Photal MCPD 1000 (Otsuka Electronics, Osaka) spectrophotometer, has been applied to the routine assessment of patients at the University Hospital of North Durham since 1999. Since then a healing rate of 94% has been achieved with a below knee to above knee amputation ratio of 9:2[2].

In a study to investigate the possible cause of the high infection rate in groin wounds following vascular bypass surgery Raza et al.[3] used the Erlangen microlightguide spectrophotometer[4] (EMPHO, BGT-Medizintechnik, Überlingen) to measure SO$_2$ in the groin skin medially and laterally to the incision sites in patients undergoing femoro-popliteal or femorodistal bypass operations prior to and at 2 and 7 days post-operatively. The equivalent contralateral sites were used as controls. The results showed a significant difference ($p < .01$) between the medial and lateral SO$_2$ values post operatively. On this basis, it was postulated that a disruption of blood supply may be responsible for the high incidence of infection in such surgical wounds.

The experience using lightguide spectrophotometry in the visible wavelength range to measure skin SO$_2$ (SSO$_2$) as a possible predictor of healing viability or infection of surgical wounds led to the study described elsewhere in this volume[5]. In addition, it was proposed that infrared spectroscopy should be used in the clinical study on the premise that muscle SO$_2$ may be a better predictor of wound infections emanating from deeper lying tissues than skin. The present study was therefore carried out in order to define the

---

[*] University Hospital of North Durham, North Road, Durham, DH1 5TW UK

[†] Northern Vascular Centre, Freeman Hospital, Freeman Road, Newcastle-upon-Tyne, NE7 7DN UK

range of values to be found at the proposed measurement site in healthy volunteers. In addition it was intended to identify some of the physiological parameters that may influence skin and muscle SO_2 (MSO_2) at these sites.

2. AIM

The aims of this study are threefold:

1. To find out if physiological variables affect the oxygen saturation of the skin and muscle.
2. To find out whether oxygen saturations of skin (or muscle) are related at different sites.
3. To find out whether oxygen saturations of the skin and muscle are related at a specific site.

3. MATERIALS

The visible range lightguide spectrophotometer (LGS) used for the study was the Whitland RM200 instrument (Spectrum Medical, Cheltenham, UK). A comparison of the properties of this instrument with the MCPD 1000 has been reported recently[6].

Briefly, a white LED light source mounted in the probe transmits light into the skin. The light scattered back from the tissue passes through a single optical fibre to the integrated diffraction grating/photodiode array. The wavelength range analysed is 500-586nm with a resolution of 3nm. Correction algorithms are applied to the measured absorption spectra for the influences of scattering and melanin[7]. A further algorithm using the theory of Kubelka and Monk[8] makes use of two reference spectra (oxygenated haemoglobin, HbO_2 and deoxygenated haemoglobin, Hb) and a least squares fitting method to calculate the SO_2 from the measured spectrum. Figure 1 illustrates the principle of the least squares fitting method. In this case a proportion of 75% Hb and 25% HbO_2 produce the best fit to the measured spectrum (shown as particularly noisy in this illustration).

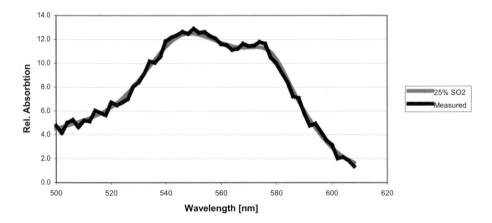

Figure 1. Illustration of the least squares fitting method.

An Inspectra® Model 325 (kindly loaned by Hutchinson Technology Inc, Hutchinson, USA) was the near infrared tissue spectrophotometer (NIS) used for the study. Again, LEDs are used as the light source and optical absorption is measured at 680, 720, 760 and 800nm. SO_2 values are derived using a second derivative technique[9]. The wavelengths used are important: 800nm is the isosbestic point (where both forms of haemoglobin absorb light equally) and is used to calculate the total haemoglobin present, 760nm is the point of maximum absorption by deoxygenated haemoglobin and minimum absorption by oxygenated haemoglobin in this range. The tissue depth achieved is that of half the probe separation[10]. For this experiment we used the 25mm probe exclusively.

4. METHODS

50 healthy volunteers were recruited by advertisement. The room temperature was kept between 21 and 23 degrees Centigrade. While the subject reached ambient temperature the procedure was explained, height and weight were measured and body mass index (BMI) calculated. The skinfold thickness over triceps (TST) was measured using Haltain callipers. Age and sex were also noted.

Using the LGS 20 measurements were taken on the right upper arm at rest. These were centred on a point 10cm below the tip of the shoulder in a grid of 4cm x 5cm. In one subject a black tattoo excluded measurements on the right and so the left arm was used instead. More than one reading was taken as it is known that vascular anomalies beneath the skin (e.g. arterio-venous shunts) can give abnormally high readings[6] and so an average reading was needed.

The NIS probe was placed at 10cm below shoulder tip over the bulk of biceps and held in place with the self-adhesive probe covers provided with the machine until the reading stabilised (usually about 10 seconds).

Readings were then taken 2cm either side of the umbilicus, 10 times each side with the LGS in a grid 2cm x 5cm and with the NIS as on the arm. The average of these two readings was then used as mean abdomen MSO_2.

5. RESULTS

For this study statistical significance was reached when $p < 0.05$.

Of the 50 subjects 32 were female and 18 male with an average age of 39 +/-11 years (mean +/- standard deviation). There were no significant differences in age or BMI between the two sexes. TST was significantly lower in males than females ($p < 0.001$). Please also refer to table 1.

Arm SSO_2 showed no difference at 95% confidence interval between male and female; however abdomen SSO_2 was significantly higher in males ($p = 0.02$). Arm MSO_2 was significantly higher in males than females ($p < 0.001$). Abdomen MSO_2 was higher in males than females but this did not reach statistical significance ($p = 0.1$).

Table 1. Results compared between the genders

Variable	Male	Female
Number in study	18	32
Age	38.8 +/- 10.4years[a]	39.4+/- 11.4 years
BMI	25.1 +/- 5.5	25.7 +/- 3.9
TST	14.1 +/-4.4	20.4+/- 6.2
Arm SSO2	47.2% +/-9.3	47.7% +/- 9.2
Abdomen SSO2	37.6% +/- 7.8	32.1% +/- 7.9
Arm MSO2	59.3% +/- 19.3	33.5% +/- 19.6
Abdomen MSO2	53.0% +/- 19.8	43.7% +/- 18.1

[a] All values in this article are mean +/- standard deviation unless otherwise stated

Using Pearson's regression there was a positive correlation (r) of 0.72 (p<0.001) between arm and abdomen MSO_2 readings, but there was no correlation between arm and abdomen SSO_2 readings. There was no correlation between muscle and skin SO_2 in either location. MSO_2 had a negative dependence on TST (r=-0.4) but this was only significant (p<0.01) in the arm. No readings correlated with age.

In 13 subjects the near-infrared spectrophotometer was unable to pick up a reading either on the arm or abdomen, henceforth known as the 'poor' readings group. These subjects consisted of 12 female and 1 male. In five of these (4 female, 1 male) neither the arm nor the abdomen yielded a result. Please refer to table 2 for differences in other variables between these groups. Skin SO_2 (as measured by LGS) in the 'poor' readings group was not significantly lower than 'good' readings group at either location.

Table 2. Differences between 'poor' and 'good' NIS readings

Variable	Good	Poor
Number (M:F ratio)	37 (17:20)	13 (1:12)
Age	39.9 +/- 11.3 years	37.2 +/- 10.2 years
BMI	24.3 +/- 4.3 kg per m^2	28.1 +/- 5.7 kg per m^2
TST	16.3 +/- 5.2	22.9 +/- 6.8
Arm SSO2	49.2% +/- 8.8	53.2% +/- 9.2
Abdomen SSO2	35.0% +/- 8.8	30.5% +/- 7.9

Age was not related to having a poor reading. However, BMI and TST were found to be significantly higher in the group with 'poor' readings. The BMI in the subjects with 'good' readings was on average 4.2 kg per m^2 lower than compared to the 'poor' group (p=0.04). Subjects with 'good' readings had an average TST of 6.6 lower compared to subjects with a 'poor' reading (p=0.01).

6. DISCUSSION

Skin SO_2 was found to vary between sites. A study by Caspary et al. has also found differences in skin oxygenation, although they measured at different sites to this study[11]. This suggests that skin blood flow is controlled locally, e.g. affected by temperature or emotion. It may be that at some locations a reading is taken directly on top of a blood vessel giving a spuriously high result; however this should be overcome by taking the average of twenty readings in the area.

As arm and abdomen MSO_2 are correlated it appears that muscle blood flow is controlled systemically.

The amount of adipose tissue present (as measured by TST) detrimentally affects MSO_2 readings. Not only do high TST readings give a low oxygenation reading, in some circumstances they can obliterate a reading all together. Van Beekvelt et al. have recorded similar findings to these[12]. This may reflect a true reduction in oxygenation, and supports the evidence that obese patients are more at risk for wound infections[13]. However it may be that it is only the circulation in the fat that is being measured which is known to be small anyway, in this case work needs to be carried out with probes that have a larger separation so the light can penetrate to the muscle. Alternative hypotheses to this are that there is more scattering of the light by adipose tissue and so less can return to the probe; however in these wavelengths lipids have minimal absorption and scattering[14].

The findings of this study could have effects on future studies into tissue SO_2.

7. REFERENCES

1. D.K. Harrison, P.T. McCollum, D.J. Newton et al. Amputation level assessment using lightguide spectrophotometry. *Prosthet Orthot Int* 1995; **19**: 139-147
2. J.M. Hanson, D.K. Harrison and I.E. Hawthorn Tissue spectrophotometry and thermographic imaging applied to routine clinical prediction of amputation level viability. In "Progress in Biomedical Optics and Imaging: Functional Monitoring and Drug-Tissue Interaction". Ed. GJ Müller and M Kessler. *Spie Proc Series* 2002; **4623**: 187-194
3. Z. Raza, D. J. Newton, D. K. Harrison et al. Disruption of skin perfusion following longitudinal groin incision for infrainguinal bypass surgery. *Eur. J. Vasc. Endovasc. Surg* 1999; **17**, 5-8
4. K.H. Frank, M. Kessler, K. Appelbaum and W. Dümmler, The Erlangen micro-lightguide spectrophotometer EMPHO I. *Phys. Med. Biol.* 1989; 34, 1883-1900
5. C.L. Ives, D. K. Harrison and G. Stansby. Prediction of surgical site infections using spectrophotometry: Preliminary results. *Adv. Exp. Med. Biol.* This Volume (2005)
6. D.K. Harrison and I.E. Hawthorn Amputation level viability in critical limb ischaemia: Setting new standards. *Adv Exp Med Biol* 2004. In Press
7. D. Parker and D.K. Harrison. Non-invasive measurement of blood analytes. *World Intellectual Property Organization*, 2000; WO 00/01294
8. P. Kubelka and F. Munk. Ein Beitrag zur Optik der Farbanstrich. Zeitschrift f. techn. *Physik.* 1931; 11a: 593-601.
9. D. Myers, C. Cooper, G. Beilman et al. Noninvasive method for measuring local hemoglobin oxygen saturation in tissue using wide gap second derivative near-infrared spectroscopy. *Adv. Exp. Med. Biol.* This volume (2005)
10. W. Cui, C. Kumar, B. Chance. Experimental study of migration depth for the photons measured at sample surface. *Proc SPIE* 1991; 1431: 180-191
11. L. Caspary, J. Thum et al. Quantitative reflection spectrophotometry: spatial and temporal variation of Hb oxygenation in human skin. *Int J Microcirc Clin Exp* 1995; **15**(3): 131-6.
12. M.C.P. van Beekvelt, M.S. Borghuis, B.G.M. van Engelen et al. Adipose tissue thickness affects in-vivo quantitative near-IR spectroscopy in human skeletal muscle. *Clin Sci* 2001;101: 21-28
13. www.surgicaltutor.org. Last accessed 25 July 2004
14. C. Elwell. A practical users guide to near infrared spectroscopy. 1995 London, UCL Reprographics.

6

FORMATION AND ROLE OF NITRIC OXIDE STORES IN ADAPTATION TO HYPOXIA

Eugenia B. Manukhina, Anatoly F. Vanin, Khristo M. Markov, and Igor Yu. Malyshev[*]

1. INTRODUCTION

Periodic hypoxic episodes can exert either detrimental or protective effects on the organism, which are defined to a considerable extent by the regimen, that is, the ratio of duration and intensity of hypoxic exposures. It was shown that acute, chronic and intermittent hypoxia induce different responses of intracellular signaling pathways for transcription of different genes; physiological responses to hypoxic exposures differ correspondingly.[1] It is well known that chronic periodic hypoxia, which is observed, for example, in sleep apnea syndrome, can result in the development of pulmonary and systemic hypertension, myocardial infarction and cognitive disorders.[2] At the same time, dosed adaptation to intermittent hypobaric or normobaric hypoxia shows a range of direct and cross-protective effects, which have been widely used for the treatment and prevention of many diseases and to increase the efficiency of exercise training.[3] This is why studying the effect of hypoxia on key physiological processes is so important. One of such universal processes is the metabolism of nitric oxide (NO) tightly regulated by environmental oxygen.[4]

The effect of NO depends on the NO level. High concentrations of NO are very toxic while comparatively low concentrations of NO provide a regulatory function through activation of cyclic guanylate cyclase and cyclic GMP production in target cells.[5]

The direction of changes in NO synthesis is related to the extent and duration of hypoxia with different mechanisms of these changes. Hypoxic inhibition of endothelial NOS (eNOS) is evident as attenuation of endothelium-dependent vascular relaxation. After 48 hours of exposing rats to hypoxia, endothelium-dependent relaxation of the isolated aorta was reduced twofold.[6]

The decrease in NO production in acute hypoxia is reversible. For example, after 90-min hypoxia, plasma levels of the NO metabolites nitrite and nitrate in the blood from the aorta and renal artery were decreased to 48% and 73% of the initial level, respectively. After 1 hour of reoxygenation, these parameters returned to baseline.[7]

[*] Eugenia B. Manukhina, Institute of General Pathology and Pathophysiology, Baltijskaya 8, Moscow 125315, Russia. Anatoly F. Vanin, Institute of Chemical Physics, Kosygin 4, Moscow 117977, Russia. Khristo M. Markov, Institute of Pediatrics, Lomonosovsky 2/62, Moscow 117296, Russia. Igor Yu. Malyshev, Institute of General Pathology and Pathophysiology, Baltijskaya 8, Moscow 125315, Russia.

Activity of NOS directly depends on the level of synthesized NO since NO regulates the NOS activity by the negative feedback mechanism. The synthesized NO binds to iron of NOS heme thereby inhibiting NOS; simultaneously apparent Kм for O_2 increases.[8] Since O_2 and NO compete for NOS heme iron, the extent of NOS dependence on O_2 is related to the decomposition rate of the NO/heme iron complex, which depends, in turn, on O_2 concentration.[9] The level of O_2 as a substrate can limit NO production in hypoxia because the apparent sensitivity of NOS isoforms to O_2 is within the range of normal tissue O_2 concentrations.[10]

Another possible mechanism for hypoxic inhibition of NO production is the decrease in heat shock proteins HSP90 characteristic of hypoxia.[11] These proteins bind to eNOS and provide its activation in response to NO-stimulating agonists.[12]

Sensitivity of eNOS to O_2 shortage increases in the presence of inducible NOS (iNOS) induction. Indeed, a moderate decrease in O_2 tension in a cell culture, which generally does not induce any attenuation of NO synthesis, suppresses eNOS activity following the iNOS gene expression stimulated by bacterial lipopolysaccharide.[4]

The decreased NO production in acute hypoxia is due to inhibition of eNOS activity but not downregulation of the enzyme protein synthesis.[11] However chronic hypoxia can suppress both eNOS activity and eNOS gene expression in vessels of the systemic circulation simultaneously stimulating the latter in the pulmonary circulation. The amount of eNOS protein falls sharply in rat aorta while it increases in lungs as early as after 12 hours of hypoxia.[6] Percentage of pulmonary arterioles expressing eNOS begins to increase after 1 day of hypoxia. After that, this parameter increases for several days and then remains unchanged for 4 weeks. The NO overproduction in pulmonary arterioles is not always beneficial and may even contribute to the development of pulmonary hypertension depending on the relationship between NO and free oxygen species.[13]

Dosed intermittent hypoxia may be an efficient stimulator of NO synthesis and eNOS expression. However, intermittent hypoxia stimulates both activity and expression of all three NOS isoforms. Potentiation of constitutive NO production is generally beneficial for the organism because it protects the organism against hypertension, thromboses, vasospasm, oxidative stress, etc. At the same time, expression of iNOS results in NO overproduction and toxic effects of excessive NO. During a hypoxic exposure, expression of iNOS begins only after 6 hours of continuous hypoxia while the constitutive isoforms eNOS and nNOS become activated almost immediately.[14,15] Therefore the duration of an adapting session of hypoxia should not exceed 6 hours.

In our experiments, adaptation to intermittent hypobaric hypoxia was used for stimulation of endogenous NO production.[16] Adaptation was performed in the altitude chamber at the simulated altitude corresponding to 4000 m above sea level. The duration of hypoxic exposure was gradually increased from 10 minutes to 5 hours. The complete course of adaptation consisted of 40 daily sessions. Nitric oxide production was evaluated by total plasma level of nitrite and nitrate. Adaptation induced a gradual increase in plasma nitrite plus nitrate, and by the end of the adaptation course this parameter was twice that of the control level.

Protective effect of 8-day adaptation to hypoxia was evident as prevention of NO overproduction in brain and a significant increase in survival of rats exposed to acute hypoxia at a simulated altitude of 11,000 m.[17] A course of the NOS inhibitor N^ω-nitro-L-arginine with the last dose administered 24 hours prior to the acute hypoxia exposure slightly reduced the resistance of rats to acute hypoxia and abolished the protective effect of adaptation. At the same time a single dose of the NO inhibitor L-NNA or the NO trap diethyldithiocarbamate administered immediately prior to the acute hypoxic exposure significantly increased the survival of rats. The experiment supported the role of the adaptive potentiation of NO production in protective effects of adaptation to hypoxia.

Obviously, an adapting organism, which increases the NO production, should possess some additional mechanisms of protection against NO overproduction. Indeed NO can be stabilized by binding to the complexes which form NO stores. NO stores can gradually release NO and therefore serve as an additional non-enzymic source of free NO. Formation of NO stores can be induced by any increase in NO level irrespective of the reason for this increase. For instance, an increase in NO level can result from both stimulation of endogenous NO synthesis and injection of exogenous NO donor. In basal conditions, NO stores are undetectable.[18]

S-nitrosothiols and dinitrosyl iron complexes (DNIC) are two major forms of NO storage and transportation in the organism. These two types of compounds can transform into each other depending on the intracellular level of iron, low molecular weight thiols and NO. S-nitrosothiols and DNICs exist in two forms, a high molecular weight, that is, a protein-bound form and a low molecular weight form containing low molecular weight thiol ligands such as cysteine or reduced glutathione. Protein-bound complexes are more stable than low molecular weight ones and they are regarded as intracellular NO stores.[19]

The NO stores in blood vessels have been detected using the electron paramagnetic resonance (EPR) assay,[20] histochemical staining for bivalent iron,[21] immunostaining for cysteine-NO residues[22] or photorelaxation response of isolated blood vessels.[23] The most common method for detecting NO stores in the vascular wall is the reaction of N-acetylcysteine or diethyldithiocarbamate (DETC) with DNIC and S-nitrosothiols, which results in the release of vasorelaxing products. Magnitude of DETC-induced vaso-relaxation reflects the volume of NO store.[24,25]

Adaptation to hypoxia resulted in formation of NO stores and a progressive increase in their size. The size of NO stores is significantly positively correlated with plasma levels of nitrite and nitrate.[16]

A possible mechanism of increased NO synthesis in adaptation to intermittent hypoxia involves alternations of hypoxia and reoxygenation. Hypoxia activates the so-called hypoxia-inducible factor-1 (HIF-1), a transcriptional regulator of certain genes including the NOS gene. HIF-1 is a heterodimer consisting of two components, HIF-1α and HIF-1β.[26] Expression of HIF-1α dose-dependently increases at O_2 tension below 20% whereas HIF-1β is expressed constitutively. HIF-1 decomposes after less than one minute of reoxygenation.[27] Such rapid dynamics of HIF-1 may provide adaptation to periodic hypoxia at the transcriptional level, with hypoxic stimulation not only of NO synthesis but also erythropoiesis, angiogenesis and glycolysis.[2] Due to activation of HIF-1, expression of the NOS gene progresses during adaptation to make adaptation more reliable and long term.

Intensity of NO production is predetermined at the genetic level. For example, in spontaneously hypertensive rats of SHR and SHRSP strains endothelial NO synthesis is decreased,[28] while in rats of the August strain NO synthesis is increased.[29] We studied the relationship between NO production and storage in rats of different genetic strains in the condition of increased NO synthesis.

Adaptation to hypoxia significantly increased total NO production as measured by urinary nitrate and nitrite, in both SHRSP and their genetic normotensive control Wistar-Kyoto (WKY). In addition, adaptation slowed down the development of hypertension in SHRSP and left blood pressure (BP) unchanged in WKY. Finally, adaptation induced formation of NO stores in the vascular wall of both rat strains. However, the volume of NO stores was twice as large in WKY than in SHRSP.[30]

A tentative relationship between the NO level, NO stores and BP in normotensive and spontaneously hypertensive rats can be represented as follows (Fig. 1).

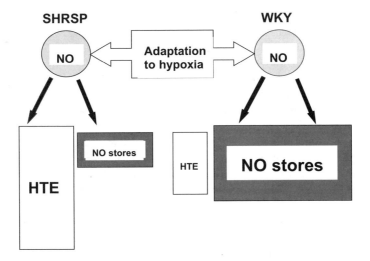

Figure 1. A tentative relationship between the NO level, NO stores and blood pressure in normotensive and spontaneously hypertensive rats in adaptation to hypoxia. SHRSP, stroke-prone spontaneously hypertensive rats; WKY, Wistar-Kyoto rats; HTE, hypotensive effect; NO, nitric oxide.

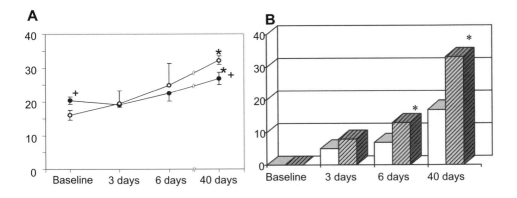

Figure 2. Dynamics of plasma nitrite/nitrate (A) and formation of NO stores (B) in Wistar and August rats during adaptation to hypoxia. A: abscissa, plasma nitrite/nitrate level, µM; ordinate, duration of adaptation. Open circles, Wistar; dark circles, August. *Significant difference from baseline, $p < 0.05$; +Significant difference between Wistar and August, $p < 0.05$. B: ordinate, DETC-induced vasorelaxation, % of contraction. Empty bars, Wistar; dashed bars, August. *Significant difference between Wistar and August, $p < 0.05$.

When adaptation to hypoxia stimulates NO synthesis, a part of the excessive NO binds to the NO store whereas the unbound part of NO exerts its biological effect, that is, induces vasodilation and a decrease in BP. When the efficiency of NO storage is high, the volume of NO stores is large while the hypotensive effect is small. This situation apparently takes place in normotensive rats. On the contrary, when the efficiency of NO storage is low, much NO remains unbound and exerts a more pronounced hypotensive effect. This is the case in spontaneously hypertensive rats.

As August rats display higher basal and stimulated production of NO than Wistar rats,[29] it was suggested that August rats might possess a more potent system of NO binding to stores for protection of the body against the detrimental effect of excessive NO.

Adaptation to hypoxia consistently increased the plasma level of nitrite and nitrate in both Wistar and August rats (Fig. 2). However, this increase proceeded slower in August rats, apparently due to the increased binding of NO to NO store. By the end of adaptation, the level of nitrite and nitrate was lower in August rats even though it had been higher than in Wistar rats at baseline. Indeed, the formation of NO stores proceeded more rapidly in August than in Wistar rats and by the end of adaptation, the volume of NO stores in August rats was almost twice as large as in Wistar rats.[31]

Therefore, volume of NO stores is larger in August rats which have a higher level of NO production than Wistar rats, and smaller in SHRSP which have a lower level of NO metabolites than their normotensive control WKY. Both plasma and urinary levels of NO metabolites are similar in Wistar and WKY (unpublished data). We suggest that the formation of NO stores is an adaptive mechanism for protection of the cardiovascular system against the detrimental effects of excessive NO synthesized during repeated hypoxic exposures. On the other hand, NO stores may serve as a non-enzyme source of free NO to compensate for decreased production of NO by endothelial cells.

2. ACKNOWLEDGEMENTS

The study was supported by NWO (grant 047.011.2001.010) and the Russian Foundation for Basic Research (grant 03-04-49065).

3. REFERENCES

1. Prabhakar N. R., Fields R. D., Baker T., and Fletcher E. C., Intermittent hypoxia: cell to system, *Am. J. Physiol.* **281**, L524-L528 (2001).
2. Neubauer J. A., Physiological and pathophysiological responses to intermittent hypoxia. *J. Appl. Physiol.* **90**, 1593-1599 (2001).
3. F.Z. Meerson, *Essentials of Adaptive Medicine: Protective Effects of Adaptation* (Hypoxia Medical LTD, Moscow, 1994)
4. Hong Y., Suzuki S., Yatoh S., Mitzutani M., Nakajima T., Bannai S., Sato H., Soma M., Okuda Y., and Yamada N., Effect of hypoxia on nitric oxide production and its synthase gene expression in rat smooth muscle cells, *Biochem. Biophys. Res. Commun.* **268**, 329-332 (2000).
5. Bredt D. S., Endogenous nitric oxide synthesis: biological functions and pathophysiology, *Free Rad. Res.* **31**, 577-596 (1999).
6. Toporsian M., Govindaraju K., Nagi M., Eidelman D., Thibault G., and Ward M. E., Downregulation of endothelial nitric oxide synthase in rat aorta after prolonged hypoxia in vivo, *Circ. Res.* **86**, 671-675 (2000).

7. Pearl J. M., Nelson D. P., Wellmann S. A., Raake J. L., Wagner C. J., McNamara J. L., and Duffy J. Y., Acute hypoxia and reoxygenation impairs exhaled nitric oxide release and pulmonary mechanics, *J. Cardiovasc. Surg.* **119**, 931-938 (2000).

8. Abu-Soud H. M., Ichimori K., Presta A., and Stuehr D. J., Electron transfer, oxygen binding and nitric oxide feedback inhibition in endothelial nitric-oxide synthase, *Biol. Chem.* **275**, 17349-17357 (2000).

9. Abu-Soud H. M., Rousseau D. L., and Stuehr D. J., Nitric oxide binding to the heme of neuronal nitric-oxide synthase links its activity to change in oxygen tension, *J. Biol. Chem.* **271**, 32515-32518 (1996).

10. Rengasamy A., and Johns R. A., Determination of Km for oxygen of nitric oxide synthase isoforms, *J. Pharmacol. Exp. Ther.* **276**, 30-33 (1996).

11. Su Y., and Block E. R., Role of calpain in hypoxic inhibition of nitric oxide synthase activity in pulmonary endothelial cells, *Am. J. Physiol.* **278**, L1204-L1212 (2000).

12. Shi Y., Baker J. E , Zhang C., Tweddell J. S., Su J., and Pritchard K. A:, Chronic hypoxia increases endothelial nitric oxide synthase generation of nitric oxide by increasing heat shock protein 90 association and serin phosphorylation, *Circ. Res.* **91**, 300-306 (2002).

13. Hampl V., and Herget J., Role of nitric oxide in the pathogenesis of chronic pulmonary hypertension, *Physiol. Rev.* **80**, 1337-1372 (2000).

14. Gess B., Schricker K., Pfeifer M., and Kurtz A., Acute hypoxia upregulates NOS gene expression in rats, *Am. J. Physiol.* **273**, R905-R910 (1997).

15. Ferreiro C. R., Chagas A. C., Carvalho M. H., Dantas A. P., Jatene M. B., Bento de Souza L.C., and Lemos da Luz P., Influence of hypoxia on nitric oxide synthase activity and gene expression in children with congenital heart disease: a novel pathophysiological adaptive mechanism, *Circulation* **103**, 2272-2276 (2001).

16. Manukhina E. B., Malyshev I. Yu., Smirin B. V., Mashina S. Yu., Saltykova V. A., and Vanin A. F., Production and storage of nitric oxide in adaptation to hypoxia, *Nitric Oxide* **3**, 393-401 (1999).

17. Malyshev I. Yu., Zenina T. A., Golubeva L. Yu., Saltykova V. A., Manukhina E. B., Mikoyan V. D., Kubrina L. N., and Vanin, A. F., NO-dependent mechanisms of adaptation to hypoxia, *Nitric Oxide* **3**, 105-113 (1999).

18. Manukhina E. B., Smirin B. V., Malyshev I. Yu., Stoclet J.-C., Muller B., Solodkov A. P., Shebeko V. I., and Vanin A. F., Nitric oxide storage in the cardiovascular system, *Biology Bulletin*, **29**, 477-486 (2002).

19. Vanin A.F., Dinitrosyl iron complexes and S-nitrosothiols: two possible forms of nitric oxide stabilization and transport in biological systems, *Biochemistry (Moscow)* **63**, 782-796 (1998).

20. A.F. Vanin, and A.L Kleschyov, in: *Nitric Oxide in Transplant Rejection and Anti-Tumor Defense*, edited by S. J. Lukiewicz, and J. L. Zweier (Kluwer Academic Publ., Norwell, MA, 1998) pp. 49-82.

21. Flitney F. W., Megson I. L., Flitney D. E., and Butler A. R., Iron-sulphur cluster nitrosyls, a novel class of nitric oxide generator: mechanism of vasodilator action on rat isolated tail artery, *Br. J. Pharmacol.* **107**, 842-848 (1992).

22. Alencar J. L., Lobysheva I., Geffard M., Sarr M., Schott C., Schini-Kerth V. B., Nepveu O., Stoclet J.-C., and Muller B., Role of S-nitrosation of cysteine residues in long-lasting inhibitory effect of nitric oxide on arterial tone, *Mol. Pharmacol.* **63**, 1148-1158 (2003).

23. Megson I. L., Holme S. A., and Magid K. S., Selective modifiers of glutathione biosynthesis and 'repriming' of vascular smooth muscle photorelaxation, *Br. J. Pharmacol.* **130**, 1575-1580 (2000).

24. Muller B., Kleschyov A. L., and Stoclet J.-C., Evidence for N-acetylcysteine-sensitive nitric oxide storage as dinitrosyl iron complexes in lipopolysaccharide-treated rat aorta, *Brit. J. Pharmacol.* **119**, 1281-1285 (1996).

25. Smirin B. V., Vanin A. F., Malyshev I. Yu., Pokidyshev D. A., and Manukhina E. B., Nitric oxide storage in blood vessels in vivo, *Biull. Eksp. Biol. Med.* **127**, 629-632 (1999) (Russ).

26. Semenza G. L., HIF-1 and mechanisms of hypoxia sensing, *Curr. Opin. Cell Biol.* **13**, 167-171 (2001).

27. Yu A. Y., Frid M. G., Shimoda L. A., Wiener C. M., Stenmark K., and Semenza G. L., Temporal, spatial, and oxygen-regulated expression of hypoxia-inducible factor-1 in the lung, *Am. J. Physiol.* **275**, L818-L826 (1998).

28. Wu C.-C., Yen M.-H. Nitric oxide synthase in spontaneously hypertensive rats, *Biomed. Sci.* **4**, 249-255 (1997).

29. Mikoyan V. D., Kubrina L. N., Manukhina E. B., Malysheva E. V., Malyshev I. Yu., and Vanin A. F., Differences in stimulation of NO synthesis by heat shock in rats of genetically different populations, *Biull. Eksp. Biol. Med.* **121**, 634-637 (1996) (Russ).

30. Mashina S. Yu., Smirin B. V., Malyshev I. Yu., Lyamina N. P., Senchikhin V. N., Pokidyshev D. A., and Manukhina E. B., Correction of NO-dependent cardiovascular disorders by adaptation to hypoxia, *Ross. Fiziol. Zh. Im. I. M. Sechenova* **87**, 110-117 (2001) (Russ).

31. Pshennikova M. G., Smirin B. V., Bondarenko O. N., Malyshev I. Yu., and Manukhina E. B., Nitric oxide storage in rats of different strains and its role in the antistress effect of adaptation to hypoxia, *Ross. Fiziol. Zh. Im. I. M. Sechenova* **86**, 174-181 (2000) (Russ).

7.

MASS LAW PREDICTS HYPERBOLIC HYPOXIC VENTILATORY RESPONSE

John W. Severinghaus[*]

1. INTRODUCTION

Opening the Oxford Arterial Chemoreceptor symposium in 1966, R. W. Torrance, paraphrased Churchill's radar praise, if memory serves, "Never have so many labored so long over so little to so great an end".

In this hypothetical and speculative analysis of hypoxia's effects on the carotid body and ventilation, I have relied primarily on the extensive research of S. Lahiri, D. F. Wilson and their associates at the University of Pennsylvania.[1-6] In working with polarographic oxygen electrodes in which the electron-oxygen reaction is so clearly regulated by the law of mass action, I came to recognize the similarity to carotid body function. Several years ago I asked Britton Chance whether the mass action law applies to the electron-oxygen reaction at cytochrome oxidase. He immediately replied, "It has to". These hypotheses derive from his caveat.

2. CAROTID BODY PHYSIOLOGY AND BIOCHEMISTRY

HVRp is the relationship of $\dot{V}_E$(expired ventilatory rate, liters per min) to acute stable isocapnic hypoxia plotted as PaO_2 (arterial oxygen tension, torr). HVRp is generally reported as fitting a hyperbola with an estimated ventilatory asymptote at about 32 torr PaO_2 at which $\dot{V}_E \Rightarrow \infty$ (Fig. 1). The commonly used equation describing it is:

$$\dot{V}_E = A/(PaO_2 - 32) + B \qquad \text{(Eq. 1)}$$

where A is the CB sensitivity (about 200 liters/min at $PaO_2=33$ torr) and B is the non-CB component of ventilatory drive at rest, typically 4 L/min.[7]

Carotid body metabolic rate is very high ($\dot{V}O_2 \approx 9$ ml/dl/min), about 2.5 times $\dot{V}O_2$ of cerebral cortex). Its blood flow through enlarged capillaries or sinusoids is about 20 ml/g/min, much higher than expected for that metabolic rate, such that its capillary and venous blood PO_2 had been expected to closely track arterial PO_2. However Lahiri et al in 1993 reported direct optical measurement of CB microvascular (CBM) PO_2 (by fluorescence quenching). Their arterial to CBM SO_2 (oxygen saturation) difference,

[*] John W. Severinghaus, UCSF, San Francisco CA 94143

computed (by me) from PO_2, was 8% in normoxia, 27% at $PaO_2=40$ torr and 31% at $PaO_2=30$ torr. This implies either a 70% fall of blood flow or greatly increased O_2 consumption in hypoxia, which I have not attempted to include in these attempts to predict the effect of the mass law. For simplicity, I here assume negligible A-V SO_2 differences, assigning all the diffusion gradient to tissue. The conclusions are about the same if some gradient is in blood.

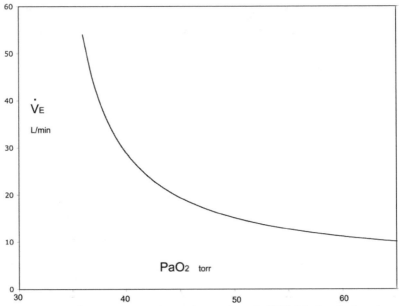

Figure 1. The human ventilatory response to isocapnic acute hypoxia (HVRp) is described as a hyperbola with an estimated asymptote of infinite ventilation at $PaO_2 = 32$ torr.

The high metabolic rate creates a diffusion gradient for oxygen between arterial blood and the glomus cell mitochondria. A variety of values have been reported for tissue or cytochrome (PcO_2) from a calculated value of less than 10 torr[8] and oxygen micro-electrode values of 0-15 torr[9] up to 70 torr.[10] I chose a gradient of 32 torr (arterial to oxidase) both as a mean of the widely discrepant values. I thus assume that that $PcO_2 = 0$ at the observed 32 torr isocapnic HVRp asymptote.

An important but questionable assumption is that tissue metabolic rate is independent of its PO_2 above the critical threshold of oxygen delivery. This critical threshold in dilute vigorously stirred isolated mitochondria was shown to be <1 torr PO_2.[11] O_2 consumption **begins** to fall at the critical threshold. The critical threshold PaO_2 for the carotid body is assumed here to be 32 torr (Fig. 2), below the range of PO_2 used in measurement of HVRp. That is to say that CB hypoxic stimulation does not imply decreased CB oxygen consumption.

3. MASS LAW APPLIED TO CYTOCHROME OXYGEN REDUCTION

I assume that the CB tissue oxygen diffusion gradient is independent of PaO_2 over the range at which we test ventilatory response. I assume that mitochondrial redox becomes fully reduced at the critical threshold at which $\dot{V}O_2$ begins to fall. PcO_2 (PO_2 at the copper atom in cytochrome oxidase) should then be about 32 torr less than PaO_2 (Fig. 3) and approach zero as PaO_2 approaches 32 torr.

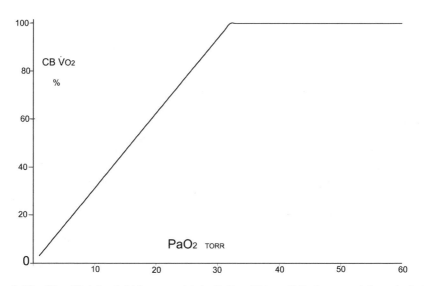

Figure 2. The CB critical threshold is assumed to be $PaO_2 = 32$ torr. Critical oxygen delivery is the PO_2 at which oxygen consumption begins to fall. Above this level $\dot{V}O_2$ is independent of PO_2.

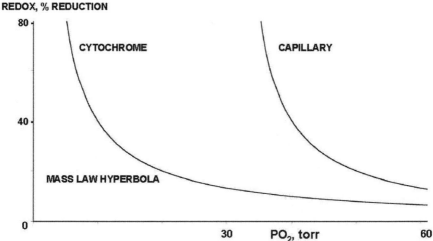

Figure 3. Predicted cytochrome redox state rises hyperbolically in isocapnic hypoxia.

The mass law must apply to the electron-oxygen reaction at the aa_3 copper atom in mitochondria. As O_2 supply falls, I assume by analogy that redox reduction increases the (negative) electron pressure (voltage, Ec), maintaining O_2 consumption by the negative feedback system matching glycolysis to energy need.[12] The potential drop associated with each citric acid cycle step is unchanged but the absolute potential shifts negatively equal to the Ec shift. The concentrations of pyruvate and lactate are similarly elevated. This response is so effective that moderate hypoxia has no detectable effect on $\dot{V}O_2$. With constant metabolic rate, Eq. 2 describes a hyperbola, with a units-adjusting constant k.

$$\dot{V}O_2 = k(Ec \times PcO_2) \qquad\qquad (Eq.\ 2)$$

The primary oxygen sensor of the CB is cytochrome aa$_3$ in type I cells.[5, 13] $\dot{V}$E is assumed to be proportional to redox reduction during isocapnic hypoxia.

4. ANALOGY OF CAROTID BODY TO POLAROGRAPHIC ELECTRODE

I think of CB physiology and biochemistry in the light of the physical chemistry of the reduction of oxygen by electrons at a negatively charged platinum metal surface. Leland Clark invented the membrane covered oxygen polarographic electrode 50 years ago this October and publicly disclosed it at the FASEB meeting in April, 1956. His invention is incorporated in most blood gas analytic apparatus. Clark's electrode consumes oxygen at a rate precisely proportional to the PO$_2$ at the outside surface of its semi-permeable membrane, usually polypropylene. The current resulting from the electrons that reduce oxygen is therefore an exact measure of membrane surface PO$_2$. A polarogram is a plot of the current–voltage relationship of a platinum cathode (Fig. 4).

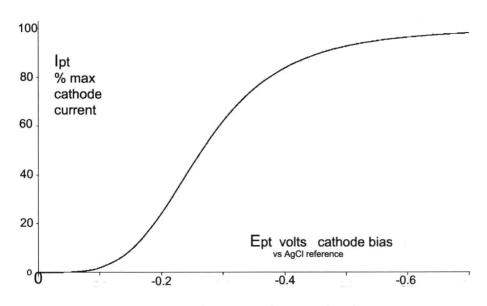

Figure 4. Schematic polarogram of an oxygen electrode.

As the bias voltage increases (negatively) from zero, increasing amounts of the dissolved oxygen are reduced resulting in an S shaped plot. For oxygen measurement, the cathode is negatively charged to about –0.65v, a level at which the oxygen current is nearly maximum. Increasing the bias above –0.7v begins to cause other reactions, first with hydrogen ions. At this 0.65v plateau, the PO$_2$ at the cathode surface is nearly zero. This is because all O$_2$ molecules diffusing within the capture radius of the negatively charged cathode are captured and reduced, leaving none to escape which, by definition, means PO_2 = zero torr. The O_2 diffusion gradient from the outer membrane surface to the cathode surface and the resulting electrode current reach a maximum. The absolute potential of an oxygen cathode vs a hydrogen electrode (instead of AgCl) is about –0.35v.

At a constant current, the mass law applied to a polarographic cathode should yield a hyperbolic relationship of the potential E to the available oxygen, aO_2 , an unknown but linear function of the PO_2 at the outside of the semi-permeable electrode membrane. This is schematically shown in Fig. 5, choosing a current that requires maximum cathode bias at a membrane surface PO_2 of 32 torr as in the putative CB scheme.

Another useful analogy between polarography and CB response relates to the "stirring" effect. With Clark's large cathode (2mm diam.), O_2 consumption depleted the dissolved O_2 in unstirred liquid adjacent to the membrane, making the reading far less than with gas of the same PO_2 outside the membrane. To accurately measure blood PO_2, rapid stirring of the sample and calibration with equilibrated blood were required. Norman Staub (U. Pennsylvania) and Richard Shepard (John Hopkins U.) showed that this stirring error could almost be eliminated by reducing cathode diameter to about 10 microns. This is relevant to the high $\dot{V} O_2$ of the CB. Because blood flow velocity is very high, the plasma layer on CB endothelial surfaces is rapidly "stirred", making the CB into a PO_2 sensor, almost independent of hematocrit and O_2 content of blood. [14]

$$I = k[E \times f(PO2-32)] \tag{Eq.3}$$

5. LINKING HVRs TO HVRp

HVRs refers to the slope of the relationship of ventilation to arterial oxyhemoglobin desaturation, $\Delta\dot{V}E /\Delta SaO_2$. The normal value is about -1 l/min/%. This relative linearity does not imply that the CB measures oxygen saturation. I here calculate a hypothetic HVRs by combining two relationships, that of the normal human oxygen dissociation curve (ODC) used to compute PcO_2 from SaO_2 (Fig. 6) and that of aa_3 redox (as a proxy for $\dot{V}E$) vs PcO_2 (Fig. 3).

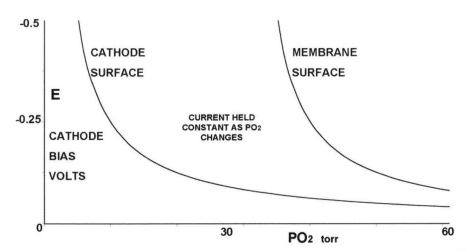

Figure 5. Hypothetical polarographic oxygen electrode cathode EMF vs PO_2 at a current set to reach a bias maximum at about 32 torr analogous to the hypoxic asymptote of HPRp.

In Fig. 6, the following equation of the ODC relates SaO_2 to PaO_2:[15]

$$S = 100/\{[23,400/(PaO_2^3 + 150PaO_2)] + 1\} \tag{Eq. 4}$$

From Eq. 4, PcO_2, defined as $(PaO_2\text{-}32)$ may be calculated from SaO_2 (Fig. 6).

From the mass law hyperbola (Fig. 3), $\dot{V}E$ is assumed numerically equal to the redox reduction % and computed from PcO_2:

$$\dot{V}E = 200/PcO_2 + 4 \tag{Eq. 5}$$

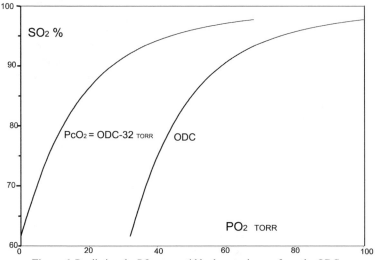

Figure 6. Predicting the PO_2 at carotid body cytochrome from the ODC.

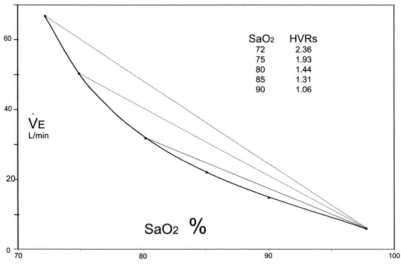

Figure 7. HVRs predicted from the ODC combined with the Fig. 3 hyperbola.

To predict HVRs from HVRp, I eliminate PcO_2 by combining these equations (Fig. 7). This computed HVRs is not linear, the slope increasing from 1.06 at 90% to 2.36 at 72% SaO_2. The relationship of $\dot{V}_E$ to SaO_2 is reported by many investigators has been more linear than in Fig. 7, although a trend of increasing slope was seen in a test of linearity in subjects at 3180 m altitude.[16] This difference between predicted and actual HVRs can be examined in the opposite way by predicting the shape of HVRp if HVRs is assumed to be linear down to 60% SaO_2 (Fig. 8). This cannot be forced to fit a hyperbola. I conclude that the HVRs slope is not linear but will be found to increase, especially in the range below 80% SaO_2. The experimental evidence for relative linearity of HVRs also implies that HVRp will not accurately fit a hyperbola when low saturations are carefully examined.

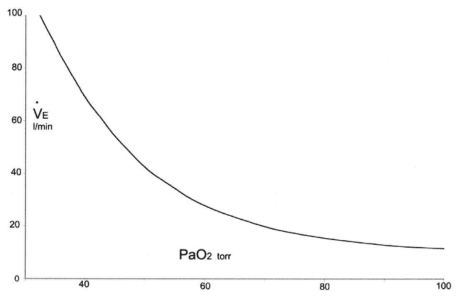

Fig. 8. Hypothetic HVRp if HVRs is linear. This shape fails to represent observed responses at low PO_2.

6. CONCLUSION

The hyperbolic hypoxic ventilatory response vs PaO2, HVRp, is interpreted as relecting a mass hyperbolic relationship of cytochrome PcO2 to cytochrome potential Ec, offset 32 torr by the constant diffusion gradient between arterial blood and cytochrome in CB at its constant metabolic rate $\dot{V}O_2$. Ec is taken to be a linear function of redox reduction and CB ventilatory drive. As Ec rises in hypoxia, the absolute potentials of each step in the citric acid cycle rises equally while the potential drop across each step remains constant because flux rate remains constant. A hypothetic HVRs ($\dot{V}_E$ vs SaO_2) response curve computed from these assumptions is strikingly non linear. A hypothetic HVRp calculated from an assumed linear HVRs cannot be fit to the observed hyperbolic increase of ventilation in response to isocapnic hypoxia at PO_2 less than 40 torr. The incompatibility of these results suggest that in future studies HVRs will not be found to be linear, especially below 80% SaO_2 and HVRp will fail to be accurately hyperbolic.

7. REFERENCES

1. D. F. Wilson, W. L. Rumsey, T. J. Green, J. M. Vanderkooi. The oxygen dependence of mitochondrial oxidative phosphorylation measured by a new optical method for measuring oxygen concentration. *J Biol Chem* **263**, 2712-2718 (1988).
2. D. F. Wilson, A. Mokashi, S. Lahiri, S. A. Vinogradov. Tissue PO₂ and mitochondrial enzymes. Cytochrome C oxidase as O₂ sensor. *Adv Exp Med Biol* **475**, 259-264 (2000).
3. D. F. Wilson, K. M. Laughlin, C. Rozanov, et al. Tissue oxygen sensing and the carotid body. *Adv Exp Med Biol* **454**, 447-454 (1998).
4. S. Lahiri, D. F. Wilson, R. Iturriaga, W. L. Rumsey. Microvascular PO2 regulation and chemoreception in the cat carotid body. *Adv Exp Med Biol* **345**, 129-135 (1994).
5. S. Lahiri, D. K. Chugh, A. Mokashi, S. Vinogradov, S. Osanai, D. F. Wilson. Cytochrome oxidase is the primary oxygen sensor in the cat carotid body. *Adv Exp Med Biol* **388**, 213-217 (1996).
6. S. Lahiri, C. Rozanov, A. Roy, B. Storey, D. G. Buerk. Regulation of oxygen sensing in peripheral arterial chemoreceptors. *Int J Biochem Cell Biol* **33**, 755-774 (2001).
7. J. V. Weil. Hypoxic drive: assessment and interpretation. *Proc R Soc Med* 68, 239 (1975).
8. S. Lahiri, R. E. n. Forster. CO₂/H(+) sensing: peripheral and central chemoreception. *Int J Biochem Cell Biol* **35**, 1413-1435 (2003).
9. H. Acker. PO₂ chemoreception in arterial chemoreceptors. *Annu Rev Physiol* **51**, 835-844 (1989).
10. W. J. Whalen, J. Savoca, P. Nair. Oxygen tension measurement in the carotid body of the cat. *Am J Physiol* **225**, 986-999 (1973).
11. D. F. Wilson, C. S. Owen, M. Erecinska. Quantitative dependence of mitochondrial oxidative phosphorylation on oxygen concentration: a mathematical model. *Arch Biochem Biophys* **195**, 494-504 (1979).
12. M. Erecinska, D. F. Wilson, K. Nishiki. Homeostatic regulation of cellular energy metabolism: experimental characterization in vivo and fit to a model. *Am J Physiol* **234**, C82-89 (1978).
13. D. F. Wilson, A. Mokashi, D. Chugh, S. Vinogradov, S. Osanai, S. Lahiri. The primary oxygen sensor of the cat carotid body is cytochrome a3 of the mitochondrial respiratory chain. *FEBS Lett* **351**, 370-374 (1994).
14. S. Lahiri, W. L. Rumsey, D. F. Wilson, R. Iturriaga. Contribution of in vivo microvascular PO₂ in the cat carotid body chemotransduction. *J Appl Physiol* **75**, 1035-1043 (1993).
15. J. W. Severinghaus. Simple, accurate equations for human blood O₂ dissociation computations. *J Appl Physiol* **46**, 599-602 (1979).
16. M. Sato, J. W. Severinghaus, F. L. Powell, F. Xu, M. J. J. Spellman. Augmented hypoxic ventilatory response in man at altitude. *J Appl Physiol* **73**, 101-107 (1992)

8

COST EFFECTIVE METAL AFFINITY CHROMATOGRAPHY FOR PROTEIN C SPECIFIC, MINI-ANTIBODY PURIFICATION

Samin Rezania, Doh G. Ahn, and Kyung A. Kang[*]

1. INTRODUCTION

Protein C (PC) is a vitamin K dependent (VKD) anticoagulant, antithrombotic, and anti-inflammatory protein in blood plasma[1]. PC deficiency causes abnormal blood clot formation that hinders oxygen transport to vitals organs. Therefore, inexpensive PC supply is important for patients with various thromboembolic disorders, including the PC deficient patients. All VKD proteins are synthesized in the liver. Molecular weights and structures of some VKD proteins are very similar to PC but most of them are clotting factors[1]. Therefore, only highly specific purification methods such as immuno-affinity chromatography can separate PC from these homologues. Since monoclonal antibodies (Mabs) used in affinity chromatography are expensive to produce, recombinant *E. coli* strains producing single chain variable fragments (ScFv, mini-Mabs) against PC has been developed[2]. Currently, the mini-Mab production level using the protocol developed by our research group is approximately 450 mg/L per day.

Because the purification yield for our mini-Mab using protein A was relatively low[3] and protein A is rather costly, immobilized metal affinity chromatography (IMAC) was investigated as an alternative to protein A chromatography. Histidine in the protein is a common target amino acid for IMAC[4, 5].

This paper reports the effect of pH on the adsorption of the mini-Mabs against PC, for the IMAC purification using Cu^{+2} and IDA chelator combination. Utilizing the purified mini-Mab, PC purification performances of four commercially available matrices were investigated. After the selection of the best performing matrix, purification of PC from the mixture of PC and four PC homologues, factors II, VII, IX, and X, was also performed.

---

[*] Samin Rezania, Doh G. Ahn and Kyung A. Kang, Department of Chemical Engineering, University of Louisville, Louisville, KY 40292.

2. MATERIALS AND METHODS

2.1. Mini-Antibody Purification Using IMAC

Materials and methods for the *E. coli* growth, the mini-Mab production, and its purification using Protein A (Sigma, St. Louis, MO) are as described by Korah, et al[3]. IMAC purification of the mini-Mab was done using HiTrap™ Chelating Sepharose (Amersham Biosciences; Piscataway, NJ), following the manufacturer's instructions. For this feasibility study, pure PC mini-Mab was loaded to the column (binding capacity, 12 mg of histidine tagged proteins) and the non-adsorbed material was washed with the equilibrium/washing (E/W) buffer (0.02 M phosphate buffer). Five column volumes (CV) of 0.5 M imidazole in 0.02 M phosphate buffer was used to elute the bound protein. Five tenths (0.5) M NaCl was added to all buffers to minimize non-specific electrostatic binding. In the experiment to study the efficacy of a lower concentration of imidazole for removing the weakly retained impurities, 5 mM imidazole was employed in E/W buffer. After the purification process, Cu^{+2} was stripped off the resin using 50 mM EDTA (5 CV) and then the resin was washed with DI water (5 CV). The mini-Mab quantification was performed by the ELISA, using anti-c-myc antibody, 9E10[3] (Santa Cruz Biotechnology; Santa Cruz, CA).

2.2. PC Purification

The four affinity chromatography resins tested for the PC purification performance are CNBr-activated Sepharose™ 4B (CNBr; Amersham Biosciences), NHS-activated Sepharose™ 4 Fast Flow (NHS; Amersham Biosciences), Actigel ALD™ (Actigel; Sterogene; Carlsbad, CA), and Epoxy Activated Ultraflow 4™ (Epoxy; Sterogene). PC was provided by the American Red Cross (Rockville, MD) and factors II, VII, IX, and X were purchased from Innovative Research, Inc. (Southfield, MI). Rabbit polyclonal factor II antibody was from Biomeda Corporation (Foster city, CA) and mouse monoclonal factor II antibody, from Enzyme Research Laboratories (South Bend, IN). Rabbit polyclonal factor VII antibody was obtained from Novus Biologicals, Inc. (Littleton, CO), and goat polyclonal factor X antibody, from US biological (Swampscott, MA). Rabbit polyclonal antibodies for PC and factor IX, mouse monoclonal antibodies against PC, and factors VII, IX, and X were purchased from Sigma (St. Louis, Mo).

For PC purification, purified PC mini-Mab was immobilized on the selected matrix. The PC purification was performed by the method described by Kang, et al[6]. The amounts of the mini-Mab and PC were quantified by ELISA.

3. RESULTS AND DISCUSSION

3.1. Mini-Antibody Purification Using IMAC

IMAC has two main components: a chelator and an immobilized metal ion. The chelator is linked to a matrix via a spacer arm by covalent bonds and the metal ion is immobilized to the chelator by coordinate bonds[7]. In IMAC, the binding force between the metal ion and the chelator, and the force between the protein to be purified and the metal ions are strongly dependent on pH[8]. With IDA chelator, the force between Cu^{+2}

and the histidine residue is usually the strongest at pH 7.4. In the PC mini-Mab purification study performed by Korah, et al.[3], the leaching of Cu^{+2} at pH 7.4 was a serious problem, indicating that Cu^{+2} binds to the mini-Mab much stronger than to the chelator. Since metal ion leaching may yield less adsorption capacity for the product, therefore, the focus of the study here was to determine the appropriate pH for the mini-Mab adsorption to the matrix with the minimal metal ion leaching. One (1) ml of HiTrap Chelating resin is reacted with Cu^{+2} (0.1 M) and then 1 mg of the pure mini- Mab in E/W buffer was applied to the resin at pH 5.4, 6.4, 7.4 or 8.4 (Figure 1a). At pH 7.4, as in the previous study[3], extensive metal ion leaching was observed, but still a 40% purification yield was obtained. At pH 6.4, much less metal ion leaching was observed and the yield increased to 55%, higher than that by protein A (33%). For the other pHs (5.4 and 8.4), yields were very small (2%), with extensive metal ion leaching. The same study was performed at pH 6.0, but the purification yield was similar to that at pH 5.4 (data not shown).

At a physiological pH, histidine binds to the metal ion by sharing electron density with the electron-deficient orbital of the metal ion. Elution occurs when the nitrogen group of the histidine is protonated. This can be induced by a competitive electron donor agent that is similar to the histidine residues. Imidazole is one of the commonly used competitive agents[5]. The above study was performed using the pure mini-Mab as a preliminary study, but, eventually, the mini-Mab needs to be purified from the supernatant of the reactor broth, which requires the non-specific binding of other bio-molecules to Cu^{+2}. Imidazole at 0.5 M is used to elute the bound mini-Mab in our protocol. Researchers have also reported that adding a small amount of imidazole in the E/W buffer can remove weakly retained impurities and metal ions and, therefore, increasing the purity of eluted protein[9].

Five (5) mM imidazole, 1/100 of the imidazole amount used in the elution buffer, was added to the E/W buffer and its effect on the mini-Mab purification was studied (Figure 1b). Imidazole in the E/W buffer caused a significant amount of the mini-Mab loss during the adsorption and washing steps, indicating that 5 mM is too high for this purpose. At pH 6.4, a yield of only 3% was obtained. Nevertheless, imidazole in E/W buffer prevented copper ion from leaching. Currently, more studies are being performed to minimize metal leaching and to maximize the mini-Mab yield with little impurity.

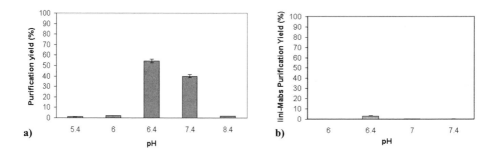

Figure 1. a) Effect of pH and b) 5mM imidazole in the E/W buffer on the mini-Mab purification yield.

3.2. PC Purification Using the Mini-Mab

Four commercially available affinity chromatography gel matrices, CNBr, NHS, Actigel, and Epoxy were tested for their PC purification efficiency. The CNBr reacts with amine residues of proteins. Actigel uses the monoaldehyde to bind with the amine group via a 5-atom spacer. NHS has 6-amino-hexanoic acid forming a 10-atom spacer. The terminal carboxyl group reacts with the amine group. Epoxy binds to the amine residue via the 12-atom spacer.

Approximately 2 mg of purified PC mini-Mab was reacted with 1 ml of each of the four selected matrices and the immobilization efficiency was quantified by ELISA (Table 1). NHS showed the highest (95%) and Epoxy had the lowest immobilization efficiency (70%).

The mini-Mab immobilized gel matrices were packed in a column of 0.7 cm diameter and their PC purification performance was tested using 20 μg of pure PC as a preliminary study. PC in the various chromatography steps was quantified by ELISA (Table 1). CNBr showed the lowest PC purification efficiency (6%) and NHS showed the best performance (54%), possibly indicating that the longer spacer of the NHS matrix better retained the mini-Mab 3-D configuration. Although Epoxy has the longest spacer (12-atom), only 30% of the purification yield was obtained, showing that there is an optimal spacer length for the best purification performance.

As a next step, PC purification performance of the mini-Mab from PC homologues was studied. The cross reactivity test between the mini-Mab and other PC homologues was performed by ELISA and factor IX showed some cross reactivity with the mini-Mab but other homologues showed very little any cross reactivity (data not shown). The source material was composed of PC and the homologous proteins at their ratio in human plasma (20 μg, PC; 450 μg, factor II; 2.5 μg, factor VII, 20 μg, factor IX; and 32 μg, factor X). The chromatography purification was performed using the NHS matrix immobilized with 2 or 4 mg of the mini-Mab. PC and PC homologues at various purification steps were quantified by ELISA (Figure 3).

Table 1. Mini-Mab immobilization efficiency and PC purification performance of the four matrices.

	CNBr	Actigel	NHS	Epoxy
Length of the spacer arm (atom)	1	5	10	12
Mini-Mab Immobilization efficiency (%)	85	85 ± 2.6	95 ± 2.3	70
Purification yield (%)	6±1	48 ± 18.2	54 ± 11	30 ± 7

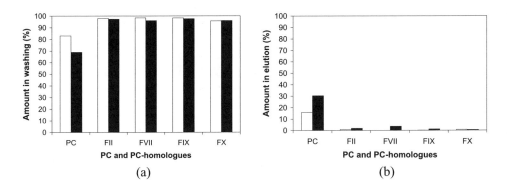

Figure 3. PC purification efficiency of the NHS matrix. The source material is the mixture of PC and four PC homologues: The amount of the factors in (a) the washing and (b) the elution steps, when 2 mg (□) or 4 mg (■) of mini-Mab was immobilized in 1 ml of the matrix.

The NHS matrix immobilized with 2 mg of the mini-Mab showed that approximately 80% of PC was washed away, indicating that large amounts of the PC were not adsorbed [Figure 3 (a)]. However, more than 95% of PC homologues were washed away. For the elution, 16% of the PC purification yield was obtained [Figure 3 (b)], which shows much less yield than that when the pure PC was used as the source material. For the elution in the same matrix, 4% of PC homologues were included, as expected.

For the NHS matrix with 4 mg of mini-Mab, approximately 70% of PC was washed away and these data show 10% less washing than that with 2 mg of mini-Mab. More than 95% of PC homologues were washed away [Figure 3 (a)] as in the 2 mg case. During elution, 30% of PC purification was obtained [Figure 3 (b)].

The results indicate that the larger amount of the immobilized mini-Mab in the matrix provides a higher PC yield.

4. CONCLUSIONS

Purified PC can be used for the treatment of various thrombotic complications, especially for PC deficiency. Inexpensive mini-Mab against PC was developed to be produced in *E. coli*. IMAC was tested for the mini-Mab purification as a less expensive alternative to protein A. Since the adsorption of the mini-Mab to the metal is strongly dependent on pH, the effect of pH on mini-Mab purification was investigated. At pH 6.4, the yield was the best, showing the yield of 55%, which is comparable to or better than protein A chromatography. The effect of 5 mM imidazole in the E/W buffer was studied to decrease non specific binding during the adsorption step. At pH 6.4, a yield of only 3% was obtained, possibly too high imidazole concentration. However, it prevented copper ion from leaching. More study is in progress to optimize the imidazole concentration.

Four commercially available immuno affinity matrices were tested to purify PC using the purified PC mini-Mab to select the best performing matrix. When pure PC was used as a source material, NHS matrix showed the best PC purification efficiency (54%). Using the mini-Mab immobilized NHS matrix, PC purification from PC homologues was studied. Sixteen (16) and 30% of PC purification yields were obtained by the NHS matrices immobilized with 2 and 4 mg of the mini-Mab, respectively. Less than 5% of homologues in the source material were present in the eluate. Currently, the NHS matrices immobilized with the mini-Mab greater than 4 mg/ml-matrixes are tested for the PC purification performance from the PC homologues. Purification of PC from blood Cohn Fraction IV-1, which is an inexpensive PC source, is also being investigated.

5. ACKNOWLEDGEMENT

Authors thank the American Red Cross for their protein C supply and Dr. Michael Sierks in the Arizona State University for providing *E. coli* colonies producing PC mini-Mab.

6. REFERENCES

1. D. Josic, L. Hoffer, and A. Buchacher , Preparation of vitamin K-dependent proteins, such as clotting factors II, VII, IX and X and clotting inhibitor Protein C, *J. Chromatogr. B* **790**, 183-197 (2003).
2. H. Wu, G. N. Goud, and M. R. Sierks, Artificial Antibodies for Affinity Chromatography of Homologous Proteins: application to blood proteins, *Biotechnol. Progr.* **14**, 496-499 (1998).
3. L. K. Korah and K. A. Kang, in: *Advances in Experimental Medicine and Biology*, edited by Thorniley, Harrison, and James (Kluwer Academic/ Plenum Publishers, 2003), pp. 171-176.
4. J. Porath, Immobilized Metal-Ion Affinity Chromatography, *Protein Express. Purif.* **3**, 263-281 (1992).
5. J. Porath, J. Carlsson, L. Olsson, and G. Belfrage, Metal Chelate Affinity Chromatography, A new Approach to Protein Fractionation, *Nature (London)* **258**, 598-599 (1975).
6. K. A. Kang, D. Ryu, W. M. Drohan and C. L. Orthner, Effect of matrices on affinity purification of protein C. *Biotechnol Bioeng.* **39**, 1086-1096 (1992).
7. E. S. Hemdan, and J. Porath, Development of Immobilized Metal Affinity-Chromatography .2. Interaction of Amino-Acids with Immobilized Nickel Iminodiacetate, *J. Chromatogr.* **323**, 255-264 (1985).
8. G. Chaga, J. Hopp, and P. Nelson, Immobilized Metal Ion Affinity Chromatography on Co^{+2} Carboxymethylaspartate-Agarose Super flow, as Demonstrated By One-Step Purification of Lactate Dehydrogenase from Chicken Breast Muscle, *Biotechnol. Appl. Biochem.* **29**, 19-24 (1999).
9. P. Verheesen, C. T. Verrips, and J. W. Haard, Beneficial Properties of Single-domain Antibody Fragments for Application in Immuno affinity Purification and Immuno-perfusion Chromatography, *Biochimicaet Biophysica. Acta.* **1624**, 21-28 (2003).

9

PRELIMINARY RESULTS OF OXYGEN SATURATION WITH A PROTOTYPE OF CONTINUOUS WAVE LASER OXIMETER

R. Tommasi[1,2], M. G. Leo[3], G. Cicco[2], T. Cassano[3], L. Nitti[1,2], and P. M. Lugarà[2,3]

1. INTRODUCTION

The state of oxygenation of biological tissues is measured by the oxygen saturation (*Y*) defined as the percentage of oxy-hemoglobin with respect to the total hemoglobin concentration. The oximetry techniques currently in use can be divided into two categories: invasive methods, that require the use of catheters, or provide in-vitro analysis of a blood or tissue sample and non-invasive methods such as the gas-transcutaneous oximetry and the pulse oximetry[1,2,3]. The gas-transcutaneous oximetry uses a Clark electrode to measure the weak current which is related to the partial pressure of oxygen in the underlying capillaries. Although this method is widely used, it presents many disadvantages: the measurement is indirect, it is limited to the subclavicular or dorsalis pedis area, where perfusion is more consistent, and the result of the analysis is strongly affected by numerous parameters, which are difficult to control, such as the room temperature and humidity. Moreover it does not permit long lasting detection.

Near-infrared spectroscopy provides a non-invasive, real time measurement of the tissue optical parameters (absorption coefficient μ_a and reduced scattering coefficient μ'_s), which allows the calculation of the concentrations of oxy-, deoxy- and total hemoglobin in tissues[4,5].

The pulse oximetry is an optical technique which exploits the measurement of near infrared (NIR) radiation transmitted by the tissue and can therefore only be used to monitor thin layers of tissue such as earlobe or fingertips[1,3].

Moreover, different NIR optical techniques exist, namely time-resolved, phase-resolved and continuous wave (CW) oximetry, which measure *Y* by detecting the radiation backscattered by the tissue and can be applied to any part of the body giving

[1] Dipartimento di Biochimica Medica, Biologia Medica e Fisica Medica, Università degli Studi di Bari
[2] CEMOT, Centro Interdisciplinare di Ricerca in Emoreologia, Microcircolazione, trasporto di Ossigeno e Tecnologie ottiche non invasive, Università degli Studi di Bari
[3] Dipartimento Interateneo di Fisica, Università degli Studi e Politecnico di Bari

readings independent of external conditions[5] (temperature and humidity). Although time- and phase-resolved optical oximetry give very precise Y estimation, CW oximetry, which gives a fairly accurate estimation of Y, has the advantage to be a low-cost technique which can be implemented in a non cumbersome instrument.

We developed a prototype of quasi-CW tissue oximeter employing two laser diodes at 750 and 810 nm. The radiation is delivered to the tissue by means of two plastic optical fibres and the backscattered radiation is collected using four plastic optical fibres and detected by a large-area silicon photodiode.

2. THEORY

As known, optical spectroscopy in the NIR region, ranging from 700 to 900 nm, achieves sufficient photon penetration depth for non invasive probing of macroscopic tissue volume[6].

The light propagation in biological tissue is governed mainly by absorption and diffusion mechanisms. The latter must be ascribed to the local mismatches in the index of refraction that occur in correspondence to membrane interfaces or to organelles and particles that act as diffusion centres. In biological tissues this process is anisotropic, so that in a single event, photons are scattered preferentially in the forward direction[3,5,7].

In the NIR region of the spectrum the only species that consistently contribute to absorption are water, hemoglobin, and lipids; in particular in the wavelength range between 700 and 900 nm, the absorption is mostly due to oxy- and deoxy-hemoglobin[8].

Thus, at two near-infrared wavelengths λ_1 and λ_2, the absorption coefficient can be written in terms of the Lambert-Beer relationship

$$\mu_a(\lambda_1) = \varepsilon_{Hb}(\lambda_1)[Hb] + \varepsilon_{HbO_2}(\lambda_1)[HbO_2] \tag{1}$$

$$\mu_a(\lambda_2) = \varepsilon_{Hb}(\lambda_2)[Hb] + \varepsilon_{HbO_2}(\lambda_2)[HbO_2] \tag{2}$$

where ε_{Hb} (ε_{HbO_2}) and [Hb] ([HbO$_2$]) are the extinction coefficient and tissue concentration of deoxy- (oxy-) form of hemoglobin, respectively.

The ratio of absorption coefficients can be written as

$$\frac{\mu_a(\lambda_1)}{\mu_a(\lambda_2)} = \frac{\varepsilon_{Hb}(\lambda_1) + Y\left[\varepsilon_{HbO_2}(\lambda_1) - \varepsilon_{Hb}(\lambda_1)\right]}{\varepsilon_{Hb}(\lambda_2) + Y\left[\varepsilon_{HbO_2}(\lambda_2) - \varepsilon_{Hb}(\lambda_2)\right]} \tag{3}$$

where

$$Y = \frac{[HbO_2]}{[Hb] + [HbO_2]}$$

is the oxygen saturation.

Furthermore, if λ_2 is the isosbestic wavelength of hemoglobin (810 nm), which is the wavelength at which the extinction coefficients of oxy- and deoxyhemoglobin are equal, the saturation Y can be written as

$$Y = \frac{\varepsilon_{Hb}(\lambda_2)}{\Delta\varepsilon(\lambda_1)}\left[\frac{\mu_a(\lambda_1)}{\mu_a(\lambda_2)} - \frac{\varepsilon_{Hb}(\lambda_1)}{\varepsilon_{Hb}(\lambda_2)}\right] \qquad (4)$$

where

$$\Delta\varepsilon(\lambda_1) = \varepsilon_{HbO_2}(\lambda_1) - \varepsilon_{Hb}(\lambda_1)$$

Since the values of the extinction coefficients are available in literature[9], from Equation 4, Y can be simply calculated if the ratio $\mu_a(\lambda_1)/\mu_a(\lambda_2)$ is measured. It is possible to obtain this ratio considering the relation between the input light intensity I_0 and the transmitted intensity I expressed as a modified Lambert-Beer equation[10]

$$I(\lambda) = I_0(\lambda)\{\exp[-\mu_a(\lambda)\cdot d \cdot DPF(\lambda)]\} \qquad (5)$$

In tissues, light is scattered at different angles and for this reason, pathlength L becomes longer than the distance d between the emission and the detection points. The differential pathlength factor DPF accounts for the difference between the geometric distance source-detector and the effective pathlength.

$$DPF(\lambda) = \frac{\sqrt{3}}{2}\sqrt{\frac{(1-g)\mu_s(\lambda)}{\mu_a(\lambda)}}$$

where $g = <cos\ \theta>$ is a single anisotropy parameter, which is the average cosine of the scattering angle.

Considering Equation 5 at two different wavelengths, the ratio of the absorption coefficients can be written as a function of the intensity I_0 of the incident light on the tissue and the intensity I of the backscattered radiation

$$\frac{\mu_a(\lambda_1)}{\mu_a(\lambda_2)} = \left[\frac{\ln(I_0(\lambda_1)/I(\lambda_1))}{\ln(I_0(\lambda_2)/I(\lambda_2))}\right]^2$$

and the expression for the saturation Y becomes

$$Y = \frac{\varepsilon_{Hb}(\lambda_2)}{\Delta\varepsilon(\lambda_1)}\left\{\left[\frac{\ln(I_0(\lambda_1)/I(\lambda_1))}{\ln(I_0(\lambda_2)/I(\lambda_2))}\right]^2 - \frac{\varepsilon_{Hb}(\lambda_1)}{\varepsilon_{Hb}(\lambda_2)}\right\} \qquad (6)$$

3. MATERIALS AND METHODS

The CW oximetry is an optical technique based on the detection of the radiation backscattered by the tissue; continuous wave, monochromatic sources at two different wavelengths are used.

We developed a new prototype of CW oximeter modifying substantially the emission-detection system of the quasi-continuous oximeter Oxyraf[11]. This is a dual wavelength, battery operated instrument, which employs two LEDs as light sources. The emission characteristics of the LEDs, e.g. wavelengths and bandwidth, do not fulfil the requirements of the above mentioned theoretical model; therefore raw data must be suitably processed to get a correct evaluation of the saturation Y.

In order to directly apply the theoretical model, the new prototype of continuous wave optical oximeter employs two laser diodes, emitting at 750 and 810 nm (isosbestic point) with a bandwidth of 1 nm. The sources are coupled to plastic optical fibres (1 mm core diameter), which convey the radiation to the patient's tissue. The backscattered radiation is detected by a silicon photodiode coupled to four plastic optical fibres to increase the signal-to-noise ratio; the resulting signal is analyzed to evaluate the tissue oxygen saturation.

The ends of the two optical fibres, connected to the sources, are positioned in the middle of the optical probe and four optical collecting fibres are placed in a symmetrical position with respect to the emission fibres. The use of a fibre optic probe provides a flexible mechanical connection between the spectroscopic device and the patient, cutting any electrical contact.

The two laser sources are operated alternately by an appropriate timing circuit, so that the detector collects separately, and in sequence, the radiation which, after interaction with the tissue underlying the measuring probe, partially emerges through the skin, in backscattered configuration.

In order to evaluate its safety, the system has been calibrated following the CEI EN 60825-1 standard in relation to wavelength and emission duration of the lasers used; then it was possible to classify it as a Class 1 system[12].

We performed some very preliminary tests on volunteers (smokers, non-smokers and non-smoking healthy carriers of thalassemia), following the venous occlusion protocol, to evaluate the ability of the instrument to non-invasively, continuously monitor the hemoglobin saturation of a limb[1,13].

The optical probe was located on the subject's forearm and a pneumatic cuff was placed immediately above the optical probe. After the acquisition of two minutes of baseline data, the pneumatic cuff was manually inflated by bulb to a pressure of 60 mmHg, and kept constant for three minutes. Then the pressure was removed by opening the deflation valve, and the parameters were monitored for another two minutes.

4. RESULTS AND DISCUSSION

An external pressure of 60 mmHg is generally sufficient to interrupt the venous flow from the limb to the heart, while it will not affect the arterial outflow of oxygenated blood from the heart to the limb. As a consequence, an increase in the blood volume, and therefore of oxy-, deoxy- and total hemoglobin, is expected in the region below the pressure cuff, monitored by the optical probe[13].

In a state of venous occlusion the blood pools in the veins and microvasculature and the average saturation goes down because the amount of venous blood increases, compared to arterial blood. The increase in blood volume is due to the distension of the vascular space due to the hydrostatic back pressure caused by the cuff[13].

In the curves in Fig. 1 the values of saturation for three types of volunteers go down during the procedure of venous occlusion and the different slopes in the curves are ascribed to the different affinity between the hemoglobin and the oxygen that influences the oxygen-tissue exchange process. In particular, the smoker subject presents an initial value of saturation smaller because a part of molecules of hemoglobin is linked with CO instead of the oxygen.

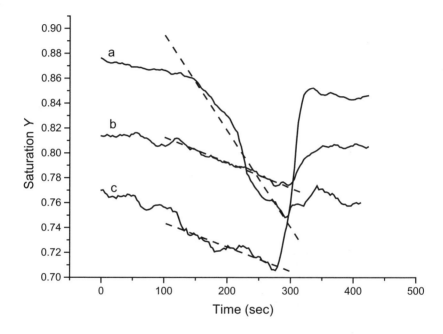

Figure 1. Saturation curves for three types of volunteers during the procedure of venous occlusion (a: non-smoker, b: non-smoker healthy carrier of thalassemia, c: smoker).

5. CONCLUSIONS

Near-infrared tissue spectroscopy is a non invasive, innocuous technique to measure oxygen saturation in tissues. NIR oximetry can be used in almost any area of the body, according to the type of analysis to be done.

The development of a laser oximeter, more selective and closer to the requirements of the diffusion theoretical model allows to better evaluate the oxygen saturation from the intensity of the backscattered light at two different wavelengths; the use of a fibre optic probe provides a flexible mechanical connection between the spectroscopic device and the patient, cutting any electrical contact.

From the safety point of view, the system has been calibrated following the CEI EN 60825-1 standard and classified as a Class 1 system.

Preliminary tests performed on volunteers have shown that CW laser oximeter can easily real time follow variations of the biological tissue oxygenation.

6. REFERENCES

1. C. Casavola, L. A. Paunescu, S. Fantini, M. A. Franceschini, P. M. Lugarà, and E. Gratton, Application of near-infrared tissue oxymetry to the diagnosis of peripheral vascular disease, *Clinical Hemorheology and Microcirculation* **21**, 389-393 (1999).
2. G. Cicco, A non invasive optical oximetry in humans: preliminary data, *Clinical Hemorheology and Microcirculation* **21**, 311-314 (1999).
3. P. M. Lugarà, Current approaches to non-invasive optical oxymetry, *Clinical Hemorheology and Microcirculation* **21**, 307-310 (1999).
4. B. C. Wilson and S. Jacques, Optical Reflectance and Transmittance of Tissue: Principles and Applications, *IEEE Journal of Quantum Electronics* **26**(12), 2186-2199 (1990).
5. E. M. Sevick, B. Chance, J. Leigh, S. Nioka, and M. Maris, Quantitation of Time- and Frequency-Resolved Optical Spectra for the Determination of Tissue Oxygenation, *Analytical Biochemistry* **195**, 330-351 (1991).
6. M. A. Franceschini, E. Gratton and S. Fantini, Noninvasive optical method of measuring tissue and arterial saturation: an application to absolute pulse oximetry of the brain, *Optics Letters* **24**(12), 829-831 (1999).
7. J. R. Mourant, J. P. Freyer, A. H. Hielscher, A. A. Eick, D. Shen, and T. M. Johnson, Mechanisms of light scattering from biological cells relevant to noninvasive optical-tissue diagnostics, *Applied Optics* **37**(16), 3586-3593 (1998).
8. S. Fantini, M. A. Franceschini-Fantini, J. S. Maier, S. A. Walker, B. Barbieri, and E. Gratton, Frequency-domain multichannel optical detector for noninvasive tissue spectroscopy and oximetry, *Optical Engineering* **34**(1), 32-42 (1995).
9. S. Wray, M. Cope, D. T. Delpy, J. S. Wyatt, and Reynolds, Characterization of the Near Infrared Absorption Spectra of Cytochrome aa$_3$ and haemoglobin for the non-invasive Monitoring of Cerebral Oxygenation, *Biochimica and Biophysica Acta* **933**, 184-192, 1988.
10. D. T. Delpy, M. Cope, P. van der Zee, S. Arridge, S. Wray, and J. Wyatt, Estimation of optical pathlength through tissue from direct time of flight measurement, *Phys. Med. Biol.* **33**(12), 1433-1442 (1988).
11. C. Casavola, G. Cicco, A. Pirelli, and P. M. Lugarà, Preliminary tests on a new, near infra-red, continuous-wave tissue oximeter, *Proc. SPIE* **4160**, 170-176 (2000).
12. Norma CEI EN 60825-1 Class. CEI 76-2: Sicurezza degli apparecchi laser. Parte 1: Classificazione delle apparecchiature, prescrizioni e guida per l'utilizzatore. Milano: Ed. CEI, 1998.
13. J. S. Maier, B. Barbieri, A. Chervu, I. Chervu, S. Fantini, M. A. Franceschini, M. Levi, W. E. Mantulin, A. Rosenberg, S. A Walker, and E. Gratton, In vivo study of human tissues with a portable near-infrared tissue spectrometer, *Proc. SPIE* **2387**, 240-248 (1995).

10

RELATIONSHIP BETWEEN CARBON DIOXIDE ELIMINATION KINETICS AND METABOLIC CORRELATES OF OXYGEN DEBT IN SEPTIC PATIENTS

Renzo Zatelli [*]

1. INTRODUCTION

Many clinical situations cause inadequate tissue oxygenation: sepsis is the most serious of them, since an imbalance between oxygen delivery to peripheral tissue and oxygen need is the most important factor in the development of multiple system organ failure and mortality [1].

Oxygen debt develops in the presence of an abnormal ratio between O_2 supply and O_2 demand, inducing increased anaerobic metabolism and tissue and blood lactate concentration. Since O_2 demand is neither measurable, nor calculable [2], in order to evaluate O_2 debt we must use markers of impaired O_2 utilization: elevated arterial base deficit and abnormal lactate concentration [3].

O_2 kinetics in sepsis have been widely studied; in contrast, CO_2 kinetics in the same condition is poorly investigated. The purpose of our study was to verify the hypothesis that CO_2 elimination kinetics is different in sepsis versus in non septic conditions, and that in these circumstances this parameter could have a relationship with lactate increase.

This relationship would indicate CO_2 kinetics as an useful parameter to assess the wearing off of the septic condition during clinical course, because the latter would indeed be the first parameter to get back to normal, due to CO_2 large diffusibility.

[*] Renzo Zatelli, Dept. of Anesthesia and Intensive Care, University of Ferrara, 44100 Ferrara, Italy

2. PATIENTS AND METHODS

We studied 55 patients admitted during six months to our Intensive Care Unit suffering from different diseases and monitored with a pulmonary artery catheter. Measurements of arterial and mixed venous acid-base and hemodynamic parameters, arterial lactate and base deficit were serially performed, as well as O_2 consumption, CO_2 production and respiratory quotient by means of an indirect calorimetric device. We then determined CO_2 elimination kinetics: exhaled CO_2, mixed venous CO_2 content, CO_2 clearance and calculated O_2 debt according to clinical techniques [4]. We then divided patients into a "non septic" group and a "septic" group.

Eleven patients met sepsis criteria with at least five of the following: body temperature > 38.5° or < 36°C, white blood cell count > 12,000 or < 3,500 / mL, heart rate > 100 beats/min, respiration rate > 28 breath/min or FIO_2 > 0.21, mean arterial pressure < 75 torr, cardiac index > 4.5 L/min/m^2 , thrombocytes < 100,000 cells/mL, positive blood culture, systemic vascular resistance < 0.08 kPa· s· cm^{-3}, clinical evidence for sepsis (surgical or invasive procedure during the preceding 48 h or presence of an obvious primary septic site) [5].

After an adequate resuscitation, when the hemodynamic and ventilatory situation was stable, from these patients fifty-seven data sets were obtained and compared with the data of non septic patients, using an unpaired t - test. Correlations were performed using Pearson correlation with Bonferroni correction. A p< 0.05 was considered statistically significant. All values were reported as mean ± SEM.

3. RESULTS

There was no significant difference between septic and non septic patients in age (62.6 ± 3.3 vs. 53.7 ± 2.9 y), weight (69.4 ± 3.8 vs. 69.8 ± 2.0 Kg), sex distribution (M:F 7/4 vs. 27/17) and APACHE II score (22.1 ± 0.9 vs. 21.1 ± 0.4).

Table 1. Main hemodynamic and O_2 / CO_2 transport data

variable	septic	non septic	P value
O_2 consumption index (mmol/min/m^2)	5.58 ± 0.21	5.36 ± 0.11	0.3256
base excess (mEq/L)	-6.15 ± 0.05	2.04 ± 0.18	<.0001
arterial lactate (mmol/L)	5.18 ± 0.38	1.42 ± 0.05	<.0001
O_2 debt (mmol/min/m^2)	0.96 ± 0.22	1.00 ± 0.11	0.8456
cardiac index (L/min/m^2)	4.21 ± 0.20	4.31 ± 0.08	0.6049
syst. vascular resistance (kPa· s· cm^{-3})	0.1272 ± 0.0094	0.1590 ± 0.0046	0.0015
mean pulm. arterial pressure (kPa)	3.76 ± 0.13	3.19 ± 0.05	<.0001
pulm. vascular resistance (kPa· s· cm^{-3})	0.0215± 0.0018	0.0183 ± 0.0006	0.0252
P $_x$ (kPa)	5.77 ± 0.28	6.44 ± 0.12	0.0120
CO_2 elimination (mmol/min/m^2)	5.72 ± 0.23	6.05 ± 0.13	0.2126
mean venous CO_2 concentr. (mmol/L)	20.3 ± 0.46	26.7 ± 0.24	<.0001
CO_2 clearance (L/min/m^2)	0.30 ± 0.02	0.23 ± 0.01	<.0001

All patients were ventilated for similar periods of time.

The hospital mortality rate in the septic group was 64% compared with 41% in the non septic group.

VO_2I (mmol/min/m^2) was similar in the two groups.

On the contrary, arterial lactate (5.18 ± 0.38 versus 1.42 ± 0.05 mmol / L) and base excess (- 6.15 ± 0.05 versus 2.04 ± 0.18 mEq/L) concentrations were significantly different in septic versus non septic patients.

O_2 debt was the same in the two groups, independently of the method used to quantify it and showed no correlation with lactate.

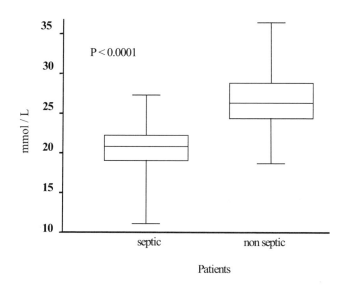

Figure 1. Mixed venous CO_2 concentration.

There was no difference in cardiac index between groups, while systemic (0.1272 ± 0.0094 kPa's'cm^{-3} versus 0.1590±0.0046) and pulmonary (0.0215±0.0018 versus 0.0183 ± 0.0006) vascular resistances were different between septic and non septic patients.

The arterial-venous O_2 saturation difference was moderately different between the two groups (0.24 ± 0.01 in septic and 0.22 ± 0.01 in non septic with a p = 0.051).

The respiratory quotient was similar in the two groups and had no correlation with lactate.

The CO_2 elimination showed no significant difference in septic patients (5.72 ± 0.23 mmol/min/m^2) and non septic patients (6.05 ± 0.13) (p = 0.213). In the septic group there was a slight positive correlation of CO_2 elimination with the arterial lactate.

The mixed venous CO_2 concentration was significantly lower in septic patients (20.3 ± 0.5 versus 26.7 ± 0.2 mmol/L) (Fig 1) and in this group it had a close correlation (r^2 = 8626, p=<0.0001) with lactate (Fig 2).

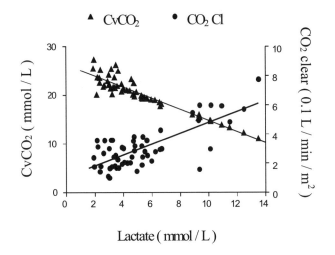

Figure 2. Mixed venous CO_2 and CO_2 clearance in septic patients.

Due to a normal CO_2 elimination and a reduced mixed venous CO_2 concentration, the CO_2 clearance appeared greater in septic patients (0.30 ± 0.02 L/min/m^2) than in non septic patients (0.23 ± 0.01).

The correlation between CO_2 clearance and the lactate concentration was greater in septic patients ($r^2 = 0.5616$, p<0.0001) than in non septic patients.

4. DISCUSSION AND CONCLUSION

At the time of data collection in our patients a normal cardiac index and an arterial-venous oxygen saturation difference provided evidence of a stable critical situation. In septic patients the arterial pCO_2 was kept normal by means of an adequate ventilation, while there was a mixed venous CO_2 content lower than in non septic patients. Therefore it is unlikely that the acid-base disturbances associated with sepsis were due to a venous hypercarbia [6].

The methods used to quantify the O_2 debt are of little interest in clinical practice, even if they are largely employed in kinesiologic, nutritional and experimental studies [7]. Not even P_x (the blood pO_2 after 2.3 mmol/L of O_2 are removed) could detect tissue hypoxia in our patients [8]. At present metabolic correlates of O_2 debt (arterial lactate and base deficit) seem the best method to quantify tissue hypoxia for clinical purposes [9], although in sepsis, serum lactate concentrations may be increased due to a reduction of pyruvate dehydrogenase activity without evidence for tissue hypoxia.

In our patients the normalization of mixed venous CO_2 concentration was often associated with a mitigation of the septic state, but the connection between the two elements did not reach statistical significance.

The capillary shunting and the different behaviour of the pulmonary vascular resistances can explain the CO_2 elimination kinetics in septic patients: a next step of research would be to verify if, besides a mean tissue pO_2 lower than seven torr [10], in patients with sepsis there is an increased tissue CO_2 in spite of its larger volatility.

5. REFERENCES

1. U. Haglund and R.G. Fiddian-Green, Assessment of adequate tissue oxygenation in shock and critical illness: oxygen transport in sepsis *Intensive Care Med* **15** (7), 475 – 477 (1989).
2. C. Chopin and B. Vallet, in: *Oxygenation tissulaire*, edited by Ch. Richard and J.L. Teboul (Masson, Paris, 1995) pp. 38-55.
3. E.H. Kincaid, P.R. Miller, J.W. Meredith, N. Rahman and M.C. Chang, Elevated arterial base deficit in trauma patients: a marker of impaired oxygen utilization, *J Am Coll Surg* **187** (4), 384 – 392 (1998).
4. W.C. Shoemaker, P.L. Appel and H.B. Kram, Tissue oxygen debt as a determinant of lethal and non lethal postoperative organ failure, *Crit Care Med* **16** (11), 1117 – 1120 (1988).
5. P. Boekstegers, S. Weidenhofer, T. Kapsner and K. Werdan, Skeletal muscle partial pressure of oxygen in patients with sepsis, *Crit Care Med* **22** (4), 640 – 650 (1994).
6. C.E. Mecher, E.C. Rackov, M.E. Astiz and M.A. Weil, Venous hypercarbia associated with severe sepsis and systemic hypoperfusion, *Crit Care Med* **18** (6), 585 – 589 (1990).
7. D.W. Hill, Determination of accumulated O_2 deficit in exhaustive short-duration exercise, *Can J Appl Physiol* **21** (1), 63 – 74 (1996).
8. O. Siggaard – Andersen, I.H. Gothgen, P.D. Wimberley, N. Fogh-Andersen, The oxygen status of the arterial blood revised: relevant oxygen parameters for monitoring the arterial oxygen availability, *Scand J Clin Lab Invest* **203** (1), 17 – 28 (1990).
9. D. Rixen and J.H. Siegel, Metabolic correlates of oxygen debt predict posttrauma early acute respiratory distress syndrome and the related cytokine response, *J Trauma* **49** (3), 392 – 403 (2000).
10. K. Reinhart, F. Bloos, F. Konig, L. Hanneman and B. Kuss, Oxygen transport and muscle tissue oxygenation in hyperdynamic septic shock, *Anesthesiology* **71** (3) A 1200 (1998).

11

ASSAYING ATP SYNTHASE ROTOR ACTIVITY

D. Maguire, J. Shah and M. McCabe[*]

1. INTRODUCTION

   Current focus during non-invasive investigations into disturbances of oxygen utilization is on either the level of NADH, using techniques such as near infra-red spectroscopy (NIRS), or on ATP/ADP,Pi ratios, by techniques such as NMR or MRI. These tools bracket the process of oxidative phosphorylation. In order to investigate molecular disturbances further, it is necessary to resort to destructive techniques, such as electron transport chain complex analysis by enzyme assay following tissue homogenisation, or mutation analysis following DNA extraction. Although these investigations are extremely informative in many instances, the enzyme assays are technically challenging even in competent hands. There is also always doubt about the integrity of the tissue before and during assay. Therefore those assays are routinely normalized against a mitochondrial 'housekeeping' enzyme, citrate synthase. Furthermore one of these enzymes, ATP synthase, can only be measured as its reverse reaction, as an ATP hydrolase (ATPase). So unreliable are the techniques available for this assay, however, that values for this enzyme are generally not reported in routine laboratory diagnosis. One published exception to that statement is the report by Baracca et al.[1] who studied ATP synthesis in isolated sub-mitochondrial particles prepared from platelets of two individuals with mitochondrial DNA T8993G mutations in the ATPase 6 gene encoding subunit a. By contrast, the mechanism of this enzyme complex and its role in the fundamental process by which mitochondria convert chemical energy in the form of food into the usable energy currency of the cell, namely ATP, is now known down to the molecular level[2-4]. Several reports have demonstrated the mechanism of ATPase function in vitro[2-4]. It is the purpose of the present work to assess whether it might be possible to harness this knowledge to design cellular probes to assay the activity of complex V in its synthetic mode in vivo.

---

[*] Genomics Research Centre, Griffith University, Nathan, Brisbane, Australia, 4111

2. STRUCTURAL AND MECHANISTIC STUDIES ON ATP SYNTHASE

The structure of the complexes of the electron transport chain (ETC) in humans and other mammals has been largely predicted from comparative studies using bacterial or yeast models. The tools in those investigations include DNA and protein sequencing and comparative genomics and proteomics, coupled with secondary structure studies. Confirmation of those predictions has been made possible by X–ray diffraction and solution NMR[5]. From the results of such approaches, it has been possible to build representations of complex V such as that shown in Figure 1. The human complex V is believed to consist of at least eight different subunit components in total assembled into the two sectors identified by classical biochemical techniques and electron microscopy as F_1 and F_0. These sectors have subunit composition $\alpha_3\beta_3\gamma_1\delta_1\epsilon_1$ and $a_1b_2c_{12}$ respectively. The mechanism of ATP generation in this model involves a rotor (of stoichiometric composition $c_{12}\gamma_1\epsilon_1$) rotating clockwise when viewed from the mitochondrial matrix side in a stator that consists of $a_1b_2\alpha_3\beta_3\delta_1$ subunits. Gavin, Devenish and Prescott[6] have implicated two other subunits in anchoring the stator within the membrane, namely OSCP and subunit b. In recent years, several investigators [2,3], have turned their attention to the physical demonstration of the mechanism of the ATP synthase rotor.

Such approaches generally have used non-mammalian ATP synthase as the in-vitro study material and laser microscopy as the investigating tool. The techniques used in those studies have necessarily involved analysis of the reverse reaction, i.e. ATPase analysis. A typical such investigation involves the attachment of some component of the structure to a solid surface and the attachment of a signalling molecule to another part of the structure. Using such approaches, it has been convincingly demonstrated that ATPase activity is indeed accompanied by rotation at a rotary torque of 40 pN nm [2,7].

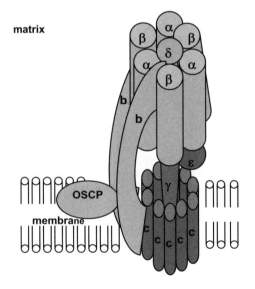

Figure 1. Model of ATP synthase (Complex V of oxidative phosphorylation). Rotation of the rotor ($c_{12}\gamma_1\epsilon_1$) within the stator is driven by proton binding with c subunits through a channel in the stator ($a_1b_2\alpha_3\beta_3\delta_1$). An additional component to anchor the stator within the membrane (OSCP) is shown although precise positioning within the complex is speculative.

Table 1. The three isoforms of the beta subunit of ATP synthase, with the aromatic amino acid residues, phenylalanine (P), tyrosine (Y) and tryptophan (W) highlighted in capitals. Square brackets in isoform 1 delineate segments that are combined to form isoform 3 and underlined residues are those present in isoform 2

- **(ISOFORM 1 256 RESIDUES)**
- mlsrvvlsaa ata] [apslkna aFlgpgvlqa trtfhtgqph lvpvpplpeY ggkvrYglip eeFFqFlYpk tgvt] [gpYvlg tglvlYalsk eiYvisaetF talsvlgv<u>mv YgikkyYgpFv adFadklneq klaqleeakq asiqhiqnai dteksqqalv qkrhYlFdvq rnniamalev tYrerlYrvY kevknrldYh isvqnmmrrk eqehminWve khvvqsistq qeketiakci adlkllakka qaqpvm</u>
-
- **(ISOFORM 2 148 RESIDUES)**
- mvYgikkYgp FvadFadkln eqklaqleea kqasiqhiqn aidteksqqa lvqkrhYlFd vqrnniamal evtYrerlYr vYkevknrld Yhisvqnmmr rkeqehminW vekhvvqsis tqqeketiak ciadlkllak kaqaqpvm
-
- **(ISOFORM 3 195 RESIDUES)**
- <u>mlsrvvlsaa ata</u>][][gpYvlgt glilYalske iYvisaetFt alsvlgvmvY gikkYgpFva dfadklneqk laqleeakqa siqhiqnaid teksqqalvq krhYlFdvqr nniamalevt YrerlYrvYk evknrldYhi svqnmmrrke qehminWvek hvvqsistqq eketiakcia dlkllakkaq aqpvm

3. HUMAN ATP SYNTHASES

The active sites of ATP synthase resides in each of the three β-subunits of the F_1 sector. There are three known b-subunits of ATP synthases in humans, namely isoforms 1,2 and 3. These are all encoded by the same gene (ATP5B) at position p 13-qter on chromosome 12. A mitochondrial (signal peptide) is cleaved during the process of delivery and assembly into the mitochondrial membrane. The different isoforms are created by differential splicing of the original DNA transcript.

4. METHODS AND RESULTS

The sequences of the human α-and β-subunits were accessed via the NCBI website and potentially fluorescent residues (tyrosine and phenylalanine) were highlighted (in capital letters). The single tryptophan residue in each of these molecules has also been highlighted. The sequences of the three isoforms of the β-subunit are shown in Table 1 with segments in each isoform delineated in square brackets. The sequences are those reported on the NCBI website. 3-D models of β-subunits were generated using software freely available on the internet. One view of a 3-D model of the b-subunit is shown with the tyrosine residues highlighted. Tagging of the three-dimensional structure at the potentially fluorescent sites will enable the maximum radius of signal to be estimated.

It can be seen from Figure 2 that the β-subunits of ATP synthase have a significant number of residues that are capable of fluorescence in their native forms and/or are also able to be tagged with fluorescent dyes. It is also obvious that some of these may exhibit movement under the stereochemical alterations known to be induced during the binding of ADP and synthesis of ATP. This is particularly true of the β-subunits of ATP synthase.

A diagram of the experimental assembly that might be used in an *in-vivo* ATP synthase analysis using laser excitation and fluorescent detection is shown (Fig 3). This assembly consists of two laser sources, one to anchor the complex and the other to generate a light source to probe for movement within the complex.

5. DISCUSSION

In order to develop a technique for analysis of ATP synthase activity *in-vivo*, it is necessary to (a) anchor one component of the complex and (b) probe for movement in another part of the complex. Attempts to anchor the stator components of the synthase complex by cross-linking have been shown to have drastic effects upon mitochondrial morphology [8]

The most likely components that will exhibit any movement are the β-subunits, parts of which are expected to undergo stereochemical alterations during the binding of ADP and ATP synthesis and release. From our modelling of the β-subunit structure, there appear to be ample native signals that might undergo such changes distributed throughout the molecule. The α-subunit also has a considerable number of native fluorescent residues but it is doubtful whether these subunits undergo significant movement.

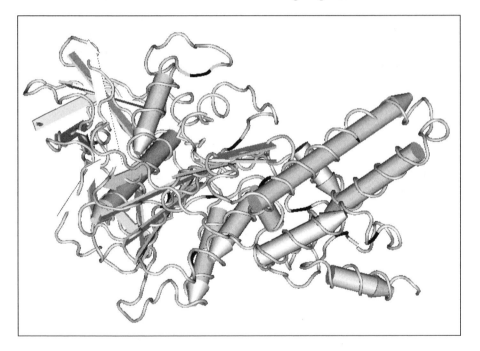

Figure 2. Structure of the β-subunit of ATP synthase, with tyrosine residues high-lighted in black.

A more dramatic signal change might be seen from a rotating fluorophore, such has been obtained by the *in vitro* examination of preparations of chemically modified F_0 sectors. The symmetrical nature of the native c_{12}- and γ- subunits of the F_0 rotor preclude the use of such signals in any attempt to follow ATP synthesis *in vivo*. However, the ε-subunit might be a potentially useful source of signal as it has been suggested that this subunit is asymmetrically situated in the rotor assembly.

Other approaches to investigating rotor movement require lengthy molecular genetics manipulations that may affect the activity of the complex. Such approaches include genetic modification to the γ-subunit such that a wider fluorescent signal arc would be created[9], introduction of bio-engineered tRNA molecules which might introduce modified amino acid residues into one or more components of the complex. This latter approach may lead to the further exploitation of other detection techniques such as the fluorescent resonant energy transfer (FRET)

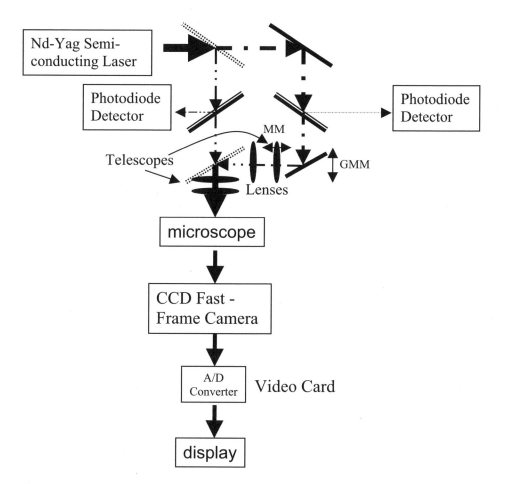

Figure 3. Proposed instrumental layout, based on instruments previously used, but with additional optics to anchor complex components

principle. Such an approach to the study of ATP synthase rotor movement has been taken by Borsch et al.[10] and Diez al.[11]. Techniques that change the basic structure of the ATP synthase complex have the disadvantage that they might alter the binding or reaction characteristics of the complex. In addition, those techniques, while being of potential use in basic science applications, would not have the capacity to be applied to a clinical situation *in vivo*.

6. CONCLUSIONS

With the development of laser excitation and detection systems, it has become possible to consider analyses that have otherwise been beyond the capacity of experimentalists. In this investigation, a case is made for the development of a device to analyse ATP synthase activity *in-vivo*. Such analyses ultimately have applications in clinical practice, particularly in the field of mitochondrial disorders, of both nuclear and mitochondrial DNA origin.

7. REFERENCES

1. A. Baracca, S. Barogi, V. Carelli, G. Lenaz, and G. Solaini, Catalytic activities of mitochondrial ATP synthase in patients with mitochondrial DNA T8993G mutation in the ATPase 6 gene encoding subunit a, *J. Biol. Chem.,* **275**, 4177-82 (2000).
2. Y. Wada, Y. Sambogni, and M. Futai, Biological nano motor, ATP synthase F_0F_1: from catalysis to $\gamma\varepsilon c_{10-12}$ subunit assembly rotation. *Biochem. Biophys. Acta,* **1459**, 499-505 (2000).
3. P. Dimroth, U. Matthey, and G. Kaim, Critical evaluation of the one- versus the two-channel model for the operation of the ATP synthase's F_0 motor. *Biochem. Biophys. Acta,* **1459**, 506-513 (2002)
4. R. H. Fillingame, W. Jiang, O. Y. Dmitriev, and P. C. Jones, Structural interpretations of F(0) rotary function in the Escherichia coli F(1)F(0) ATP synthase..*Biochim. Biophys. Acta.* 1458(2-3), 387-403 (2000)
5. J. Weber and A. E. Senior, Features of F(1)-ATPase catalytic and noncatalytic sites revealed by fluorescence lifetimes and acrylamide quenching of specifically inserted tryptophan residues. *Biochemistry* **39**:5287-94. (2000)
6. P. D. Gavin, R. J. Devenish, and M. Prescott, FRET reveals changes in the F1-stator stalk interaction during activity of F1F0-ATP synthase. *Biochim. Biophys. Acta.* **1607**, 167-79 (2003)
7. H. Omote, N. Sambonmatsu, K. Sito, Y. Sambongi, A. Iwamoto-Kihara, T. Yanagida, Y. Wada, and M. Futai, The gamma-subunit rotation and torque generation in F1-ATPase from wild-type or uncoupled mutant Escherichia coli. *Proc. Nat. Acad. Sci.,* USA, **96**, 7780-7784 (1999)
8. P. D. Gavin, M. Prescott, S. E. Luff, and R. J. Devenish, Cross-linking ATP synthase complexes in vivo eliminates mitochondrial cristae. *J. Cell. Sci.* **117**, 2333-43 (2004)
9. P. Turina, Structural changes during ATP hydrolysis activity of the ATP synthase from Escherichia coli as revealed by fluorescent probes. *J Bioenerg. Biomembr.* **32**:373-81, (2000)
10. M.Borsch, M. Diez, B. Zimmermann, R. Reuter, and P. Graber, Stepwise rotation of the gamma-subunit of EF(0)F(1)-ATP synthase observed by intramolecular single-molecule fluorescence resonance energy transfer. *FEBS Lett.* **527**, 147-52 (2002)
11. M. Diez, B. Zimmermann, M. Borsch, M. Konig, E. Schweinberger, S. Steigmiller, R. Reuter, S. Felekyan, V. Kudryavtsev, C. A. Seidel, and P. Graber, Proton-powered subunit rotation in single membrane-bound F0F1-ATP synthase. *Nat. Struct. Mol. Biol.* **11**, 135-41 (2004)

12.

ALU SEQUENCES IN THE HUMAN RESPIROME

David J Maguire, Harald Oey and Michael McCabe [*]

1. INTRODUCTION

The concept of the 'ome' has dominated some scientific literature in recent years since the introduction of the various 'genome' projects began in the late eighties. In a presentation at a recent conference on inborn errors of metabolism (ISIEM 2003), almost fifty such examples were cited. The concept is a useful one in so far as it serves to amalgamate and expedite access to a body of specific information within an area of scientific interest. The application of the concept is perhaps best illustrated by the published outcomes of the human genome projects in both the printed versions and by the easily accessible web-based version. The latter is particularly convenient in terms of ease of revision of information and also in terms of ability to link to related sites. It is the purpose of the current paper to highlight the need to develop and maintain a web-site that encompasses the information of interest to the members of the International Society on Oxygen Transport to Tissues (ISOTT). The structure of the 'respirome' and the specific components of this concept is presented in broad terms. As an example of the facility of such a resource, an example is given that encompasses the search for particular DNA sequence motifs (Alu sequences) within the human respirome. Analysis of the results of such searches can often provide insights into fundamental concepts. In this instance the search highlights the relative paucity of Alu repeats and an almost complete lack of inverted Alu repeats in exonic regions of some components of the human respirome elements investigated.

2. CASCADE MODEL OF OXYGEN FLOW FROM AIR TO CELL; A SCAFFOLD FOR THE HUMAN RESPIROME

In 1992, Erdmann presented a model that encompassed most of the information that represents the concept of the human respirome. He described the flow of oxygen from air to cell as a general model that was applicable to creatures as diverse in environments and mechanism of respiration as fish and humans. This model adequately describes the first part of the oxygen transport process. However, it fails to account for some of the other important aspects of the role of oxygen in sustaining life, namely the processes of oxygen transport and respiration within cells and the subsequent steps of synthesis and

---

[*] Genomics Research Centre, Griffith University, Nathan, Brisbane, Australia, 4111

detoxification of injurious oxygen metabolites. It is also considered important to include the processes by which inadequate control of those toxic components lead to cell death. In this context, processes that protect cells against those toxic products are also relevant. A number of molecules that impact upon the processes described above have only been recognized since the human genome publications at the end of the last millennium and rate special mention. A modified version of Erdmann's model[1] is presented elsewhere in this volume[2]. The modules that compose this expanded model are; oxygen transport within blood, oxygen exchange between blood and cells, oxygen transport within cells, cellular respiration, antioxidant processes, free radical generation and programmed cell death (apoptosis).

3. APPLICATION OF THE HUMAN RESPIROME; SEARCH FOR ALU REPEAT SEQUENCES

In order to illustrate the potential utility of the respirome, an investigation is reported into the Alu content of a subset of the genes encoding members of the respirome. Alu sequences are a sub-group of the Lines (long interspersed nuclear elements) that are almost exclusive to humans and their very close ancestral relatives. The ability of direct inverted Alu repeat sequences to form stable secondary RNA structures is believed to be an important step in the process of RNA editing by enzymes such as adenosine deaminases acting against RNA (ADARs)[3]. The action of these enzymes against pre-mRNA leads to the conversion of adenosine (A) to inosine (I). A recent genome-wide investigation into A to I editing sites has increased by two orders of magnitude the known incidence of such sites [4].

4. METHODS

A bioinformatics approach was used to search for Alu sequences and Alu repeat sequences in the set of genes specified in Table 1. The gene-coding sequences were obtained from the NCBI web-site online database of reference sequences (Refseq)[5]. When splice variants with differing 5' or 3' untranslated regions (UTRs) were available for a gene, the longest form of the gene was used. The mRNAs for the genes were aligned with the corresponding genomic DNA sequences obtained from NCBI (build 35.1). The sequences were analysed for presence of Alu repeats only, using the repeatmask program[6], specified to only mask Alus, and data were extracted from the output of these analyses. Alu fraction was calculated as the percentage of total length of Alu repeats in a gene relative to total gene length. After Alu repeat sequences were highlighted by this technique they were assigned to either intronic or exonic regions of the genes analysed. Further analysis of these results enabled predictions to be made regarding susceptibility to the editing enzymes known as ADARs. Total Alu length and total gene length in all the components of the human respirome examined in this study was calculated and expressed as a percentage.

5. RESULTS

The results of the bioinformatics approach to examination of the respirome are shown in Table 1 which shows that there is huge variability in gene length, mRNA length, exon content, Alu content and total Alu length. Some members of the respirome gene assemblage contain no such sequences, significantly the two human globin gene clusters found on chromosomes 11 and 16. Other genes in the respirome that do not contain any Alu sequences are those coding for caspase 1 and for the thioredoxin reductases 1, 3 and 5. In the table Alu content is expressed as a percentage of total gene length. The highest such content is seen in peroxiredoxin 1 at almost 40%. Genes with relatively low Alu content include myoglobin, neuroglobin, peroxiredoxin 6, caspases 4 and 5 and superoxide dismutase 3. Total gene length in the components of the human respirome examined in this study was 796479 base pairs, of which 112566 base pairs were assigned to Alu repeats. This represents approximately 14% of the total length of gene regions studied.

6. DISCUSSION

A proposal is made for the development of a web-resource to be known as the 'human respirome' web-site. A number of other sites exist with focus on areas close to the proposed one, but do not exclusively concentrate on the human situation. The minimal requirement for each molecule in the modules represented in the respirome should include details of or links to the following; gene location, gene sequence and shorthand exonic structure, gene transcript, protein sequence, protein 3D structure (if available), protein function, common polymorphisms and disease associations. To this bare scaffold, active links to important physicochemical features for each protein could be added. For example, these might include details such as molecular weight, pK, Vmax, Km, PO_2, as well as information regarding tissue specificity, developmental expression and post-translational modification of the proteins of the respirome.

As an example of the utility of such a resource, the genomic DNA sequences of the components of some of the genes that constitute the respirome were examined for their Alu sequence content. The nuclear-encoded sub-units of the electron transport chain were included in this analysis whereas the mitochondrial genome is known to have no Alu repeat sequences although a number of small direct repeat sequences have been shown to be present in that genome. It is significant that none of the members of the alpha or beta globin gene cluster contain such sequences, reflecting perhaps some protective mechanism against retro-insertion events in this region of the human genome. Similarly, Alu repeat sequences are absent from the genes for caspase 1, as well as glutathione reductase 1, 3 and 5. This finding would indicate that these genes are not substrates for ADARs and that they are unlikely to undergo editing at the pre-mRNA level. It is expected that lack of editing at the pre-mRNA level would provide phenotypic stability to the processes of oxygen transport, and parts of the apoptotic and endogenous antioxidant pathways. However, only two genes, namely caspase 6 and caspase 8, have Alu repeats within an exonic sequence. Morse et al.[3] demonstrated ADAR editing processes specifically in Human pre-mRNA. Such events could lead to mutations in exonic regions of mRNA and subsequent disparity between expressed protein sequence and that predicted from genomic sequencing. In the present study, caspase 6 and caspase 8 are the only genes in which there is a possibility of genomic-phenomic discordance. A detailed examination of what sort of mutations might be introduced into these two genes in the time between transcription of pre-mRNA and translation of protein sequences has not been undertaken. Such an approach might define whether such editing would produce significant residue changes in the expressed proteins. For the high levels of Alu sequences in other genes of the respirome, any base-changes induced within intronic regions would need to exert their effect via translational control mechanisms.

Table 1. Alu content in the respirome assemblage. Note the absence of Alu sequences in some members of this group, particularly the alpha and beta globin clusters. Legend: hb; hemoglobin, TXN; thioredoxin, PRDX; peroxoredoxin, CASP; caspase, SOD; superoxide dismutase, GPX; glutathione peroxidase, EPX; eosin peroxidase, TPO; thyroid peroxidase, NDUF; NADH dehydrogenase (ubiquinone), SDH; succinate dehydrogenase, CYC; cytochrome c, HSPCO; ubiquinol-cytochrome c reductase complex, QP-C; low molecular mass ubiquinone-binding protein, UQCR; ubiquinol cytochrome c reductase

Gene name	gene length (bp)	mRNA length	Number of exons	Number of Alus	Total Alu length (bp)	Alu fraction of gene (%)
HBB_Hb, beta	1606	626	3	0	0	0.0
HBD_Hb, delta	1650	624	3	0	0	0.0
HBG1_Hb, gamma A	1586	584	3	0	0	0.0
HBG2_Hb, gamma G	1591	583	3	0	0	0.0
HBE1_Hb, epsilon 1	1794	816	3	0	0	0.0
HBZ_Hb, zeta	1651	589	3	0	0	0.0
HBA2_Hb, alpha 2	831	575	3	0	0	0.0
HBA1_Hb, alpha 1	842	576	3	0	0	0.0
HBQ1_Hb, theta 1	844	651	3	0	0	0.0
Myoglobin	16591	1170	4	4	997	6.0
Neuroglobin	5828	1909	4	1	311	5.3
Cytoglobin	10343	1951	4	4	1225	11.8
NDUFA1	4885	479	3	9	1702	34.8
NDUFA2	2283	600	3	1	310	13.6
NDUFA3	4109	360	4	6	1606	39.1
NDUFA4	6999	823	4	2	599	8.6
NDUFA5	16876	1550	5	5	1696	10.0
NDUFA6	5359	1202	3	7	1806	33.7
NDUFA7	10023	531	4	20	4246	42.4
NDUFA8	15762	859	4	11	2943	18.7
NDUFA9	38117	1334	11	6	1278	3.4
NDUFA10	64643	1557	10	10	2721	4.2
NDUFA11	9343	819	4	8	2433	26.0
NDUFAB1	15305	804	5	16	4281	28.0
NDUFB1	5687	433	3	9	2396	42.1
NDUFB2	9966	509	4	11	2892	29.0
NDUFB3	13932	693	3	24	6121	43.9
NDUFB4	6047	541	3	1	303	5.0
NDUFB5	19714	1088	6	21	4664	23.7
NDUFB6	19660	862	4	14	3836	19.5
NDUFB7	5997	570	3	9	2246	37.5
NDUFB8	6140	686	5	3	924	15.0
NDUFB9	10864	740	4	7	1916	17.6
NDUFB10	2458	736	3	4	304	12.4
NDUFB11	2819	953	3	1	135	4.8
NDUFC1	5867	418	4	8	2020	34.4
NDUFC2	11536	2159	3	13	3767	32.7
NDUFS1	36385	3417	19	55	13126	36.1

Table 1. (continued)

NDUFS2	15081	2061	15	13	3559	23.6
NDUFS3	5483	899	7	3	869	15.8
NDUFS4	122684	668	5	49	13838	11.3
NDUFS5	8282	540	3	15	4064	49.1
NDUFS6	14660	750	4	2	602	4.1
NDUFS7	11677	787	8	8	2278	19.5
NDUFS8	6006	779	7	5	1431	23.8
NDUFV1	5599	1566	10	1	294	5.3
NDUFV2	31610	827	8	16	4061	12.8
NDUFV3	16396	2151	4	31	7211	44.0
SDHA	38346	2277	15	18	5241	13.7
SDHB	35394	1100	8	44	10826	30.6
SDHC	48808	1315	6	68	18037	37.0
SDHD	8896	1313	4	8	2127	23.9
CYC1	2430	1273	6	0	0	0.0
HSPC051	3045	934	2	2	595	19.5
QP-C	817	388	2	0	0	0.0
UQCR	8261	1305	3	13	3623	43.9
UQCRB	5078	965	4	2	396	7.8
UQCRC1	10667	1636	13	8	1967	18.4
UQCRC2	30060	1674	14	23	5925	19.7
UQCRFS1	5948	1203	2	0	0	0.0
UQCRH	13041	515	4	26	6299	48.3
COX4I1	7413	802	5	2	615	8.3
COX5A	17754	645	5	38	9342	52.6
COX5B	2137	523	4	1	299	14.0
COX6A1	2626	548	3	2	466	17.7
COX6B1	10530	578	4	20	5010	47.6
COX6C	15524	444	4	15	3011	19.4
COX7A1	1948	783	4	0	0	0.0
COX7B	5819	456	3	7	1819	31.3
COX7C	2800	448	3	1	297	10.6
COX8A	1937	521	2	2	623	32.2
ATP5A1	20090	1950	13	30	8054	40.1
ATP5B	7894	1857	10	11	2575	32.6
ATP5C1	19672	1162	10	12	3048	15.5
ATP5C2			PSEUDO	GENE		
ATP5D	3076	1005	4	5	852	27.7
ATP5E	3690	498	3	0	0	0.0
ATP5G1	3086	663	5	1	272	8.8
ATP5G2	11481	1094	5	17	4028	35.1
ATP5G3	3847	1014	5	1	315	8.2
ATP5H	8120	628	6	12	3003	37.0
ATP5I	1898	368	4	3	589	31.0
ATP5J	11175	1303	4	8	2181	19.5

Table 1. (continued)

ATP5L	8459	1343	3	6	1768	20.9
ATP5O	12402	815	7	9	2640	21.3
ATP5S	13622	1567	5		2422	17.8
TXNRD1	63336	3923	15	46	13763	21.7
TXNRD2	66470	2180	18	43	10759	16.2
TXN	12469	508	5	8	2044	16.4
TXN2	14595	1342	4	15	4191	28.7
PRDX1	10903	1262	6	16	4326	39.7
PRDX2	5061	886	5	8	2108	41.7
PRDX3	11131	1591	7	11	3108	27.9
PRDX4	18870	921	7	18	5162	27.4
PRDX5	3715	959	6	2	625	16.8
PRDX6	11461	1715	5	2	445	3.9
CASP1_caspase_1	9623	1373	10	0	0	0.0
CASP2	19387	4145	11	13	3544	18.3
CASP3	21751	2646	8	10	2857	13.1
CASP4	25699	1286	9	6	1495	5.8
CASP5	14730	1400	9	2	579	3.9
CASP6	14845	1661	4	14	3828	25.8
CASP7	51235	2586	7	15	4386	8.6
CASP8	54220	2903	10	41	10932	20.2
CASP9	32004	2034	5	34	8059	25.2
CASP10	46256	2054	10	44	11091	24.0
SOD1	9310	981	5	6	1596	17.1
SOD2	11032	1026	5	9	2555	23.2
SOD3	6410	1984	3	1	322	5.0
Catalase	33115	2279	13	12	3396	10.3
GPX1	1179	921	1or2	0	0	0.0
GPX2	3660	1024	2	2	454	12.4
GPX3	8631	1856	5	0	0	0.0
GPX4	2813	896	7	0	0	0.0
GPX5	8156	668	5	3	601	7.4
GPX6	12498	1712	5	3	924	7.4
GPX7	6679	1228	3	2	598	9.0
EPX	11577	2148	12	3	930	8.0
TPO	129261	3135	17	20	5355	4.1

The total Alu repeat length in the human respirome as a percentage of the total gene length examined (18.1%) is above that of the equivalent calculation for the whole human genome, reported by the International Human Gene Sequencing Consortium[5] to be 10.6%. We showed a paucity of or total absence of Alu sequences in particular components of the human respirome. That finding, coupled with the overall elevated level of Alu repeats in this group of genes would indicate that Alu sequences are significantly over-represented in other members of this group.

There is considerable sequence homology between some members of the respirome at both the gene and protein level, particularly within the globins, reflecting a common ancestral lineage. Common evolutionary pathways might be evoked as a mechanism for the observed low level of Alu insertion in those genes. However, most of the evolution of members of these genes took place before the separation of the human clade. The low level of Alu sequence incorporation into this group of genes compared to the genome-wide level of such incorporation might therefore reflect some protective mechanism against insertion.

7. REFERENCES

1. W. Erdmann, The oxygen molecule and its course from air to cell. *Adv Exp Med Biol.* **317**, 7-17 (1992).
2. D.Maguire. Cancer and the respirome. *Adv Exp Med Biol.* (2005) In press.
3. D. P. Morse, P. J. Aruscavage, and B. L. Bass, RNA hairpins in non-coding regions of human brain and *Caenorhabditis elegans* are edited by adenosine deaminases that act on RNA, PNAS **99**, 7906–7911 (2002).
4. E. Y. Levanon, E. Eisenberg, R. Yerlin, S. Nemzer, M. Halleger, R. Shemesh, Z. Y. Fligelman, S. R. Pollock, D. Sztybel, M. Olshansky, G. Rechavi, and M. F. Jantsch, Systematic identification of abundant A-to-I editing sites in the human transcriptome. *Nature Biotech,* **22**, 1001-1005(2004).
5. NCBI website; (http://www.ncbi.nlm.nih.gov)
6. Repeatmasker website; (http://repeatmasker.org)
7. International Human Genome Consortium, Initial sequencing and analysis of the human genome, *Nature* **409**, 860-921 (2001).

13.

ROLE OF DIFFERENCES IN MICROCIRCULATORY BLOOD FLOW VELOCITY IN OPTIMIZING PARAMETERS OF THE SKELETAL MUSCLE OXYGEN MODEL

Katherine G. Lyabakh & Irina N. Mankovskaya[*]

1. INTRODUCTION

Oxygen is the major substrate oxidizer in muscle fiber respiration. Its shortage induces functional-metabolic disturbances, namely, alteration of energy metabolism in cells. However, oxygen excess is also undesirable, being as a source of free radical oxidation reactions. The working range of oxygen concentration in skeletal muscle is limited and it is possible to speak of optimal pO_2 values. An optimal oxygen regimen (OR) in muscle tissue implies a reduction to minimum of areas with high and low tissue pO_2 values, irrespective of muscle motor activity. It was shown that the tissue O_2 supply is always sufficient for oxygen demand satisfaction[1]. At the same time, temporal-spatial heterogeneity of muscle pO_2, blood flow rate F (volume), or v (linear velocity) with changeable local capillarity is typical for the resting state. It creates oxygen mixing in tissues and the movement of hypoxic zones inside muscle if they occur. The action of local hypoxia depends on its degree and its duration. Thus the dynamics and heterogeneity of the microcirculation in specific cases may create either tissue pO_2 equalization or tissue damage, or training of the subcellular systems for hypoxic effects. It is suggested that blood flow heterogeneity is an important biophysical regulator of resting muscle OR on the level of the entire organ, creating the specific peculiarities of oxygen transport in muscle. The notion of blood flow heterogeneity within the microcirculation involves different blood flow velocities along muscle capillaries, and a variety of blood flow rates along vessels with different diameters, passing through muscle thickness and involved in oxygen transport. The first source of the pO_2 heterogeneity (and consequently the blood flow velocities) is the different vessel diameters and vessel interaction between each other due to their close proximity to each other.

[*] Katherine G. Lyabakh, Institute of Cybernetics, Irina N. Mankovskaya, Institute of Physiology, Kiev Ukraine 01024

As far back as the 1970s, it was shown that at the wall of arterioles with a diameter of 25μm the pO$_2$ value was 60-70 mm Hg, and at arterioles with a diameter of 10μm it was 30-50 mm Hg[2,3]. Thus, the greatest leakage of oxygen on the way to the capillaries occurred through resistance vessels, the arterioles, and the pO$_2$ value in capillaries might be less than in veins. The intensity of extracapillary oxygen flux in tissue, as evidenced by data from experiments performed on muscle at resting conditions[3] may achieve approximately half of the overall muscle gas exchange values [3,4]. The aim of this study was to investigate the different typical microcirculatory situations in skeletal muscle and estimate the influence of blood flow heterogeneity in creating and maintaining optimal OR in skeletal muscle.

2. METHODS

The OR was studied using the mathematical model of convective and diffusion transport of O$_2$ to muscle tissue at steady state [5,6]. The model describes the muscle fiber amidst similar fibers, adjacent to it. This model operates with the values of the muscle blood flow velocity F, size and location of vessels, intervessel distances d_{AV}, intercapillary distance d_C, oxygen demand qO_2, the blood oxygen capacity and pH, oxygen content in the arterial blood, position and form of the oxyhemoglobin dissociation curve, the blood and tissue oxygen diffusion and solubility coefficients, as well as Km – the apparent Michaelis constant for oxygen. We used a biochemical criterion of hypoxia - the state when oxygen consumption rate $VO_2(x, y,z)$ becomes less than oxygen demand qO_2 because of the lack of oxygen at a point of tissue (x,y,z). That provides an opportunity to get a graphical image of tissue hypoxia. The calculated value of ratio $S(x,y,z)=VO_2(x,y,z)/qO_2 \times 100(\%)$ describes the degree of hypoxia at (x,y,z). The surfaces $Si(x,y,z)$ represent constant values of VO_2 (where $VO_2 < qO_2$) and map hypoxia in the muscle fiber (Figure1).

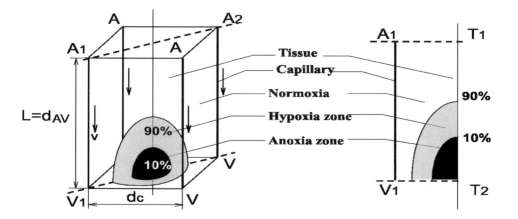

Figure 1. Model construction of muscle fiber with hypoxic zone inside; the model incorporates parts of the arteriole and venule, (shown by dotted lines) and capillaries. L- capillary length, A and V –capillary arterial and venous ends.

For simplicity, we show a symmetrical cross-section which passes through the axis and a capillary, thus obtaining a flat picture of the hypoxic zone. We present the simplest case: concurrent capillary blood flow with equal values of red blood cells velocity. This provides a symmetrical distribution of lines representing values of S. Therefore we need show only half of the picture. The model served as a basis for the information technology for the computation of oxygen consumption rates – the VO_2 field (Figure1), tissue and blood pO_2 values, and pO_2 distribution in fiber - pO_2 histogram.

Taking into consideration the observations[3] illustrated in Figure 2, we have considered the variants of local O_2 transport in microzones of muscle fiber at rest during transarteriolar and transvenular O_2 diffusion with and without participation of active capillaries in O_2 transport. Extracapillary O_2 transport was investigated using the configuration shown in Figure 1 with parts of arteriole (A_1A_2 dotted), venule (V_1V_2 dotted) and capillaries (AV). Along the lines A_1A_2 and V_1V_2, the values PaO_2 and PvO_2 respectively, were chosen according to experimental data presented in Figure2, from[2].

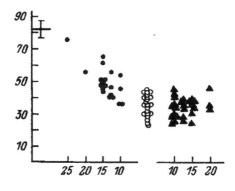

Figure 2. Experimental values of oxygen tension at the surface of muscle capillaries (squares), arterioles (circles), and venules (triangles). Ordinate - pO_2, mm Hg, abscissa - microvessel diameter,μ m.

3. RESULTS AND DISCUSSION

Since after leaving the arterioles, oxygen contributes to tissue oxygenation, it is worth paying attention to the specifics of its transport, namely: reciprocal positioning of arterioles and venules capable of interacting between each other, blood flow rate in the vessels, size and qO_2 of supply domain.

Mathematical modeling allows the investigation of the different cases of O_2 transport to tissue: oxygen is supplied from capillaries only, or by extracapillary means only in local zones, or by both ways: extra- and transcapillary. Fiure 3 shows the results of the calculation of local muscle pO_2 values between an arteriole and venule for different distances between them (d_{AV}) and for different values of qO_2 (for the case of extra-capillary O_2 transport). It is evident that between arteriole and venule an O_2 diffusive shunt can appear and its intensity is the greater the smaller are d_{AV} and qO_2: $qO_2 = 0.15$ ml/min/100g, $d_{AV} = 200\mu$m (Figure 3, curve 1). In the presence of arteriolar-venular shunt, pO_2 in vessels gradually decreased.

At d_{AV}=200μm and 0.2<qO_2<0.3 ml/min/100g, oxygen is transported predominantly from arterioles (Fig.3, curves 2,3). With increasing d_{AV} and qO_2 (200< d_{AV}< 600 μm and qO_2 >0.3 ml/ min/100g) venules can also participate in the process of muscle oxygenation (Figure 3, curves 4-6).

If qO_2=0.6 ml/min/100g and $d_{AV} \geq 300$ μm, PaO_2 and PvO will be equal to 50 and 30 mm Hg, respectively; this situation led to the appearance of hypoxic zones in muscle fiber (Figure 3, curves 5,6). Our calculations in part agree with the results of others using other models[7].

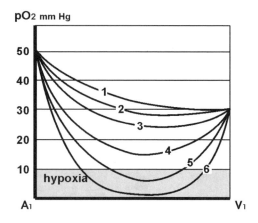

Figure 3. Tissue pO_2 between arteriole and venule at different qO_2 and d_{AV}=200μm: qO_2=0.15 ml/min/100g, (1), qO_2=0.20 ml/min/100g (2), qO_2=0.30 ml/min/100g (3): and at different d_{AV} and qO2=0.60 ml/min/100g,: d_{AV}=200μm (4), d_{AV}=300μm (5), d_{AV}=600μm (6).

We have investigated the case when $d_{AV} \leq 700$ μm, qO_2=0.15 ml/min/100g, PaO_2 =100mm Hg, and PvO_2=30 mm Hg. Calculations showed that in the case of such a large value of P(a-v) O_2 gradient, an oxygen diffusive shunt can exist. So, the greater the d_{AV} value the greater might be the P(a-v) O_2 gradient which led to the appearance of a diffusive O_2 shunt.

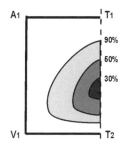

Figure 4. Hypoxic image in muscle fiber at rest (a half of the cross section, passing through the capillary AV, arteriole A_1T_1 and venule V_1T_2). Muscle fiber is fed by O_2 from capillaries, arteriole and venule. % denotes the degree of hypoxia at the curves.

With increasing qO_2, oxygen shunting diminishes and can disappear completely, especially during exercise when increasing oxygen demand is accompanied by increasing volume and linear blood flow velocities. When muscle fiber is simultaneously fed by O_2 from capillaries, arterioles, and venules, the muscle pO_2 in most cases would be equalized. But hypoxic zones may appear in muscle fiber at rest with slow F and long capillaries, e.g. F=1.6ml/min/100g) and L=1500μm,. PaO_2=60mm Hg, PvO_2=30mm Hg, F=1.6 ml/min/100g, d_{AV}=L=1500 μ m, qO_2=0.6 ml/min/100g, shown in Figure 4.

The period of arteriolar blood contact with muscle tissue is sufficient for oxyhemoglobin unload to attain half-saturation value in a resting muscle at a relatively slow muscle blood flow[3,4]. But if flow increases in a resistive vessel bed, erythrocytes may pass along the arterioles in less time than the minimum time required for O_2 exchange with the surrounding muscle tissue. Consequently, the ratio between volumes of extra- and transcapillary O_2 transport will be decreased. So at different blood flow velocities the sites of very intensive oxygen discharge may have a different spatial length and location. During exercise the diffusion oxygen transport to tissue will take place mainly through capillaries. Owing to the high linear blood flow velocity, the oxygen transfer by blood becomes maximally efficient. With changes in nutritive vessel length the diffusive resistance to oxygen flux will be significantly changed too.

It is known that tissue pO_2 is regulated mainly by changes in blood flow velocities v and diffusive resistance. The latter is related to O_2 tissue path length d. We tried to compare the efficiency of these two regulatory influences d and v on tissue OR.

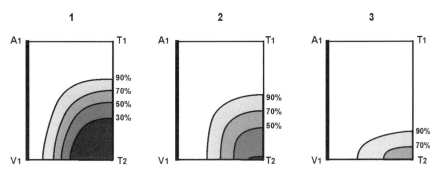

Figure 5. Comparison of v and d influences on a hypoxic zone in muscle fiber. % denotes the degree of hypoxia at the corresponding curves.

Figure 5 presents three variants of a calculated hypoxic image in resting muscle. Image 1 is the starting regimen with the vast zone of anoxia and hypoxia shown as a shaded area at qO_2=0.66ml/min/100g, F=4ml/min/100g, v=0.1mm/s and an intercapillary distance d_C=130 μm. Images 2 and 3 depict a reduction of the hypoxic zone owing to a 0.33-fold increase in the blood flow F=6ml/min/100g.

Image 2 shows to what extent the hypoxic zone is reduced with an increasing blood flow velocity F at the expense of an increase in v from 0.1 mm/s to 0.15 mm/s. Image 3 characterizes the hypoxic zone with a 0.33-fold blood flow increment (compared to initial F value) accompanied by an opening of reserve capillaries (d_C=105μm) but with the v value remaining unchanged (0.1 mm/s). The hypoxic zone becomes much smaller. Obviously, a change in linear blood flow velocity can serve as a significant regulator of oxygen transport in skeletal muscle. However, a change in intervessel distances d_C (as seen in Figure 5) exerts the most powerful influence on pO_2 and hypoxia image.

A logical continuation of the above will be to bring all these facts into one picture of oxygen transport regulation within a microcirculatory system of skeletal muscle: each time the oxygen supply by blood will differ depending on muscle functional activity, blood flow and distribution of its linear velocities within a microcirculatory system. The site of the system in which the oxygen passes into tissue may encompass various lengths of the arterial bed: arterioles, capillaries, and venules, or concentrate predominantly in capillaries. The resultant oxygen distribution in tissue will depend on the extracapillary/transcapillary oxygen flux ratio. Considering the high sensitivity of tissue pO_2 to alteration of the diffusion distance, one can suggest that regulation of oxygen tension in tissue is accomplished not only by a change in general oxygen delivery but also by change in diffusion parameters of the microcirculatory bed coupled with a change in F (apart from recruitment or cut-off of capillaries). This sharply increases the sensitivity of tissue pO_2 distribution to the peculiarities of oxygen delivery. The nutritive bed within a skeletal muscle microvasculature is very dynamic, and it changes its own structure and oxygen-transport properties with each change in blood flow rate in vessels.

4. CONCLUSIONS

Changes in extracapillary oxygen transport to tissue due to variation in microcirculatory blood flow velocities have an impact on parameters of the nutritive vessel bed normalizing the muscle oxygen regimen of skeletal muscle under changes in its functional activity.

5. REFERENCES

1. D.K.Harrison, S. Birkenhake, S.K.Knauf and M. Kessler, Oxygen supply and blood flow regulation in contracting muscle in dogs and rabbits, J. Physiol., 227-243 (1990).
2. B. Duling, R.Berne, Longitudinal gradient in periarteriolar pO_2, Circ. Res., 669-678 (1970).
3. K.P.Ivanov, E.P. Vovenko and A.N Dery, Oxygen pressure in muscle capillaries and mechanisms of capillary gas exchange, DAN SSSR, 265, 494-497 (1982) (in Russian).
4. V. Moshizuki, Study on the oxyhemoglobin velocity, Jap. J. Physiol., 635-648 (1966).
5. K.G. Lyabakh, Mathematical modeling of oxygen transport in skeletal muscle during exercise: hypoxia and VO_2 max, Adv. Exper. Med. Biol., 471, 585-593 (1999).
6. K.G. Lyabakh, I.N., Mankovskaya, Oxygen transport to skeletal muscle working at VO2max at acute hypoxia: theoretical prediction, Comparative Biochemistry and Physiology (A), 53-60 (2002).
7. A. Popel, Oxygen diffusive shunt under heterogeneous oxygen delivery, J.Theor. Biol., 533-541 (1982).

14.

CHRONIC HYPOXIA MODULATES ENDOTHELIUM-DEPENDENT VASORELAXATION THROUGH MULTIPLE INDEPENDENT MECHANISMS IN OVINE CRANIAL ARTERIES

William J. Pearce, James M. Williams, Mohammad W. Hamade, Melody M. Chang, and Charles R. White[*]

1. INTRODUCTION

Of the many external physiological perturbations that may alter blood O_2 transport, chronic hypoxia is probably the most widely studied. Chronic hypoxia is a challenge not only for mountain climbers and pilots, but also for patients with respiratory disease and fetuses compromised by placental insufficiency. Motivated by this clinical relevance, numerous studies have detailed the effects of hypoxic adaptation. Because physiological stresses that influence O_2 delivery stimulate homeostatic responses not only in blood composition, but also in the blood vessels that serve at the interface between oxygen supply and demand, many studies have examined hypoxic modulation of vascular composition and function[1-6]. For example, high altitude acclimatization has been shown to alter vascular protein content, contractility, receptor profile, and perivascular nerve function[7-9]. Aside from the intense attention focused on the effects of hypoxia on vascular contractility, however, comparatively little scrutiny has been directed toward the effects of chronic hypoxia on mechanisms of vasorelaxation. This is somewhat surprising, given that vasodilatation is a key initial response to acute hypoxia in most vascular beds[10, 11]. In particular, the effects of chronic hypoxia on endothelial vasodilator function remain largely unexplored, even though the endothelium may significantly augment vasodilator responses to acute hypoxia[12]. The present study was designed to address this deficit.

---

[*] The mailing address for all authors is: Center for Perinatal Biology, Loma Linda University School of Medicine, Loma Linda, CA, 92350. For correspondence, e-mail William Pearce at wpearce@som.llu.edu.

2. MECHANISMS OF ENDOTHELIUM-DEPENDENT VASODILATATION

To examine the effects of chronic hypoxia on endothelial vasodilator function, we used common carotid arteries taken from young non-pregnant adult sheep that had been maintained for 110 days at an altitude of 3820 m. The arteries were mounted in tissue baths for measurements of contractile responses, as previously described in detail[7, 12]. To enable the study of relaxation responses, the arteries were first contracted with 1 μM serotonin, which is approximately the EC_{50} concentration in this preparation[13]. To stimulate endothelium-mediated vasorelaxation, we treated the arteries with 1 μM A23187. This calcium ionophore facilitates calcium entry into endothelial cells in a receptor-independent manner, and is thus highly useful for maximally stimulating endothelial NO release[14]. To verify that all A23187-induced relaxation was due to activation of the enzyme eNOS (endothelial nitric oxide synthase), we also verified that treatment with 100 μM L-Nitro-Arginine Methyl Ester (L-NAME) could completely block all responses to A23187 (see Figure 1).

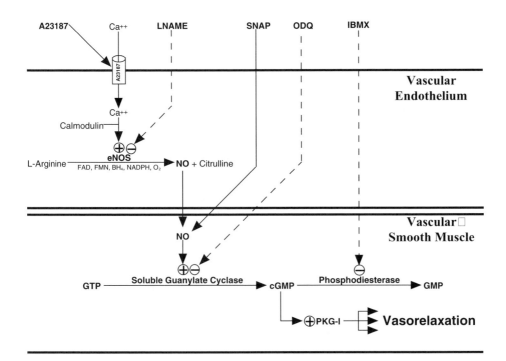

Figure 1. Endothelium-dependent vasodilatation is initiated by a rise in endothelial calcium concentration, which in turn stimulates the enzyme eNOS to synthesize and release NO. The NO is synthesized from L-Arginine and requires the cofactors FAD, FMN, BH4, NAPDPH, and O_2. This NO then diffuses into adjacent smooth muscle where it binds to the heme moiety on soluble guanylate cyclase and activates the synthesis of cGMP from GTP. This cGMP then combines with protein kinase G (PKG) to promote vasorelaxation through multiple mechanisms until is it hydrolyzed by phosphodiesterase. Please see text for additional details.

In response to A23187, carotid arteries from normoxic animals relaxed 65±5% whereas responses from hypoxic animals averaged 71±5%; this difference was not significant. In contrast, when exogenous NO was administered via the donor S-nitroso-N-acetyl-penacilamine (SNAP, see Figure 1), magnitudes of relaxation averaged 70±4 and 87±3% in normoxic and hypoxic arteries, respectively. This significant difference revealed that hypoxia enhanced reactivity to NO. This latter finding, together with the parallel finding that the magnitude of endothelium-dependent relaxation was unchanged by hypoxia, and the finding that NO is the main endothelium-dependent vasodilator in this preparation[15], suggests that hypoxia must correspondingly depress endothelial synthesis and release of NO.

3. CHRONIC HYPOXIA DEPRESSES NO RELEASE AND eNOS SPECIFIC ACTIVITY

To directly test the hypothesis that chronic hypoxia depresses endothelial NO release, we perfused 4-cm lengths of carotid arteries at 4 ml/min with a 20 mM Hepes buffer at a physiological transmural pressure (60 mm Hg). The perfusate collected was analyzed for NO content using a flurometric assay based on the nitrate reductase method[16]. In response to 1 μM A23187 added to the perfusate, NO release was significantly greater in normoxic than in hypoxic arteries (Figure 2). To further understand the basis for this effect, we measured the abundance of eNOS in the perfused arteries using standard Western blotting, as previously described[15]. Values of eNOS abundance were then divided into the corresponding rates of NO release to estimate in situ rates of maximal eNOS activity. Chronic hypoxia had no significant effect on eNOS abundance, but significantly depressed eNOS specific activity (Figure 2), suggesting that chronic hypoxia may alter one or more of the many mechanisms now recognized to govern posttranslational modification of eNOS[17-19].

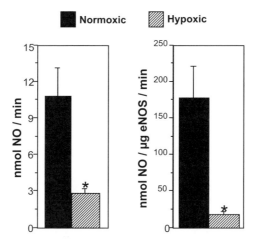

Figure 2. In ovine carotid arteries, chronic hypoxia significantly depressed the maximal rates of NO release induced by A23187 (left panel). Hypoxic depression of NO release was attributable primarily to reductions in the specific activity of the eNOS enzyme (right panel), and did not involve significant changes in eNOS abundance. Vertical error bars indicate standard errors for carotid arteries from 6-8 sheep in each group. Asterisks indicate significant effects of hypoxia at the P<0.05 level.

4. CHRONIC HYPOXIA DEPRESSES SOLUBLE GUANYLATE CYCLASE ABUNDANCE AND V_{MAX}

Our finding that NO-induced relaxation was enhanced in arteries from chronically hypoxic animals suggests that chronic hypoxia acts within vascular smooth muscle cells to upregulate mechanisms mediating cGMP-dependent vasodilatation (Figure 1). To test this hypothesis, we repeated our measurements of NO-induced relaxation in the presence of 10 μM ODQ, a selective and specific inhibitor of soluble guanylate cyclase[20]. Treatment with ODQ completely eliminated all relaxation responses to exogenous NO, thus verifying that soluble guanylate cyclase (sGC) was the predominant "NO receptor" in this tissue. Using a broken cell preparation as previously described[21], we next determined substrate-velocity relations for sGC. Although chronic hypoxia had no significant effect on K_m values, it significantly depressed maximal rates of cGMP synthesis (Figure 3, left panel). To further understand the basis for this effect, we also measured the abundance of sGC via Western blot as previously described[21]. Values of soluble guanylate cyclase abundance were then divided into the corresponding rates of cGMP synthesis to estimate in situ rates of maximal soluble guanylate cyclase activity. Chronic hypoxia significantly depressed abundance in ovine carotid arteries (Figure 3, center panel), suggesting that chronic hypoxia must depress rates of transcription of the genes for one or more sGC subunits[22], or must depress either mRNA stability or efficiency of translation. These are clearly testable hypotheses that offer promise in furthering understanding of how chronic hypoxia modulates mechanisms of vasorelaxation. In contrast, chronic hypoxia had no significant effect on in situ estimates of sGC specific activity (Figure 3, right panel). This absence of physiological regulation of sGC specific activity is consistent with the view that post-translational regulation of sGC activity is rare and has not been elucidated despite considerable investigative effort[23, 24].

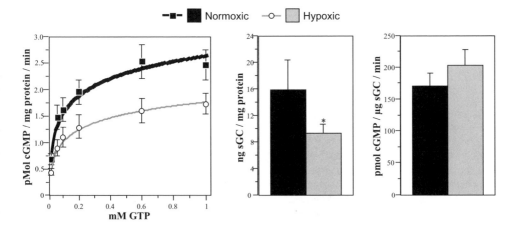

Figure 3. Substrate-velocity measurements for soluble guanylate cyclase (left panel) revealed that chronic hypoxia depressed the Vmax without significantly affecting Km values. Chronic hypoxia also significantly depressed the abundance of soluble guanylate cyclase, as measured via quantitative Western blots and expressed relative to total soluble protein. In situ estimates of soluble guanylate cyclase specific activity (right panel) revealed no significant effect of chronic hypoxia. Vertical error bars indicate standard errors for carotid arteries from 6-8 sheep in each group. The asterisk indicates a significant effect of hypoxia at the P<0.05 level.

5. CHRONIC HYPOXIA DEPRESSES PHOSPHODIESTERASE ACTIVITY

Given that relaxation responses to exogenous NO were enhanced by chronic hypoxia, we expected that chronic hypoxia would upregulate sGC activity and rates of cGMP synthesis. Because this is the opposite of what we found, it is clear that hypoxia upregulates some other component of NO-dependent vasorelaxation, downstream from cGMP synthesis. One possible candidate would be rates of cGMP degradation by phosphodiesterase (Figure 1). A variety of phosphodiesterase (PDE) subtypes are abundant in vascular smooth muscle[25, 26] and are highly subject to physiological regulation[27]. To examine the effects of chronic hypoxia on PDE activity, we performed timed measurements of NO-induced cGMP accumulation in the presence and absence of the PDE inhibitor IBMX, as previously described[28]. In normoxic arteries, rates of phosphodiesterase activity (in pmol cGMP / min/ mg protein) averaged 12.3±4.0 (N=8), but in hypoxic arteries averaged only 3.4±0.5 (N=7); chronic hypoxia significantly depressed PDE activity against cGMP in ovine carotid arteries. Which isoforms were involved in this effect, and the relative importance of decreased expression and decreased enzyme activity remain unexplored but promising topics for future investigation.

6. SUMMARY

Acclimatization to chronic hypoxia involves numerous compensatory changes in many tissues, including blood vessels. The present data demonstrate that in addition to well-documented changes in contractility, chronic hypoxia also produces important changes in the mechanisms mediating endothelium-dependent vasodilatation. At the level of the endothelium, hypoxia attenuates endothelial release of NO and this appears to be mediated through reductions in eNOS specific activity; chronic hypoxia has little effect on eNOS abundance. In contrast, chronic hypoxia depresses the abundance of sGC, which functions as the downstream vascular receptor for NO released from the endothelium. The decreased abundance of sGC produced by chronic hypoxia occurs without changes in sGC specific activity and results in decreased rates of NO-induced cGMP synthesis. Nonetheless, the vasodilator efficacy of NO is enhanced in hypoxic arteries, which suggests that mechanisms downstream from sGC are upregulated by hypoxia. Consistent with this view, chronic hypoxia significantly depresses PDE activity, which serves to prolong cGMP half-life and enhance its vasodilator effects. It remains possible that chronic hypoxia may also enhance PKG activity and/or the abundance of its substrates; this possibility remains a promising topic for future investigation. Overall, it is important to recognize that the mechanisms of adaptation to chronic hypoxia identified in the present study may be somewhat unique to adult carotid arteries. Adaptive responses to chronic hypoxia can vary considerably between small and large arteries, and also between immature and adult arteries[7]. Still, the present data clearly demonstrate that both the endothelium and vascular smooth muscle of major arteries are profoundly influenced by chronic hypoxia, and thereby participate fully in whole-body adaptation to reduced oxygen availability.

7. REFERENCES

1. M. L. Fung, Hypoxia-inducible factor-1: a molecular hint of physiological changes in the carotid body during long-term hypoxemia?, *Curr Drug Targets Cardiovasc Haematol Disord* **3**(3), 254-9 (2003).

2. C. Fradette and P. Du Souich, Effect of hypoxia on cytochrome P450 activity and expression, *Curr Drug Metab* **5**(3), 257-71 (2004).

3. L. G. Moore, M. Shriver, L. Bemis, B. Hickler, M. Wilson, T. Brutsaert, E. Parra and E. Vargas, Maternal adaptation to high-altitude pregnancy: an experiment of nature--a review, *Placenta* **25**(Suppl A), S60-71 (2004).

4. C. Peers and P. J. Kemp, Ion channel regulation by chronic hypoxia in models of acute oxygen sensing, *Cell Calcium* **36**(3-4), 341-8 (2004).

5. C. A. Raguso, S. L. Guinot, J. P. Janssens, B. Kayser and C. Pichard, Chronic hypoxia: common traits between chronic obstructive pulmonary disease and altitude, *Curr Opin Clin Nutr Metab Care* **7**(4), 411-7 (2004).

6. J. T. Reeves and F. Leon-Velarde, Chronic mountain sickness: recent studies of the relationship between hemoglobin concentration and oxygen transport, *High Alt Med Biol* **5**(2), 147-55 (2004).

7. L. D. Longo, A. D. Hull, D. M. Long and W. J. Pearce, Cerebrovascular adaptations to high-altitude hypoxemia in fetal and adult sheep, *Am J Physiol* **264**(1), R65-72 (1993).

8. N. Ueno, Y. Zhao, L. Zhang and L. D. Longo, High altitude-induced changes in alpha1-adrenergic receptors and Ins(1,4,5)P3 responses in cerebral arteries, *Am J Physiol* **272**(2), R669-74 (1997).

9. J. Buchholz, K. Edwards-Teunissen and S. P. Duckles, Impact of development and chronic hypoxia on NE release from adrenergic nerves in sheep arteries, *Am J Physiol* **276**(3), R799-808 (1999).

10. W. J. Pearce, Mechanisms of hypoxic cerebral vasodilatation, *Pharmacol Ther* **65**(1), 75-91 (1995).

11. J. M. Marshall, Adenosine and muscle vasodilatation in acute systemic hypoxia, *Acta Physiol Scand* **168**(4), 561-73 (2000).

12. W. J. Pearce, S. Ashwal and J. Cuevas, Direct effects of graded hypoxia on intact and denuded rabbit cranial arteries, *Am J Physiol* **257**(3), H824-33 (1989).

13. G. Q. Teng, J. Williams, L. Zhang, R. Purdy and W. J. Pearce, Effects of maturation, artery size, and chronic hypoxia on 5-HT receptor type in ovine cranial arteries, *Am J Physiol* **275**(3), R742-53 (1998).

14. H. Taniguchi, Y. Tanaka, H. Hirano, H. Tanaka and K. Shigenobu, Evidence for a contribution of store-operated Ca2+ channels to NO-mediated endothelium-dependent relaxation of guinea-pig aorta in response to a Ca2+ ionophore, A23187, *Naunyn Schmiedebergs Arch Pharmacol* **360**(1), 69-79 (1999).

15. J. M. Williams, A. D. Hull and W. J. Pearce, Maturational Modulation of Endothelium-Dependent Vasodilatation in Ovine Cerebral Arteries, *Am J Physiol Regul Integr Comp Physiol* **288**(1), R149-57 (2004).

16. D. J. Kleinhenz, X. Fan, J. Rubin and C. M. Hart, Detection of endothelial nitric oxide release with the 2,3-diaminonapthalene assay, *Free Radic Biol Med* **34**(7), 856-61 (2003).

17. Y. C. Boo and H. Jo, Flow-dependent regulation of endothelial nitric oxide synthase: role of protein kinases, *Am J Physiol Cell Physiol* **285**(3), C499-508 (2003).

18. I. Fleming and R. Busse, Molecular mechanisms involved in the regulation of the endothelial nitric oxide synthase, *Am J Physiol Regul Integr Comp Physiol* **284**(1), R1-12 (2003).

19. R. D. Minshall, W. C. Sessa, R. V. Stan, R. G. Anderson and A. B. Malik, Caveolin regulation of endothelial function, *Am J Physiol Lung Cell Mol Physiol* **285**(6), L1179-83 (2003).

20. J. Garthwaite, E. Southam, C. L. Boulton, E. B. Nielsen, K. Schmidt and B. Mayer, Potent and selective inhibition of nitric oxide-sensitive guanylyl cyclase by 1H-[1,2,4]oxadiazolo[4,3-a]quinoxalin-1-one, *Mol Pharmacol* **48**(2), 184-8 (1995).

21. C. R. White, X. Hao and W. J. Pearce, Maturational differences in soluble guanylate cyclase activity in ovine carotid and cerebral arteries, *Pediatr Res* **47**(3), 369-75 (2000).

22. M. Russwurm and D. Koesling, Isoforms of NO-sensitive guanylyl cyclase, *Mol Cell Biochem* **230**(1-2), 159-64 (2002).

23. D. Koesling and A. Friebe, Soluble guanylyl cyclase: structure and regulation, *Rev Physiol Biochem Pharmacol* **135**(1), 41-65 (1999).

24. S. Andreopoulos and A. Papapetropoulos, Molecular aspects of soluble guanylyl cyclase regulation, *Gen Pharmacol* **34**(3), 147-57 (2000).

25. T. Matsumoto, T. Kobayashi and K. Kamata, Phosphodiesterases in the vascular system, *J Smooth Muscle Res* **39**(4), 67-86 (2003).

26. S. D. Rybalkin, C. Yan, K. E. Bornfeldt and J. A. Beavo, Cyclic GMP phosphodiesterases and regulation of smooth muscle function, *Circ Res* **93**(4), 280-91 (2003).

27. S. H. Francis, I. V. Turko and J. D. Corbin, Cyclic nucleotide phosphodiesterases: relating structure and function, *Prog Nucleic Acid Res Mol Biol* **65**(1), 1-52 (2001).

28. C. R. White and W. J. Pearce, Effects of maturation on cyclic GMP metabolism in ovine carotid arteries, *Pediatr Res* **39**(1), 25-31 (1996).

15.

THE "HEMOLYSIS MODEL" FOR THE STUDY OF CYTO-TOXICITY AND CYTO-PROTECTION BY BILE SALTS AND PHOSPHOLIPIDS

Piero Portincasa, Antonio Moschetta, Michele Petruzzelli, Michele Vacca, Marcin Krawczyk, Francesco Minerva, Vincenzo O. Palmieri and Giuseppe Palasciano[*]

1. INTRODUCTION

Hemolysis of human erythrocytes is a simple *in vitro* model to assess both toxic and protective properties of different molecules. The method was extensively used to characterize the detergent properties of bile salts. Erythrocyte lysis strongly depends on bile salt concentration.[1] We systematically characterized the protection offered by different phospholipids and elucidated molecular mechanisms underlining this phenomenon. Here we describe the hemolysis model, review its use in lipid research, and underscore the similarities with other commonly used *in vitro* models for cyto-toxicity and cyto-protection.

2. THE "HEMOLYSIS MODEL"

Aliquots of fresh blood obtained from healthy subjects with a syringe containing 100 units of heparin[2] are sedimented 3 times by centrifugation at 3000 rpm for 15 min; plasma and buffy coat are discarded and pellet reconstituted to the original blood volume in TRIS-buffer of the following composition (mM): TRIS 10, NaCl 130, glucose 10, pH 7.4 (Figure 1). Although the buffer solution provides an isotonic environment and glucose sustains erythrocytes activity, experiments must be performed within a few hours to avoid erythrocyte distress. For the incubation, washed erythrocytes are added to testing substances dissolved in the same buffer solution in a 1:4 ratio and final vol. of 1 mL, at 37°C for 30 to 45 min.[2-5] Thereafter, 7 mL of buffer are added to reduce hemolysis to negligible levels. Samples are centrifuged to sediment pellet and the absorbance of

---
[*] Piero Portincasa, Antonio Moschetta, Michele Petruzzelli, Michele Vacca, Marcin Krawczyk, Francesco Minerva, Vincenzo O. Palmieri and Giuseppe Palasciano, Clinica Medica "*A. Murri*", Department of Internal & Public Medicine, University Medical School, Piazza G. Cesare 11, 70124 Bari, Italy.

hemoglobin is assayed in the supernatant at 540 nm. Hemoglobin release (a sensitive index of hemolysis) is then normalised to maximal hemolysis of erythrocytes incubated with distilled water.[2-5]

3. BILE SALTS AND HEMOLYSIS

Bile salts are essential lipid components in bile, with phospholipids and cholesterol. Bile salts have a potent intrinsic detergent activity which allows the solubilization of cholesterol and phospholipids in the biliary tract and intestine. This step interaction is essential for intestinal lipid absorption, bile formation and cholesterol disposal.[6] The "bad" side of bile salt,[7] however, is their cellular toxicity which depends on the degree of hydrophobicity[5, 8] and on cell membrane composition.[9] Although bile salt toxicity has long been known, the mechanisms by which bile salts injure cells have been troublesome to elucidate. Since both toxicity and detergency of bile salts increase with increasing hydrophobicity, it was suggested that the toxicity of this molecule was due to plasma membrane disruption. The hemolysis model was particularly suitable to test this hypothesis: in fact, since erythrocytes lack the nucleus and are poor in intracellular organelles, they can not respond to bile salts injury by adapting some tissue-specific pathways such as bile salt uptake, transport or metabolism,[10] and most effects result from interaction between bile salts and the plasma membrane.[11] In pioneering studies, Coleman et al. [1-12] showed that the hemolytic activity of bile salts occurs exclusively at concentrations above their critical micellar concentration and strongly depends on the degree of hydrophobicity. Later, we systematically studied the effects of bile salts hydrophilic-hydrophobic balance on membrane damage.[5] The detergent effect induced by increasing concentrations of various bile salts was analysed: the most hydrophobic taurodeoxycholate caused 100% hemolysis at low concentrations (5 mM), while slightly higher concentrations were needed for the less hydrophobic taurocholate (40 mM). These bile salt concentrations are in line with those measured in the gallbladder bile under normal conditions. In contrast, hydrophilic tauroursodeoxycholate caused only moderate hemolysis, even at very high -not physiological- concentrations (see Figure 2A).[5]

The hemolysis model is also valid to study the protective role of molecules. Heuman et al.[8] demonstrated the protective role of conjugates of ursodeoxycholate on total bile salt toxicity. The protection offered by lecithins against bile salt-induced damage on plasma membrane was the object of investigation by our group.[2, 4] Velardi et al.[4] studied the effect of inclusion of phosphatydilcholine within bile salt micelles on bile salt induced hemolysis: phosphatydilcholine decreased hemolysis induced by 50 mM sodium-taurocholate in a concentration-dependent manner. These studies have a clear pathophysiological relevance. At the concentrations occurring in hepatic and gallbladder biles, bile salts could theoretically damage the hepatocyte, but the biliary phosphatidylcholine plays a complete protective role against this phenomenon. This scenario is further supported by the fact that mice with homozygous disruption of the *mdr*2 gene exhibit severe bile salt-induced hepatocyte damage in vivo. Since *mdr*2 encoded a P-glycoprotein that facilitates the phosphatidylcholine secretion in bile, there is no phosphatidylcholine protecting against bile salt-induced toxicity in the bile of these mice.[13] Interestingly, the amounts of phosphatidylcholine needed to protect against bile salt-induced cyto-toxicity positively correlated with the bile salt degree of hydrophobicity (see Figure 2B);[5] at increasing bile salt hydrophobicity index, both toxicity and amounts of protection needed increase.

Both phosphatidylcholine and sphingomyelin are the major phospholipids of the hepatocyte canalicular membrane outer leaflet.[14,15] Even if PC is the most abundant phospholipid in bile,[16] significant amounts of sphingomyelin might be present, particularly under cholestatic conditions.[17] Therefore, we compared the protective effects of phosphatidylcholine, sphingomyelin and dipalmitoyl phosphatidylcholine against taurocholate induced hemolysis. Sphingomyelin was able to strongly reduce hemolysis of human erythrocytes, while phosphatidylcholine was less effective.[2]

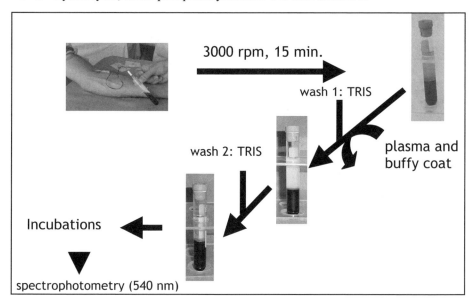

Figure 1. Steps in erythrocyte preparation to study the cell toxicity and/or protection by different molecules using the "hemolysis model". See text for details.

Several lines of evidence suggest that toxicity induced by bile salts takes place in conditions when this molecule is present as monomers and "simple" micelles, i.e. without incorporated lipids.[18-20] This so-called "intermixed micellar-vesicular bile salt concentration" may be responsible for the damaging effect on membrane bilayers. The protective effect of phospholipid inclusion within bile salt micelles is likely to occur at the level of these micelles rather than by transfer of phospholipid molecules from micelles into the plasma membrane. With sphingomyelin, in fact, there is an increased amount of bile salts associated with phospholipid in mixed micelles compared to phosphatidylcholine, and this could account for the different degree of protection offered by the two phospholipids.[2] Taking these data together, one might speculate that any water-insoluble compound that needs to be incorporated into the bile salt micelles, should be able to protect against bile salt toxicity by decreasing the free -intermixed micellar-bile salt concentration. However, the scenario is not so simple, since interactions between bile salts and other molecules might also greatly influence their micellar physical-chemical properties.[21, 22] In this respect, we have preliminary observations about some endogenous as well as exogenous molecules able to decrease the free bile salt concentration but to increase the bile salt toxicity.

4. ERYTHROCYTES vs. CACO2: HEMOLYSIS vs. LDH RELEASE

Lactate dehydrogenase (LDH) release from cell culture model has been extensively used in the literature for the study of cyto-toxicity and cyto-protection. Briefly, postconfluent cell cultures are incubated with different testing solution (i.e. bile salt or bile salt plus phospholipids mixtures) and after 30 min incubation, the medium is collected and the cells are treated with 0.4% Triton X-100.[2, 5] LDH activity, as a sensitive parameter of cell damage, is measured according to Mitchell et al.[23] in both medium and in Triton X-100 treated cells. Fat-free bovine serum albumin (final concentration 0.6%) is added to prevent interference of bile salts with the spectrophotometric assay of LDH activity. Enzyme activity in each single experiment is normalized to percentage of total LDH activity (medium plus Triton X-100 treated cells). Table 1 depicts the main studies using the hemolysis method in comparison to the CaCo-2 cell method to study bile salt cyto-toxicity and phospholipid cyto-protection. CaCo-2 represents a human neoplastic colon cell line commonly used as a model system for gastrointestinal epithelium.

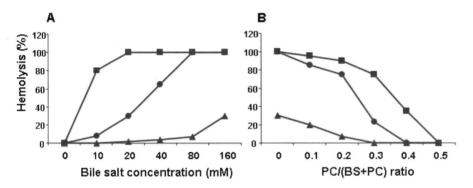

Figure 2. The "hemolysis model" for the study of cyto-toxicity by bile salts and cyto-protection by phosphatidylcholine. (**A**) Bile salt-induced hemolysis is directly related to their degree of hydrophobicity. Whereas low concentrations of hydrophobic TDC (squares) or slightly higher concentrations of intermediate hydrophobic TC (scatters) result in 100% hemolysis, this is not the case for hydrophilic TUDC (triangles) even after incubation with extremely high concentrations. (**B**) Hemolysis is completely protected at increasing PC concentrations. However, the amounts of phosphatidylcholine (PC) needed to protect against bile salts (BS)-induced hemolysis is directly related with their degree of hydrophobicity.

Heuman et al.[8] studied the relationship between bile salts hydrophobicity and toxicity on erythrocytes and on primary monolayer cultures of adult rat hepatocytes: in both systems, cytolysis increased with increasing bile salts concentration and hydrophobicity. Velardi et al.[4] compared the effect of phosphatidylcholine on bile salt-induced cyto-toxicity in human erythrocytes and cultured CaCo-2 cells. With respect to sensitivity to bile salts and protective effect of lecithin, CaCo-2 cells behaved very similar to erythrocytes: in both *in vitro* models inclusion of phosphatidylcholine within bile salt micelles resulted in increased protection against bile salt-induced cyto-toxicity. Similar to the data obtained on the "hemolysis model", the amounts of phosphatidylcholine needed to protect against bile salt-induced LDH release increases at higher bile salt hydrophobicity index.[5] Moschetta et al.[2] compared the protective effects of different phospholipids on bile salts toxicity in complementary in vitro studies: sphingomyelin was able to strongly reduce both hemolysis of human erythrocytes and LDH release by CaCo-2 cells, while phosphatidylcholine was less effective.

5. CONCLUSIONS

The hemolysis of human erythrocytes can be proposed as a valuable *in vitro* model to study the interaction of different molecules with the plasma membrane: given its simplicity and versatility, this method provides a reliable tool to screen potentially damaging or protective compounds like drugs, nutrients and endogenous molecules, with relevance to the pathophysiology and treatment of human diseases.

6. ACKNOWLEDGEMENTS

The authors thank Drs. K.J. van Erpecum, G.P. vanBerge-Henegouwen and J. DeSilvio for criticisms and helpful discussion; Mrs. R. De Venuto and P. Debenedictis for excellent technical assistance during the experiments.

Table 1. Different studies using hemolysis and cell culture methods to assess bile salt cyto-toxicity and phospholipid cyto-protection.

Authors (ref.)	Year	Model	Major finding
Coleman et al.[12]	1979	Hemolysis	Damaging properties of bile
Coleman et al.[1]	1980	Hemolysis	Bile salts induce hemolysis in a concentration-dependent manner
Heuman et al.[8]	1991	Hemolysis and cell culture	Protection by ursodeoxycholate against bile salt-induced cytolysis
Velardi et al.[4]	1991	Hemolysis and cell culture	Protection against bile salt cyto-toxicity by lecithin
Lapre et al.[3]	1992	Hemolysis and cell culture	Lytic effects of mixed micelles of fatty acids and bile acids
Puglielli et al.[24]	1994	Hemolysis	Native phospholipids protect against bile salt-induced damage.
Narain et al.[25]	1999	Hemolysis	Lecithin but not cholesterol protects against bile salt-damage
Moschetta et al.[2]	2000	Hemolysis and cell culture	Enhanced protection by sphingomyelin compared to phosphatidylcholine against bile salt-induced toxicity
Barrios et al.[26]	2000	Hemolysis and ex vivo tissue	Biliary phosphatidylcholine protects against bile salt-induced toxicity
Moschetta et al.[5]	2001	Hemolysis and cell culture	More phospholipids are needed to protect from bile salt hydrophobicity
Carubbi F al.[27]	2002	Hemolysis and cell culture	Cytotoxic and cytoprotective effects of hydrophilic bile salts
Van Gorkom et al.[28]	2002	Hemolysis	Bile salt composition in the hemolysis by human fecal water

7. REFERENCES

1. R. Coleman, P. J. Lowe, and D. Billington. Membrane lipid composition and susceptibility to bile salt damage. *Biochim Biophys Acta* **599**, 294-300 (1980).
2. A. Moschetta, G. P. van Berge-Henegouwen, P. Portincasa, G. Palasciano, A. K. Groen, and K. J. van Erpecum. Sphingomyelin exhibits greatly enhanced protection compared with egg yolk phosphatidylcholine against detergent bile salts. *J Lipid Res* **41**, 916-924 (2000).
3. J. A. Lapre, D. S. M. Termont, A. K. Groen, and R. van der Meer. Lytic effects of mixed micelles of fatty acids and bile acids. *Am J Physiol* **263** (26), G333-G337 (1992).
4. A. L. M. Velardi, A. K. Groen, R. P. Oude Elferink, R. van der Meer, G. Palasciano, and G. N. Tytgat. Cell type-dependent effect of phospholipid and cholesterol on bile salt cytotoxicity. *Gastroenterology* **101**, 457-464 (1991).
5. A. Moschetta, G. P. vanBerge-Henegouwen, P. Portincasa, W. Renooij, A. K. Groen, K. J. van Erpecum. Hydrophilic bile salts enhance differential distribution of sphingomyelin and phosphatidylcholine between micellar and vesicular phases: potential implications for their effects in vivo. *J Hepatol* **34** (4), 492-499 (2001).
6. A. F. Hofmann. The continuing importance of bile acids in liver and intestinal disease. *Arch Intern Med* **159** (22), 2647-2658 (1999).
7. A. F. Hofmann. Bile Acids: The Good, the Bad, and the Ugly. *News Physiol Sci* **14**, 24-29 (1999).
8. D. M. Heuman, W. M. Pandak, P. B. Hylemon, and Z. R. Vlahcevic. Conjugates of ursodeoxycholate protect against toxicity of more hydrophobic bile salts: In vitro studies in rat hepatocytes and human erythrocytes. *Hepatology* **14** (5), 920-926 (1991).
9. L. Amigo, H. Mendoza, S. Zanlungo, J. F. Miquel, A. Rigotti, S. Gonzalez et al. Enrichment of canalicular membrane with cholesterol and sphingomyelin prevents bile salt-induced hepatic damage. *J Lipid Res* **40**, 533-542 (1999).
10. G. A. Kullak-Ublick, B. Stieger, P. J. Meier. Enterohepatic bile salt transporters in normal physiology and liver disease. *Gastroenterology* **126** (1), 322-342 (2004).
11. D. M. Heuman. Bile salt-membrane interactions and the physico-chemical mechanisms of bile salt toxicity. *Ital J Gastroenterol* **27** (7), 372-375 (1995).
12. R. Coleman, S. Iqbal, P. P. Godfrey, and D. Billington. Membranes and bile formation. Composition of several mammalian biles and their membrane-damaging properties. *Biochem J* **178**, 201-208 (1979).
13. J. J. Smit, A. H. Schinkel, R. P. J. Oude Elferink, A. K. Groen, E. Wagenaar, L. van Deemter et al. Homozygous disruption of the murine mdr2 P-glycoprotein gene leads to a complete absence of phospholipid from bile and to liver disease. *Cell* **75** (3), 451-462 (1993).
14. .T. Kremmer, M. H. Wisher, and W. H. Evans. The lipid composition of plasma membrane subfractions originating from the three major functional domains of the rat hepatocyte cell surface. *Biochim Biophys Acta* **455** (3), 655-664 (1976).
15. J. A. Higgins, and W. H. Evans. Transverse organization of phospholipids across the bilayer of plasma membrane subfractions of rat hepatocytes. *Biochem J* **174**, 563-567 (1978).
16. D. Alvaro, A. Cantafora, A. F. Attili, S. C. Ginanni, C. De Luca, G. Minervini et al. Relationships between bile salts hydrophilicity and phospholipid composition in bile of various animal species. *Comp Biochem Physiol [B]* **83** (3), 551-554 (1986).
17. S. G. Barnwell, B. Tuchweber, and I. M. Yousef. Biliary lipid secretion in the rat during infusion of increasing doses of unconjugated bile acids. *Biochim Biophys Acta* **922** (2), 221-233 (1987).
18. J. M. Donovan, N. Timofeyeva, and M. C. Carey. Influence of total lipid concentration, bile salt:lecithin ratio, and cholesterol content on inter-mixed micellar/vesicular (non-lecithin-associated) bile salt concentrations in model bile. *J Lipid Res* **32**, 1501-1512 (1991).
19. J. M. Donovan, A. A. Jackson. Rapid determination by centrifugal ultrafiltration of inter-mixed micellar/vesicular (non-lecithin-associated) bile salt concentrations in model bile: influence of Donnan equilibrium effects. *J Lipid Res* **34**, 1121-1129 (1993).
20. J. M. Donovan, A. A. Jackson, and M. C. Carey. Molecular species composition of inter-mixed micellar/vesicular bile salt concentrations in model bile: dependence upon hydrophilic-hydrophobic balance. *J Lipid Res* **34**, 1131-1140 (1993).
21. N. A. Mazer, G. B. Benedek, and M. C. Carey. Quasielastic light-scattering studies of aqueous biliary lipid systems. Mixed micelle formation in bile salt-lecithin solutions. *Biochemistry* **19** (4), 601-615 (1980).
22. K. J. van Erpecum, and M. C. Carey. Influence of bile salts on molecular interactions between sphingomyelin and cholesterol: relevance to bile formation and stability. *Biochim Biophys Acta* **1345** (3), 269-282 (1997).
23. D. B. Mitchell, K. S. Santone, and D. Acosta. Evaluation of cytotoxicity in cultured cells by enzyme leakage. *J Tissue Cult Methods* **6**, 113-116 (1980).

24. L. Puglielli, L. Amigo, M. Arrese, L. Nunez, A. Rigotti, J. Garrido et al. Protective role of biliary cholesterol and phospholipid lamellae against bile acid-induced cell damage. *Gastroenterology* **107**, 244-254 (1994).

25. P. K. Narain, E. J. DeMaria, and D. M. Heuman. Cholesterol enhances membrane-damaging properties of model bile by increasing the intervesicular-intermixed micellar concentration of hydrophobic bile salts. *J Surg Res* **84** (1), 112-119 (1999).

26. J. M. Barrios, and L. M. Lichtenberger. Role of biliary phosphatidylcholine in bile acid protection and NSAID injury of the ileal mucosa in rats. *Gastroenterology* **118** (6), 1179-1186 (2000).

27. F. Carubbi, M. E. Guicciardi, M. Concari, P. Loria, M. Bertolotti, and N. Carulli. Comparative cytotoxic and cytoprotective effects of taurohyodeoxycholic acid (THDCA) and tauroursodeoxycholic acid (TUDCA) in HepG2 cell line. *Biochim Biophys Acta* **1580** (1), 31-39 (2002).

28. B. A. van Gorkom, M. R. van der Meer, W. Boersma-van Ek, D. S. Termont, E. G. de Vries, and J. H. Kleibeuker. Changes in bile acid composition and effect on cytolytic activity of fecal water by ursodeoxycholic acid administration: a placebo-controlled cross-over intervention trial in healthy volunteers. *Scand J Gastroenterol* **37** (8), 965-971 (2002).

16

PRELIMINARY STUDY OF FIBER OPTIC MULTI-CARDIAC-MARKER BIOSENSING SYSTEM FOR RAPID CORONARY HEART DISEASE DIAGNOSIS AND PROGNOSIS

Liang Tang and Kyung A. Kang[*]

1. INTRODUCTION

Several cardiac-specific biomarkers have emerged as strong and reliable risk predictors for coronary heart disease (CHD), the leading cause of death with an annual incidence of 6.4% and a mortality of 42% in the US[1]. Myoglobin (MG), although not very specific, is the first marker released after myocardial muscle cells are damaged. B-type natriuretic peptide (BNP), cardiac troponin I (cTnI), and C-reactive protein (CRP) are released later than MG, but they are specific markers for coronary events. BNP is useful for the emergency diagnosis of heart failure[2] and for the prognosis in patients with acute coronary syndromes (ACS)[3]. cTnI has become a standard marker for the detection of acute myocardial infarction (AMI)[4]. CRP is an important prognostic indicator of CHD and ACS[3]. Elevated concentrations of these cardiac markers in serum are associated with recurrent CHD events and higher death rates. Simultaneous quantification of these biomarkers allows clinicians to diagnose CHD quickly and/or to accurately design a patient care strategy[3]. A fast and reliable detection of these proteins will also help medical professionals differentiate diseases among those showing similar symptoms. For example, both AMI and pulmonary embolism cause chest pain[5].

The clinically significant sensing ranges of MG, BNP, cTnI, and CRP are extremely low (pM~nM), and therefore, assay methods for these biomarkers need to be highly sensitive. A frequently used method is enzyme linked immunosorbent assay (ELISA). Although very accurate, it is time-consuming, expensive, and technically complicated. Commercially available test kits for BNP (Biosite; San Diego, CA), cTnI (Roche; Basel, Switzerland), and CRP (Dade Behring; Deerfield, IL) can provide fast, easy, and point-of-care assays. However, they usually provide only qualitative single biomarker information and most of the assay kits are expensive. Therefore, a fiber-optic, multi-

[*] Liang Tang and Kyung A. Kang, Department of Chemical Engineering, University of Louisville, Louisville, KY 40292.

cardiac-marker biosensing system is currently under development in our research group, to simultaneously quantify all four markers in a rapid, accurate, cost-effective, and user-friendly way. The system performs a fluoro-mediated sandwich immunoassay within the evanescent wave field of optical fiber surface[6]. This technology has been applied for simultaneous quantification of protein C and S (anticoagulants; nM) for rapid (~5 min) deficiency diagnosis[7-10]. In this paper, as an initial effort toward realizing a multi-sensing unit, four individual BNP, cTnI, MG, and CRP sensors were developed and the results of the sensor performance and the level of their specificity are presented.

2.   MATERIALS, INSTRUMENTS, AND METHODS

2.1.   Materials and Instruments

Purified cTnI, MG, and CRP from human heart and respective murine, monoclonal antibodies were obtained from Fitzgerald (Concord, MA). Human BNP was purchased from Bachem (Torrance, CA) and two different murine, monoclonal anti-human BNP, from Strategic Biosolutions (Newark, DE). Alexa Fluor 647 reactive dye (AF647; the maximum excitation and emission at 650 and 668 nm, respectively) was from Molecular Probes (Eugene, OR) and human serum albumin (HSA), from Sigma (St. Louis, MO). Quartz optical fibers (600 μm core diameter) and Fluorometer, Analyte 2000[TM] were from the Research International (Monroe, WA).

2.2.   Methods

To emulate MG-, BNP-, cTnI-, or CRP-free human plasma, samples were prepared in HSA solution at 103 mg/ml-phosphate buffered saline[9]. The conjugation of AF647 with the respective second monoclonal antibody (AF647-2° Mab) was performed according to manufacturer's instructions. Tapered optical fibers (3-12 cm long) were chemically treated to immobilize the first monoclonal antibodies (1° Mabs), against MG, BNP, cTnI, or CRP, on the surface of respective fibers *via* avidin-biotin bridges[7, 11]. Then, the fibers were inserted into glass chambers (100-200 μl) and the two ends were hot glued to form functional sensor units[7]. Assays with static incubation (no flow) were performed as described by Spiker and Kang[7]. For the assay with convective flow, the sample and the AF647-2° Mab were injected and circulated within an enclosed sensing unit at a pre-determined velocity for a pre-determined incubation period[12].

3.   RESULTS AND DISCUSSION

3.1.   Development of BNP and cTnI Sensors

The target sensing ranges for BNP and cTnI are 0.1~1 ng/ml (26~260 pM) and 0.7~7 ng/ml (30~300 pM), respectively. The initial sensing study was performed with a 12 cm sensor and 10 minutes each for the sample and the AF647-2° Mabs incubation. With static incubation (i.e. no flow during the incubation), the signal intensity of the BNP or cTnI sensor was very low (data not shown). Our previous study results demonstrated that convective flow can improve the sensor performance significantly by increasing the analyte mass transport to the sensor surface[8, 13]. Therefore, the effect of the convective

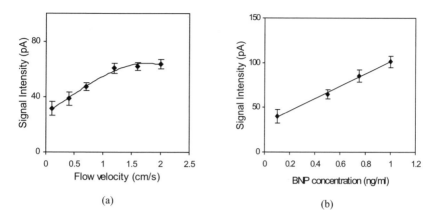

Figure 1. Development of the BNP sensor: (a) Effect of flow velocity; (b) Standard curve for BNP sensing.

flow on the sensor performance was investigated. The BNP sample at 0.5 ng/ml and AF647-2° Mab-BNP were circulated in the sensing system at a flow velocity between 0.1 and 2 cm/s during the reaction (incubation) [Fig. 1(a)]. In the range of 0.1 to 1.2 cm/s, as the velocity increases, the signal intensity rapidly increases, approximately 10 times of that without flow. At velocities higher than 1.2 cm/s, the intensity increase becomes insignificant (2%), indicating that the reaction kinetics changes from the mass-transport-limited to the reaction-limited. Therefore, 1.2 cm/s was selected to be the flow velocity for our BNP sensing.

In the clinical practice, a short assay time is especially beneficial for rapid disease diagnosis. Therefore, a study was performed to minimize the assay time. BNP sample (0.5 ng/ml) was incubated between 1 and 10 minutes while keeping the AF647-2° Mab-BNP incubation for 10 minutes, and *vice versa* (data not shown). The results indicate that the molecular mass transport and the immuno-reaction on the sensor surface for both incubations may be completed in 5 minutes. Therefore, 10 minutes is probably sufficient for the sample and reagent incubations and one assay can be completed within 15 minutes including the time for sample/reagent application and sensor regeneration. The sensitivity of the optimized BNP sensor was investigated in the target sensing range [Fig. 1(b)]. The signal intensity is linear with the analyte concentration and the sensor clearly quantifies BNP concentration at an average signal-to-noise (S/N) ratio of 25.

Similarly, the signal intensity for the cTnI (3 ng/ml) rapidly increases with the increase in the flow velocity until the velocity reaches 1.2 cm/s [Fig. 2(a)]. At this flow velocity, the incubation time optimization was also performed for the cTnI sensing and as for the BNP sensor, 15 minutes is sufficient for completing a cTnI assay (data not shown). Figure 2(b) shows the standard curve of the cTnI sensing in the targeted range with the optimized sensing protocol. The signal intensity is linear with the cTnI concentration, at an average S/N ratio of approximately 25.

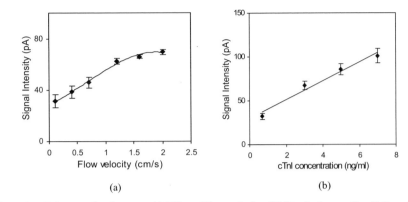

Figure 2. cTnI sensor development: (a) Effect of flow velocity; (b) Standard curve for cTnI sensing.

3.2. MG and CRP Sensors

The target sensing ranges for MG (70~700 ng/ml; 4~40 nM) and CRP (700~7000 ng/ml; 5.6~56 nM) are approximately 100 times higher than those of BNP and cTnI. Therefore, the convective flow did not improve the MG and CRP sensor performances significantly (data not shown), because the analyte transport to the sensor surface is sufficient without convection. The effect of the sample and the reagent incubation times on the MG and CRP sensor performance was also investigated and 5 minutes for each incubation period is more than sufficient for both sensors (data not shown). To minimize the amount of the sample and reagents to be used, a study was performed to reduce the sensor size. Compared to the 12-cm sensor, the signal intensities from the 3-cm sensors are about 40-50% [Fig. 3 (a) and (b)], although the sensor size is reduced by 75%. The signal of the 3-cm MG and CRP sensors is also in a linear relationship with the respective MG and CRP concentration, at an average S/N ratio of approximately 50.

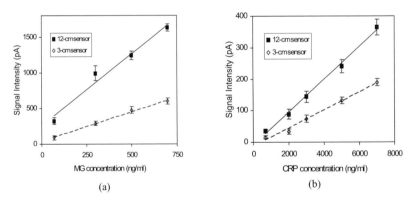

Figure 3. Sensing performance of the 3- and 12-cm sensors of (a) MG; (b) CRP.

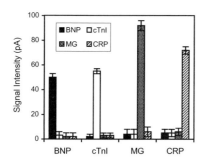

Figure 4. Cross-reactivity of the
four sensors to other markers.

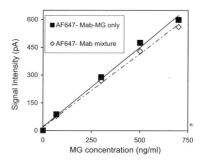

Figure 5. Effect of the mixture of
AF647-2° Mabs on the MG sensor.

3.3. Cross-reactivity of the Four Markers

To investigate the possible cross-reactivity among the BNP, cTnI, MG, and CRP
sensors, each sensor was tested to samples containing the other three analytes at their
upper limit of sensing ranges. For example, for a BNP sensor, cTnI (7 ng/ml), MG (700
ng/ml), and CRP (7000 ng/ml) were tested and the fluorescence signals were compared to
that of BNP measurement at the lower sensing limit (0.1 ng/ml). As shown in Figure 4,
all four sensors demonstrated a very high specificity with little cross-reactivity when the
other three analytes were present in the sample.

3.4. Effect of Second Antibody Mixture on the Sensor Performance

For the simultaneous quantification of these markers in a multi-sensing unit, after the
sample incubation, each AF647-2° Mab needs to be applied to the respective sensor. If
the four different AF647-2° Mabs do not cross-react, for an easier design and operation of
the sensing system, a mixture of the AF647-2° Mabs can be applied through the multi-
sensing unit.   Therefore, the effect of the second antibody mixture on each sensor
performance was studied.

Figure 5 shows the signal intensity generated by an MG sensor using only AF647-2°
Mab-MG or a mixture of four AF647-2° Mabs. The signal intensities generated by the
mixture showed a slight signal reduction (< 10%), with a similar standard deviation. For
the BNP, cTnI, and CRP measurements using the antibody mixture, similar results were
obtained (data not shown). A possible cause for the signal reduction could be that the
presence of other types of molecule causes a slightly slower mass transfer. These study
results demonstrated that the quantification of cardiac markers with the mixed AF647-2°
Mabs can be as accurate as that with the respective AF647-2° Mabs, if the measurements
are always performed with the antibody mixture. Therefore, a mixture of four different
second antibodies will be used for simultaneous four markers quantification.

4. CONCLUSIONS

A fiber-optic biosensing system for the simultaneous quantification of four cardiac
markers is currently under development, in our research group, for rapid coronary disease
diagnosis and prognosis. As an initial effort, four individual BNP, cTnI, MG, and CRP

sensors were developed. The sensors showed excellent performance in quantifying these cardiac markers in their clinically significant ranges within 15 minutes, at a S/N ratio of 25-50. The cross-reactivity of the four sensors was also found to be negligible. A mixture of four AF647-2° Mabs has shown only a slight interference to the four sensors, indicating that the mixture can be applied through a multi-sensing unit for simultaneous detection with an easier operation.

Other future work includes simultaneous quantification of these four cardiac markers using a multi-sensing unit. A microfluidic system will be incorporated in the unit for precise fluid control. Micro-electro-mechanical systems (MEMS) technology will be utilized for the development of an automatic, smaller, and more cost-effective biosensing chip.

5. ACKNOWLEDGEMENT

The authors would like to thank the National Science Foundation (BES-0330075) for the financial support of this work.

6. REFERENCES

1. American Heart Association, Heart disease and stroke statistics – 2004 update, pp. 9-11 (2004).
2. A. S. Maisel, P. Krishnaswamy, R. M. Nowak, et al. Rapid measurement of B-type natriuretic peptide in the emergency diagnosis of heart failure, *New Engl. J. Med.* **347**, 161-167 (2002).
3. M. S. Sabatine, D. A. Morrow, J. A. de Lemos, C. M. Gibson, S. A. Murphy, N. Rifai, C. McCabe, E. M. Antman, C. P. Cannon, E. Braunwald, Multimarker approach to risk stratification in non-ST elevation acute coronary syndromes: Simultaneous assessment of troponin I, c-reactive protein, and b-type natriuretic peptide, *Circulation* **105**, 1760-1763 (2002).
4. ESC/ACC. Consensus document, myocardial infarction redefined: A consensus document of the joint European society of cardiology/American college of cardiology committee for the redefinition of myocardial infarction, *Eur. Heart J.* **21**, 1502-1513 (2002).
5. J. McCord, R. M. Nowak, P. A. McCullough, et al. Ninety-minute exclusion of acute myocardial infarction by use of quantitative point-of-care testing of myoglobin and troponin I, *Circulation* **104**, 1483-1488 (2001).
6. F. S. Ligler, J. P. Golden, L. C. Shriver-Lake, R. A. Ogert, D. Wijesuria, G. P. Anderson, Fiber-optic biosensor for the detection of hazardous materials, Immunomethods **3**, 122-127 (1993).
7. J. O. Spiker and K. A. Kang, Preliminary study of real-time fiber optic based protein C biosensor, *Biotech. Bioeng.* **66 (3)**, 158-163 (1999).
8. H. I. Balcer, H. J. Kwon, and K. A. Kang, Assay procedure optimization of rapid, reusable protein immunosensor for physiological samples, *Ann. Biomed. Eng.* **30 (10)**, 141-147 (2002).
9. H. J. Kwon, H. I. Balcer, and K. A. Kang, Sensing performance of protein C immuno-biosensor for biological samples and sensor minimization, *Comparative Biochem. Phys. A* **132**, 231-238 (2002).
10. L. Tang and K. A. Kang, Preliminary study of simultaneous multi-anticoagulant deficiency diagnosis by fiber optic multi-analyte biosensor, proceedings of 2003 Annual ISOTT Meeting, Aug. 16-20, 2003, Rochester, NY, in press.
11. S. K. Bhatia, L. C. Shriver-Lake, K. J. Prior, J. H. Georger, J. M. Calvert, R. Bredehorst, F. S. Ligler, Use of thio-terminal silanes and heterobifunctional crosslinkers for immobilization of antibodies on silica surfaces, *Anal. Biochem.* **178**, 408-413 (1989).
12. L. Tang, H. J. Kwon, and K. A. Kang, Studies on effect of sample circulation on sensing performance of protein C immunosensor, proceedings of the 24th Annual International Conference of the IEEE Engineering in Medicine and Biology Society **3**, 1805-1806 (2002).
13. L. Tang and K. A. Kang, Sensing performance improvement of protein C biosensor by sample circulation, *Adv. Exp. Med. Biol.* **540**, 177-182 (2003).

17

IMPACT OF HEMORHEOLOGICAL AND ENDOTHELIAL FACTORS ON MICROCIRCULATION

Vera Turchetti[*], Letizia Boschi, Giovanni Donati, Luca Trabalzini, and Sandro Forconi

1. ABSTRACT

Previous studies showed that endothelial alterations caused by physical stress worsened the hemorheological parameters mainly in patients affected by ischemic vascular diseases: major vascular alterations have been found in patients with very high endothelial dysfunction indexes: these indexes are given by the various substances produced by the endothelium, but it is very difficult to have a value which clearly identifies the real state of the endothelial alteration. The function of the NO, an endogenous vasodilator whose synthesis is catalyzed by NOs, can be determined by the Citrulline/Arginine ratio, which represents the level of activity of the enzyme. A very good index of the endothelial dysfunction is asymmetric dimethylarginine (ADMA), a powerful endogenous inhibitor of NOs; in fact several studies have demonstrated a strong relationship between ischemic vascular disease and high levels of plasmatic ADMA. Our recent studies on heart failure and on ischemic cerebrovascular diseases evaluate endothelial dysfunctions and hemorheological parameters.

2. BACKGROUND

Microcirculation is the most important vascular system component and, with blood and interstitial tissue, can be considered as a whole functional organ. Every vascular disease has repercussions on microcirculation and on tissular nutrition[1,2]. The worsening of the hemorheological parameters affects microcirculation: a lot of studies have shown that blood viscosity increase and erythrocyte deformability decrease can be found in ischaemic vascular diseases[3,4]. Hemorheological disturbances are present in vascular diseases, both in acute and in chronic conditions[5]. A reduction of blood fluidity, due either to increase of hematocrit or of fibrinogen concentration or of the rigidity of red cells is commonly considered a condition of high risk for acute or chronic brain

[*] Vera Turchetti, Dipartimento di Medicina Interna, Cardiovascolare e Geriatrica - Università degli Studi di Siena, Policlinico Santa Maria alle Scotte, 53100 Siena, Italy – e-mail: turchetti@unisi.it

ischemia [6,7]. Many studies have demonstrated these alterations in ischemic vascular diseases: in acute and chronic cerebrovasculopathies, in ischemic cardiac vasculopathies and especially in peripheral obliterating arteriopathy [8-12].

3. PHYSICAL EXERCISE AND HEMORHEOLOGY

We studied physical exercise in healthy subjects and in patients suffering from ischemic vascular diseases with different kinds of exercise, such as the cycloergometer test, treadmill test and isotonic stress test[13-15]. In our previous research, we found that hematic hyperviscosity was also present in the resting condition in the vasculopathic patients suffering from coronary heart disease or peripheral ischemic diseases compared to healthy subjects[16-18]. However, when stronger ischemia was provoked by means of general or specific exercise, the hemorheological decline increased and persisted after a recovery period in a significant way in patients suffering from ischemic vasculopathies compared to healthy people. Subsequently, together with increased viscosity and reduced erythrocyte deformability, we found a significant increase in erythrocytic cytosolic calcium which explained the reduced erythrocyte deformability and activation of the fibrinolytic system, as in conditions of hypercoagulability[16,17,19]. After devising a new method to evaluate erythrocyte morphology, (the Zipursky-Forconi method), we observed that the EMI was altered in the vasculopathic subject, with values <1 due to a higher number of discocytes, which are normal but more rigid than the bowls which are more abundant in healthy subjects[20,21]. The exercise further decreased the EMI which did not return to basal values even after 20 minutes of recovery. In those studies, we considered that the blood, in all its cellular and plasmatic components, and the vessel were involved in the phenomena that caused the ischemia during the stress test[22,23]. The rheological worsening, secondary to the ischemic pathology, was affected by, and in turn caused, a microcirculatory deterioration. Previous studies showed that endothelial alterations caused by physical stress worsened the hemorheological parameters mainly in patients affected by ischemic vascular diseases: major vascular alterations have been found in patients with very high endothelial dysfunction indexes[3,4,24].

4. HEMORHEOLOGY AND ENDOTHELIUM

In the last 20 years, there has been great interest in the study of the vascular endothelium. Today it is considered a true organ with a surface area of about 400 square meters, the same mass as 6 normal hearts with a weight of 1500 grams and containing more than 1 trillion cells[25]. It has many functions that are important in guaranteeing vascular homeostasis: it can modulate vascular tone, the proliferation of vascular smooth muscle cells, hemostasis, thrombolysis, platelet aggregation, adhesiveness of monocytes, inflammation, the immune response and the production of free radicals[26-28]. It secretes many substances which can be divided into vasodilators, such as nitric oxide (NO), prostacyclin, bradykinin and hyperpolarizing factors, and vasoconstrictors, such as endothelin-1, thromboxane and activation of angiotensin II[27,29]. The main vasodilatory factor is NO which also has an anti-platelet action and is an inhibitor of smooth muscle cells, of ET-1 synthesis, reduces the expression of molecular adhesion for inhibition of leukocytes[30,31]. The metabolic pathway of NO starts from l-arginine catalyzed by NO syntase and produces one molecule of NO and one molecule of l-citrulline. All the atherosclerotic risk factors for example hypercholesterolemia, hyperglycemia, hyperhomocysteinemia, smoke, diabetes and local factors, can by various mechanisms, cause an alteration of endothelial function and an increase of free radicals and of superoxide anion released by the endothelium and smooth muscle cells. The superoxide anion induces a decrease of NO levels. Therefore, oxidative stress represents the common pathway of these risk factors [28,29,31,32].

There are two different hypotheses on the origin of the endothelial dysfunction: the activation of the cycle-oxigenase pathway and the consequent production of endoperoxides and superoxide anion; or the risk factors of ATS increase the uncoupled NOs with the consequent increase of superoxide anion: shear stress becomes the link between hemorheology and endothelial activation[30,31]. In the normal subject, physical exercise increases the shear stress. This causes a temporary increase of hematic viscosity but is also a stimulus for NO production and thus vasodilation which brings the hemorheological and coagulative-fibrinolytic parameters to the basal conditions. In the vasculopathic subject, the increase of shear stress in the dysfunctional endothelium increases the production of superoxide anion which inactivates NO with the formation of peroxinitrites. These molecules decompose into two potent free radicals, OH and NO_2, which can initiate lipid peroxidation. This has a cytotoxic effect on the endothelium, promotes the recruitment of inflammatory cells, reduces the levels of NOs and pushes the activity of the endothelium towards vasoconstriction, coagulation and platelet aggregation[22,24,44]. The results that we have seen in subjects with ischemic vascular pathologies, during and after the stress test, can be explained by these phenomena; in fact the increase of hematic viscosity, the increase of erythrocytic cytosolic calcium and the decrease of EMI do not regress in the recovery phase but remain highly altered with respect to the basal values. In normal subjects, the L-citrulline/L-arginine ratio decreases at peak effort but then returns to basal values in the recovery phase. In the vasculopathic patient, it progressively decreases even in the recovery phase. The assay of VCAM-1, a vasocellular adhesion molecule which reflects the degree of activation of leucocytes, shows significantly higher values in patients than in normal subjects but its trend is not uniform during the exercise and in the recovery phase[22,24,32]. From a methodological point of view, we can try to identify an index of endothelial dysfunction by measuring the various substances secreted by the endothelium but it is very difficult to find a parameter that indicates the true state of endothelial alteration. The possibility of identifying endothelial dysfunction at an early stage would be of great clinical value because it could allow us to block or delay the development of atherosclerotic lesions well in advance. However, even if we know and can measure a series of soluble substances produced by the endothelium, we are far from having a reliable marker of endothelial dysfunction[29,30,31]. A fairly new method has been proposed by Vasquez-Vivar et al.: their evaluation of endothelial function by the determination of endothelial progenitor cells in the blood. Their number is correlated with FMD (Flow-Mediated Dilation) and with the risk factors for ATS (33). The function of NO, one of the most potent endogenous vasodilators whose synthesis is catalyzed by NOs, can be determined by the ratio between the hematic concentrations of citrulline and arginine. The first is a co-product and the last is a precursor of the NO synthesis pathway which reflects the degree of activity of the enzyme. In our research, we have seen that this ratio can give us an indication of the activity of NOs and also of the production of NO, even if a quota of the citrulline produced re-enters in the arginine pool and therefore these values could be underestimates[24,34,35]. A very reliable index of endothelial dysfunction is asymmetric dimethyl arginine (ADMA), a potent endogenous inhibitor of NOs. Many studies have shown that ischemic vascular pathologies are associated with high plasma levels of ADMA. Recently it was demonstrated that renal insufficiency is associated with a large increase in the plasma values of ADMA. In addition, endothelial dysfunction that leads to increased peripheral resistance and to arterial hypertension in these patients can be attributed to these high ADMA levels[36-38]. These parameters are measured by high performance liquid chromatography (HPLC). This technique provides very reliable qualitative results, as well as the presence of substances even in very low concentrations like ADMA[39].

5. OUR RESEARCH

In a recent paper[40] we evaluated endothelium-dependent dilatation induced by an ACE-inhibitor, calcium antagonist and beta blocker in patients suffering from heart failure (NYHA class II and III). We studied 34 patients (19M, 15F, mean age 76.96 ± 8.82) in pharmacological wash-out for at least one week, divided into 3 groups: Group A (15 patients, 9M and 6F) taking ramipril (5 mg/die); Group B (10 patients, 6M and 4F) taking amlodipine (10 mg/die), Group C: (9 patients, 4M and 5F) taking carvedilole (25 mg/die). The groups were homologous for NYHA class and instrumental echographic parameters (mean EF = 22.5 ± 6.7 and mean sAPP 38.4 ± 8.7). At the beginning and after 3 weeks of therapy, we performed a clinical and instrumental assessment; we studied endothelial function and hemorheological and coagulative/fibrinolytic parameters. The results show that L-citrulline and L-arginine increase, while VCAM-1 decreases. The L-citrulline/L-arginine ratio increases in a statistically significant way. This trend is maintained in each group. These results demonstrate that the drugs used induce an improvement of endothelium-dependent dilatation. In addition, there is progressive hemorheological and fibrinolytic improvement, with a reduction of PAI-1 and blood viscosity[40,41]. Our current research on ischemic cerebral vasculopathies evaluates the levels of ADMA, the L-citrulline/L-arginine ratio, and quantity of VCAM-1 and endothelin-1 in order to correlate them with hemorheological parameters and the degree of cognitive decline. We have analyzed, in patients with chronic cerebral ischemia, hemorheological and endothelial parameters and some new atherogenic factors as homocysteina compared to the degree of cognitive decline. The results shown that these patients have a blood viscosity increase and erythrocyte deformability decrease, an alteration of endothelial dysfunction index (l-citrulline/l-arginine ratio decrease and ADMA increase). In Group A (multiple cortical and undercortical infarction), we have observed a blood viscosity increase and EMI reduction in a statistical significant way; in Group B (leucoaraiosi), in which cognitive decline is very important, we have observed an homocystemia and ADMA increase in a statistically significant way[42-44].

6. REFERENCES

1. L. Reid, B. Meyrick, Microcirculation: definition and organization at tissue level. *Ann. N. Y. Acad. Sci.* **384**, 3-20 (1982).
2. J.A. Firth, Endothelial barriers: from hypothetical pores to membrane proteins. *J. Anat.* **200**(6), 541-8 (2002).
3. OK Baskurt, O Yalcin, HJ Meiselman. Hemorheology and vascular control mechanisms. *Clin Hemorheol Microcirc.***30**(3-4),169-78 (2004).
4. AR Pries, TW Secomb,. Rheology of the microcirculation *Clin Hemorheol Microcirc.***29**(3-4), 143-8 (2003).
5. P Kowal, A Marcinkowska-Gapinska, W Elikowski, Z Chalupka, Comparison of whole blood viscosity in vascular diseases *Pol Merkuriusz Lek,***15**(90),515-7 (2003 Dec)
6. S. Forconi, V. Turchetti, R. Cappelli, M. Guerrini and M. Bicchi, Haemorheological disturbances and possibility of their correction in cerebrovascular diseases, *J. Mal. Vasc.* **24**(2), 110–6 (Review)(1999)

7. B Turczynski, K Pierzchala, L Slowinska, J Becelewski, Dynamics of the changes of blood and plasma rheological properties in ischaemic stroke *Wiad Lek.*,**55**(3-4):189-96 (2002).

8. N Antonova, I Velcheva, Hemorheological disturbances and characteristic parameters in patients with cerebrovascular disease, *Clin Hemorheol Microcirc*,**21**(3-4):405-8(1999).

9. B Turczynski, L Slowinska, S Szczesny, M Baschton, A Bartosik-Baschton, J Szygula, J Wodniecki, J. Spyra, The whole blood and plasma viscosity changes in course of acute myocardial infarction *Pol Arch Med Wewn.* Oct;**108**(4):971-8.(2002)

10. TL De Backer, M De Buyzere, P Segers, S Carlier, J De Sutter, C Van de Wiele, G De Backer, The role of whole blood viscosity in premature coronary artery disease in women, *Atherosclerosis*, **165**(2):367-73 (2002 Dec).

11. R Raddino, R Ferrari, TM Scarabelli, C Portera, G Galeazzi, L Sarasso, E Ascari, O Visioli, Clinical implications of rheology in cardiovascular diseases *Recenti Prog Med,* **93**(3):186-99. Review.(2002).

12. C. Koksal, M Ercan, AK Bozkurt, Hemorrheological variables in critical limb ischemia, *Int Angiol.* **21**(4),355-9 (2002)

13. A Margonato, O Carandente, G Vicedomini, C Rocco, L Falqui, D Cianflone, Changes in hemorheological parameters during physical exertion in coronary disease patients *Ric Clin Lab*,**15** Suppl 1:169-77 (1985).

14. K Nageswari, R Banerjee, RV Gupte, RR Puniyani, Effects of exercise on rheological and microcirculatory parameters, *Clin Hemorheol Microcirc*,**23**(2-4),243-7 (2000).

15. TS Church, CJ Lavie, RV Milani, GS Kirby, Improvements in blood rheology after cardiac rehabilitation and exercise training in patients with coronary heart disease, *Am Heart J.* **143**(2):349-55 (2002).

16. S Forconi, M Guerrini, C Rossi, S Pecchi, T Di Perri, Modificazioni della viscosità ematica sistemica durante prova da sforzo nell'angina pectoris. *Boll. S. I. C.*, **22**, 435, (1977).

17. S Forconi, M Guerrini, C Rossi, S Pecchi, T Di Perri Modificazioni della viscosità ematica sistemica durante claudicatio intermittens provocata nella arteriopatia obliterante periferica. *Boll. S. I. C., 22, 441 (1977)*

18. S Forconi, M Guerrini., D Agnusdei., F Laghi Pasini, T Di Perri, Abnormal Blood Viscosity in Raynaud's Phenomenon. *Lancet, II, 586 (1976).*

19. S. Forconi, M. Guerrini and V. Turchetti, Local haemorheological changes and Intracellular Ca ++ content of blood cells in exercising ischaemic limbs, *Clin. Hemorheol.* **12**, 527–534 (1999).

20. A Zipursky, E Brown, J Palko, EJ Brown, The erythrocyte differential count in newborn infants, *The American Journal of Pediatric Hematology/Oncology* **5**, Number 1, (1983)

21. V.Turchetti, C.De Matteis, F Leoncini, L Trabalzini, M Guerrini, S Forconi, Variations of erythrocyte morphology in different pathologies, *Clinical Hemorheology,***17**, 3, (1997).

22. V. Turchetti, D. Ricci, F. Leoncini, M. Bianconi, V. Trani, M. Bellini, F. Petri, L. Trabalzini and S. Forconi, Red cell morphology, blood viscosity and intraerythrocytic calcium after exercise test in normal subjects and in patients suffering from vascular disease, *Clin. Hemorheol. Microc.* **20**, 285(1999).

23. S. Church, C.J. Lavie, R.V. Milani and G.S. Kirby, Improvements in blood rheology after cardiac rehabilitation and exercise training in patients with coronary heart disease, *Am. Heart J.* **143**(2) , 349–55 (2002).

24. V. Turchetti, L. Boschi, G. Donati, G. Borgogni, D. Coppola, S. Dragoni, M.A. Bellini, S. Sicuro, V.M.A. Mastronuzzi and S. Forconi. Endothelium and hemorheology. *Clinical Hemorheology and Microcirculation* **30**, 289–295 (2004).

25. E.E. Anggard, The regulatory functions of the endothelium, *Jpn. J. Pharmacol.* **58** (Suppl. 2), 200P–206P (1992).

26. J.R. Vane, E.E. Anggard and R.M. Botting, Regulatory functions of the vascular endothelium, *N. Engl. J. Med.* **323**, 27–36 (1990).

27. E. Anggard, Nitric oxide: mediator, murderer and medicine, *Lancet* **343**, 1192–1206 (1994)

28. H. Drexler, Endothelial dysfunction: clinical implications, *Prog. Cardiovasc. Dis.* **39**, 287–324 (1997).

29. S. Taddei, L. Ghiadoni, A. Virdis, D. Versari and A. Salvetti, Mechanisms of endothelial dysfunction: clinical significance and preventive non-pharmacological therapeutic strategies, *Curr. Pharm. Des.* **9** (29), 2385–402 (2003).

30. J. Vasquez-Vivar, B. Kalyanaraman, P. Martasek, N. Hogg, B.S. Masters, H. Karoui, P. Tordo and K.A. Pritchard Jr., Superoxide generation by endothelial nitric oxide synthase: the influence of cofactors, *Proc. Natl. Acad. Sci. USA* **95**(16), 9220–5 (1998).

31. T. Gori and J.D. Parker, Nitrate tolerance: a unifying hypothesis, *Circulation* **106**(19), 2510–3 (2002).

32. A Maiorana, G. O'Driscoll, R. Taylor and D. Green, Exercise and the nitric oxide vasodilator system, *Sports Med.* **33**(14), 1013–35 (2003).

33. J. Vasquez-Vivar, B. Kalyanaraman, P. Martasek, N. Hogg, B.S. Masters, H. J.M. Hill, G. Zalos, J.P. Halcox, W.H. Schenke, M.A. Waclawiw, A.A. Quyyumi and T. Finkel, Circulating endothelial progenitor cells, vascular function, and cardiovascular risk, *N. Engl. J. Med.* **348**(7), 593–600 (2003).

34. J. Martens-Lobenhoffer and S.M. Bode-Boger, Simultaneous detection of arginine, asymmetric dimethylarginine, sym-metric dimethylarginine and citrulline in human plasma and urine applying liquid chromatography–mass spectrometry with very straightforward sample preparation, *J. Chromatogr. B Analyt. Technol. Biomed. Life Sci.* **798**(2), 231–9 (2003).

35. T Koga, WY Zhang, T Gotoh, S Oyadomari, H Tanihara, M Mori, Induction of citrulline-nitric oxide (NO) cycle enzymes and NO production immunostimulated rat RPE-J cells. *Exp Eye Res.* **76**(1), 15-21 (2003).

36. JT Kielstein, B Impraim, S Simmel, SM Bode-Boger, D Tsikas, JC Frolich, MM Hoeper, H Haller, D Fliser, Cardiovascular effects of systemic nitric oxide synthase inhibition with asymmetrical dimethylarginine in humans. *Circulation.* **109**(2): 172-7 (2004).

37. RH Boger, Association of asymmetric dimethylarginine and endothelial dysfunction, *Clin Chem Lab Med.* **41** (11), 1467-72 (2003).

38. K Sydow, T Munzel, ADMA and oxidative stress. *Atheroscler Suppl.* **4**(4), 41-51 (2003).

39. A Pettersson, L Uggla, V Backman., Determination of dimethylated arginines in human plasma by high-performance liquid chromatography, *J. Chromatogr. B Biomed. Sci. Appl.* **692**(2), 257–62 (1997).

40. V. Turchetti, M.A. Bellini, L. Boschi, G. Postorino, A. Pallassini, M.G. Richichi,L. Trabalzini, M. Guerrini and S. Forconi, Haemorheological and endothelial-dependent alterations in heart failure after ACE inhibitor, calcium antagonist and beta blocker. *Clin. Hemorheol. Microcirc.* **27**, 209–218 (2002).

41. V Turchetti, MA Bellini, D Ricci, A Lapi, G Donati, L Boschi, L Trabalzini, M Guerrini, S Forconi, Spontaneous echo-contrast as an in vivo indicator of rheological imbalance in dilatative cardiomyopathy. *Clin. Hemorheol. Microcirc.* **25**(3-4), 119-25 (2001).

42. G. Postorino, B. Provenzano, F. Giunti, V. Turchetti, L. Trabalzini, S. Forconi. Atherogenic risk factors and cognitive function decline in geriatric age. *Arch Gerontol Geriatr.* **33**, 307-312 (2001).

43. SR Lentz, RN Rodionov, S Dayal, Hyperhomocysteinemia, endothelial dysfunction, and cardiovascular risk: the potential role of ADMA. *Atheroscler Suppl.* **4**(4), 61-5 (2003).

44. V. Turchetti, F. Leoncini, C. De Matteis, L. Trabalzini, M. Guerrini and S. Forconi, Evaluation of erythrocyte morphologyas deformability index in patients suffering from vascular diseases, with or without diabetes mellitus: correlation with blood viscosity and intra-erythrocytic calcium. *Clin. Hemorheol. Microcirc.* **18**(2), 141–149 (1998).

18.

ARTEPILLIN C ISOPRENOMICS: DESIGN AND SYNTHESIS OF ARTEPILLIN C ANALOGUES AS ANTIATHEROGENIC ANTIOXIDANTS

Yoshihiro Uto[*], Shutaro Ae, Azusa Hotta, Junji Terao, Hideko Nagasawa, and Hitoshi Hori[*]

1. INTRODUCTION

Artepillin C is a diprenyl-4-hydroxycinnamic acid derivative first isolated from *Baccharis* species[1] as a major constituent (>5%) in Brazilian propolis[2] and first synthesized by Uto[3]. Artepillin C is an isoprenyl-containing compound, which has various medicinal effects such as antibacterial,[4] antitumor,[5] apoptosis-inducing,[6] immunomodulating,[7] and antioxidative[8] activities. Therefore, it is useful for drug discovery to study of medicinal chemistry and biology of the isoprenyl-containing compounds, which we have termed "isoprenomics".

Low-density lipoprotein (LDL) is a major cholesterol carrier in the blood, and it is suggested that the oxidative modification of LDL has an important role in the development of atherosclerosis.[9, 10] The effective antioxidants, which prevent the oxidation of LDL, are beneficial in reducing atherosclerosis and coronary heart disease.

Herein, we discuss the design and synthesis of artepillin C analogues and their structure-activity relationship in terms of their inhibitory activity of LDL oxidation as new antiatherogenic antioxidants.

2. MATERIALS AND METHODS

2.1. Chemicals

Synthesis of artepillin C and its analogues was done in our laboratory and the detail of this is currently under preparation for submission. Probucol in a chemically pure form was obtained from a tablet used for a trial (Daiichi Pharmaceutical Co., Ltd, Otsuka

---

[*] Yoshihiro Uto, Shutaro Ae, Hideko Nagasawa, and, Hitoshi Hori, Department of Biological Science & Technology, Faculty of Engineering, The University of Tokushima, Minamijosanjimacho-2, Tokushima, 770-8506 Japan. Azusa Hotta, and Junji Terao, Department of Nutrition, School of Medicine, The University of Tokushima, Kuramotocho, Tokushima, 770-8503 Japan.

Pharmaceutical Co., Ltd, and the Dow Chemical Company). 2-thiobarbituric acid (TBA) and 1,1-diphenyl-2-picrylhydrazyl (DPPH) were purchased from Wako Pure Chemical Industries Ltd. (Osaka, Japan). 2-(N-morpholino)ethanesulfonic acid (MES) was obtained from Sigma Chemical Company (St. Louis, USA).

2.2. Preparation of Human LDL

A LDL (density, 1.006-1.063 g/mL) was fractionated from human plasma by ultracentrifugation (80,000 rpm, 1.0 hr) using the CS120 ultracentrifuge equipped with a RP80AT rotor (Hitachi Koki Co., Ltd), dialyzed at 4°C against 3 changes of phosphate-buffered saline (PBS; pH 7.35). The LDL solution was flushed with N_2, stored in the dark at 4°C, and used within 5 days from the time of the preparation. Protein was measured by the Bradford method using bovine serum albumin as standard.

2.3. Determination of Lipophilicity

R_M values, defined as a logarithmic function of the R_F values, were determined as a measure of lipophilicity. Octadecyl silica (ODS) reverse phase plates (MERCK RP-18 $WF_{254\,s}$) were used as the stationary phase. A methanol/PBS (pH 7.35) mixture (90/10, v/v) was used as the mobile phase. The plates were developed in closed chromatography tanks, saturated with the mobile phase. Spots were detected under UV light (254 nm). R_F values were determined from three individual measurements. R_M values were calculated from the corresponding R_F values, using the equation: $R_M = \log (1/R_F - 1)$.[11]

2.4. Calculation of log$P$ and log$D$

The n-octanol-water partition coefficient (log$P$) and distribution coefficient (log$D$, pH 7.35) of antioxidants were calculated by the program Pallas 3.0 (CompuDrug International Inc., Arizona, USA).

2.5. Assay of Free Radical Scavenging Activity

Free radical scavenging activity was determined by using DPPH at 517 nm according to the method of Blois[12] with some modifications. To 3.0 mL of 100 µM DPPH ethanol solution (60%), containing 40 mM MES (pH 5.5), was added with 3 µL of different concentrations of antioxidants tested. The mean effective concentration of antioxidants, which was required to decrease by 0.200 in the absorption in the reaction with DPPH after 30 min ($IC_{0.200}$), was calculated from the relation between the absorbance vs. the concentration.

2.6. Assay of LDL Oxidation

The effects of antioxidants on kinetics of lipid oxidation of human LDL (50 µg protein/mL) were evaluated by spectrophotometric monitoring of conjugated diene lipid hydroperoxide formation at 234 nm,[13] during copper-induced oxidation (5 µM $CuSO_4$, 10 mM PBS) using a Hitachi U-3300 spectrophotometer at 37°C with a temperature controller SPR-10. The duration of the lag phase, that is the lag time, was calculated by extrapolating the propagation phase. Lipid peroxides were measured as thiobarbituric acid reactive substances (TBARS) by the method of Yagi[14]. The concentration (IC_{50}) leading to 50% decrease of the amount of TBARS was estimated by linear regression analyses from three individual measurements.

3. RESULTS

3.1. Design and Synthesis

We designed artepillin C analogues focusing on the prenyl group and synthesized according to our established prenylation and Mizoroki-Heck reaction.[3] Chemical structures of artepillin C, artepillin C analogues, and probucol are shown in Table 1.

Table 1. R_M values, log$P$, log$D$ (pH 7.35), and DPPH free radical scavenging activity (IC$_{0.200}$) of antioxidants

Antioxidant	R_M	Log$P$ [a]	Log$D$ (pH 7.35) [a]	IC$_{0.200}$ (µM)
Artepillin C	-0.056 ± 0.027	5.66	2.98	14.7 ± 0.7
TX-1959	-0.270 ± 0.035	3.53	0.84	54.1 ± 5.9
TX-1960	0.136 ± 0.042	6.11	6.11	16.8 ± 0.2
TX-2007	0.417 ± 0.053	9.81	7.12	15.1 ± 1.0
TX-2012	-0.047 ± 0.041	5.60	2.91	79.2 ± 4.0
TX-2013	0.139 ± 0.024	7.74	5.05	14.0 ± 0.8
TX-2020	-0.092 ± 0.032	4.89	2.27	13.1 ± 1.0
Probucol	0.438 ± 0.054	10.72	10.72	n.t.[b]

[a] Log$P$ and Log$D$ (pH 7.35) values were calculated using Pallas 3.0 software.
[b] Not tested.

3.2. Lipophilicity and DPPH Free Radical Scavenging Activity of Antioxidants

The lipophilicity of artepillin C and its analogues was evaluated by their R_M values, calculated log$P$, and log$D$ (pH 7.35) (Table 1). A positive correlation between prenyl elongation and lipophilicity was found as follows: TX-1959 (monoprenyl) < Artepillin C (diprenyl) < TX-2013 (monoprenyl and monogeranyl) < TX-2007 (digeranyl). In addition, R_M values had good linearity with the log$P$ values ($r^2 = 0.9542$), and also with the log$D$ values ($r^2 = 0.9070$).

Next, we determined the $IC_{0.200}$ value to evaluate the potency of DPPH free radical scavenging activity in aqueous solution as shown in Table 1. Disubstituted analogues including artepillin C ($IC_{0.200} = 14.7 \pm 0.7$ µM), TX-1960 ($IC_{0.200} = 16.8 \pm 0.2$ µM), TX-2007 ($IC_{0.200} = 15.1 \pm 1.0$ µM), TX-2013 ($IC_{0.200} = 14.0 \pm 0.8$ µM), and TX-2020 ($IC_{0.200} = 13.1 \pm 1.0$ µM) showed almost the same DPPH radical scavenging activities. On the other hand, monosubstituted analogues such as TX-1959 ($IC_{0.200} = 54.1 \pm 5.9$ µM) and TX-2012 ($IC_{0.200} = 79.2 \pm 4.0$ µM) showed weaker DPPH radical scavenging activity than disubstituted analogues including artepillin C.

3.4. Inhibitory Activity of Antioxidants against LDL Oxidation

Figure 2 shows the effects of artepillin C and probucol (each 1 µM) on the kinetics of conjugated diene formation during the oxidation of LDL induced by cupric ion at a concentration of 5 µM. With regard to the lag time, artepillin C (lag time = 102.0 min) was thought to be a more efficient protector of LDL oxidation than probucol (lag time = 76.9 min), which is well-known hypocholesterolemic drug. All the results, including artepillin C analogues, are summarized in Table 3. Monoprenyl analogue TX-1959 (lag time = 71.3 min) and digeranyl analogue TX-2007 (lag time = 74.8 min) showed lower activity of antioxidant than artepillin C, while di-*tert*-butyl analogue TX-2020 (lag time = 105.3 min) was slightly higher than artepillin C. Methyl ester analogue TX-1960 (lag time = 91.5 min) had a less inhibitory activity against conjugated diene formation than that (lag time = 102.0 min) of artepillin C, while TX-1960 ($IC_{50} = 7.7 \pm 0.5$ µM) had a higher inhibitory activity against TBARS production than that ($IC_{50} = 10.6 \pm 0.8$ µM) of artepillin C.

4. DISCUSSION

In this study, we discussed the structure activity relationship of antioxidant based on artepillin C isoprenomics. For the lipophilicity, we found that R_M values of artepillin C and its analogues in both the lipophilic prenyl group and hydrophilic carboxyl group were closely related to calculated log$P$ or log$D$ values. Exceptionally, R_M values showed that lipophilicity of TX-2007 was underestimated by calculation of log$D$ probably due to the high affinity for digeranyl group to ODS.

As a DPPH free radical scavenging activity and inhibitory activity for LDL oxidation, we have obtained an interesting result of no correlation between two activities. TX-2007 and TX-2013, which were lipophilic analogues of artepillin C, possessed only half inhibitory activity for LDL oxidation of artepillin C although their DPPH radical scavenging activity values were nearly equal to artepillin C. On the other hand, monogeranyl analogue TX-2012 showed weaker DPPH free radical scavenging activity than artepillin C, however, its inhibitory activity for LDL oxidation was comparable to it.

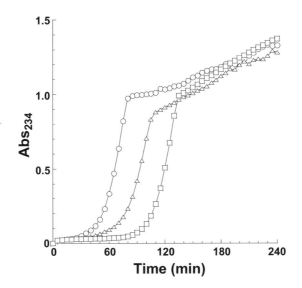

Figure 2. Kinetic of conjugated diene formation in the absence (open circle) or presence of 1 □M probucol (open triangle) or 1 □M artepillin C (open square). LDL (50 □g/mL) was oxidized in PBS at 37°C with 5 □M Cu^{2+}.

Table 3. Inhibitory activities of antioxidants against conjugated diene formation and TBARS production during LDL oxidation

Antioxidant	lag time (min)	IC_{50} (μM)
Artepillin C	102.0	10.6 ± 0.8
TX-1959	71.3	25.5 ± 4.7
TX-1960	91.5	7.7 ± 0.5
TX-2007	74.8	27.0 ± 1.4
TX-2012	89.5	8.6 ± 0.8
TX-2013	83.5	19.6 ± 1.8
TX-2020	105.3	6.3 ± 0.7
Probucol	76.9	7.5 ± 0.6

Thus, these geranyl analogues showed inverse correlation between their inhibitory activity values for LDL oxidation and lipophilicity. As compared similar lipophilic compounds, TX-2012 and artepillin C, the former shows more inhibitory activity. In this case, the degree of molecular bulkiness and linearity, rather than lipophilicity of TX-2012, can be suitable for the interaction with LDL.

We designed various prenylated artepillin C analogues such as TX-2007, TX-2012, and TX-2013 based on biosyntheses of natural products. Furthermore, we will elucidate the effect of prenylation of natural compounds for their medicinal activities involving the antiatherogenic activity through the isoprenomics.

5. CONCLUSION

In this study, we developed a promising antiatherogenic antioxidant, TX-2012 based on isoprenomics. The inhibitory potency of our artepillin C analogues for LDL oxidation depends not only on the lipophilicity and free radical scavenging activity, but also on the topological properties such as linearity and compactness required for LDL interaction.

6. REFERENCES

1. S. Huneck, C. Zdero, and F. Bohlmann, Seco-guaianolides and other constituents from *artemisia* species, *Phytochem.* **25**, 883 (1986).
2. H. Aga, T. Shibuya, T. Sugimoto, S. Nakajima, and M. Kurimoto, Isolation and identification of antimicrobial compounds in Brazilian propolis, *Biosci. Biotech. Biochem.* **58**, 945 (1994).
3. Y. Uto, A. Hirata, T. Fujita, S. Takubo, H. Nagasawa, and H. Hori, First total synthesis of artepillin C established by *o,o '*-diprenylation of *p*-halophenols in water, *J. Org. Chem.* **67**(7), 2355-2357 (2002).
4. M. C. Marcucci, F. Ferreres, C. Garcia-Viguera, V. S. Bankova, S. L. De Castro, A. P. Dantas, P. H. Valente, and N. Paulino, Phenolic compounds from Brazilian propolis with pharmacological activities, J. *Ethnopharmacol.* **74**(2), 105-112 (2001).
5. T. Kimoto, S. Arai, M. Aga, T. Hanaya, M. Kohguchi, Y. Nomura, and M. Kurimoto, Cell cycle and apoptosis in cancer induced by the artepillin C extracted from Brazilian propolis, *Gan To Kagaku Ryoho* **23**(13), 1855-1859 (1996).
6. T. Matsuno, S. K. Jung, Y. Matsumoto, M. Saito, and J. Morikawa, Preferential cytotoxicity to tumor cells of 3,5-diprenyl-4-hydroxycinnamic acid (artepillin C) isolated from propolis, *Anticancer Res.* **17**(5A), 3565-3568 (1997).
7. T. Kimoto, S. Arai, M. Kohguchi, M. Aga, Y. Nomura, M. J. Micallef, M. Kurimoto, and K. Mito, Apoptosis and suppression of tumor growth by artepillin C extracted from Brazilian propolis, *Cancer Detect. Prev.* **22**(6), 506-515 (1998).
8. K. Hayashi, S. Komura, N. Isaji, N. Ohishi, and Yagi, K, Isolation of antioxidative compounds from Brazilian propolis: 3,4-dihydroxy-5-prenylcinnamic acid, a novel potent antioxidant, *Chem. Pharm. Bull.* **47**, 1521 (1999).
9. D. Steinberg, Role of oxidized LDL and antioxidant in atherosclerosis, *Adv. Exp. Med. Biol.* **369**, 39-48 (1995).
10. J. L. Witztum, and D. Steinberg, Role of oxidized low density lipoprotein in atherogenesis, *J. Clin. Invest.* **88**(6), 1785-1792 (1991).
11. E. Tomlinson, Chromatographic hydrophobic parameters in correlation analysis of structure-activity relationships, *J. Chromatogr.* **113**(1), 1-45 (1975).
12. M. S. Blois, Antioxidant determinations by the use of a stable free radical, *Nature* **181**, 1199 (1958).
13. W. A. Pryor, and L. Castel, Chemical methods for the detection of lipid hydroperoxides, *Methods Enzymol.* **105**, 293-299 (1984).
14. K. Yagi, Simple assay for the level of total lipid peroxides in serum or plasma, *Methods Mol. Biol.* **108**, 101-106 (1998).

19

IMAGING OXYGEN PRESSURE IN THE RODENT RETINA BY PHOSPHORESCENCE LIFETIME

David F. Wilson[a], Sergei A. Vinogradov[a], Pavel Grosul[a], Newman Sund[b], M. Noel Vacarezza[a], and Jean Bennett[b#]

1. INTRODUCTION

Many diseases of the eye, especially those causing inner retinal neovascularization (diabetic retinopathy, retinopathy of prematurity, sickle cell disease, etc.) have regional hypoxia as either a primary causative or early contributory factor. This is particularly true for diabetic retinopathy, the leading cause of blindness for individuals between 20 and 74 years of age. In diabetes, multiple structures of the eye are pathologically affected by vascular changes with resultant plasma leakage and tissue disruption[1]. Currently available data suggest that the pathological growth of new vessels results from developing regions of hypoxia in the retina. These new blood vessels in the inner retina are defective and exudate from the vessels and/or bleeding due to vessel rupture are the major cause of the decreased vision/blindness associated with diabetes. Using microelectrodes, Linsenmeier and coworkers[2, 3], reported that in diabetic cats the oxygen pressures in the inner half of the retina are about one half those of normal cats, and Berkowitz et al.[4], reported decreased oxygen response in the retina of galactosemic rats. Oxygen electrodes can very effectively identify global oxygen deficiency but are invasive. Also, since they make point measurements, they are not very effective for measuring focal hypoxia induced through microvessel failure. Retinal pathology in diabetes, for example, is thought to begin with pathological changes in the intraretinal microvessels (nonproliferative diabetic retinopathy) to growth of new vessels in the extraretinal space[5]. Similar pathogenic mechanisms are thought to occur in other diseases characterized by inner retinal neovascularization. Hypoxia resulting from vascular insufficiency could be responsible for production of high levels of vascular endothelial growth factor (VEGF). Vitreous samples in humans with retinopathy have contained high levels of VEGF[5]. High levels of VEGF have also been measured in animal models of ocular neovascular disease [6, 7, 8].

[#] Departments of Biochemistry and Biophysics[a] and of Ophthalmology[b], Medical School, University of Pennsylvania, Philadelphia, PA 19104

In the present paper, we report continuing progress[9] in developing a phosphorescence lifetime imaging system capable of obtaining high resolution images of the oxygen distribution in the retina of the rodent eye. These oxygen measurements would make possible critical evaluation of the role of microcirculatory failure in vision loss using a variety of rodent models for human disease.

2. METHODS

2.1. Phosphorescence Lifetime Imaging System

The phosphorescence lifetime imaging system previously used to image oxygen in the retina of the cat eye[10, 11] was modified to allow imaging of phosphorescence lifetimes in the much smaller mouse eye following the lead of Shonat and Kight[12, 13]. A frequency domain approach [14, 15] was used in which the excitation light source was modulated in a square wave while the gate of the intensified CCD camera was similarly modulated but delayed with respect to the excitation. On axis illumination was achieved by placing a dicroic mirror in the optical path and using it to reflect the excitation light into the optical path. Long working distance (18 mm) microscope optics were used between the dicroic mirror and the eye. A Xybion ISG 750 camera (now ITT Night Vision, Roanoke, VA) with enhanced red sensitivity was used for imaging the phosphorescence. The intensifier of this camera can be turned on or off "gated" in approximately 0.10 μsec. The excitation light source is a high power LED with a response time of less than 1 microsecond. Both the camera intensifier and the LED are modulated in square waves with a 50% duty cycle at frequencies from less than 100 Hz to greater than 40,000 Hz. In order to determine the phosphorescence lifetime, phosphorescence intensity images are collected at 7 to 10 different delays over the range from 0 to 360 degrees relative to the excitation light. These are then analyzed by fitting the intensity at each pixel of the image set to a sinusoid of the frequency used for taking the images. The phase of the phosphorescence relative to the excitation is determined and from the phase shift and frequency, the phosphorescence lifetime is calculated. The quenching by oxygen follows the Stern-Volmer relationship:

$$T_0/T = 1 + k_Q * T_0 * pO_2 \tag{1}$$

where T^0 and T are the phosphorescence lifetimes at oxygen pressures (pO_2) of zero and the experimental value, respectively, and k_Q is a second order rate constant related to the frequency of collision of excited state phosphor molecules with molecular oxygen and the probability that energy transfer will occur in each collision. Equation 1 makes it possible to calculate the oxygen pressure at each pixel of the image array once the phosphorescence lifetime is known.

The imaging software calculates the best fit of the data from each pixel of the image array and generates pixel by pixel maps of: 1. phase delay (phase shift) of the emission relative to the excitation light. 2. goodness of fit to a sinusoid as given by the regression coefficient ($r = 1.0$ is a perfect fit); 3. phosphorescence lifetime; and 4. oxygen pressure as calculated from Eq. 1. In living tissue, there can be significant tissue auto-fluorescence and/or leakage of excitation light through the optical filters. These signals occur with an effective phase shift of zero relative to the excitation. If no correction is made, they add to the phosphorescence and result in a progressive decrease in phase shift with increasing fraction of the signal. In order to correct for the "in phase" signal, two additional images are taken at 0 and 180 degrees relative to excitation, but at a frequency of 36 kHz. At such a high frequency the phosphorescence signal is highly demodulated and the remaining signal is phase shifted, relative to excitation, by more than 80 degrees. Thus the difference in image intensity between 0 and 180 degrees approximates the "in phase" signal due to tissue fluorescence and excitation light leak. The phosphorescence signal at each phase delay is corrected for the "in phase" signal and the phosphorescence lifetimes calculated. This correction has been tested with samples containing mixtures of fluorophor and phosphor and shown to effectively remove the effect of the fluorescence as long as it accounted for less than about 50% of the total intensity.

2.2. Experimental Protocol

The pigmented mice were anesthetized by i.p. injection of 0.2 ml of a solution with ketamine (25 mg/ml) and xylazine (25 mg /ml) dissolved in phosphate buffered saline. A drop of 1% tropicamide (Mydriacyl®, Alcon, Ft. Worth, TX) was placed on the eyes to dilate the pupils and the Oxyphor G2 (1.6 mg/ml in unbuffered saline, pH 7.5) given by i.v. injection of 0.15 ml into the tail vein. Approximately 4 min after the Mydriacyl®, a drop of hydroxypropyl methylcellulose (Goniosol®, CIBAVision Ophthalmics, Atlanta, GA) was placed on the eye and then a small piece of clear plastic sheet was gently placed on the Goniosol®. The retina was then imaged through the cover slip. Phosphorescence imaging began as early as 3 min after injection of the phosphor and on occasion continued for periods as long as 1.5 hrs.

Oxyphor G2, recently synthesized by Vinogradov and coworkers (16,17) was excited using 450 nm light and the phosphorescence (peak 790 nm) was collected through a 2mm 695 nm long pass Schott glass filter.

3. RESULTS AND DISCUSSION

3.1. Imaging Phosphorescence and Oxygen in the Mouse Retina

Sets of images, suitable for calculation of lifetime and oxygen maps, were taken of phosphorescence from Oxyphor G2 in the blood plasma of the retina. The phosphorescence intensity images show good resolution of the vasculature of the retina. Figure 1 shows phosphorescence images at 30 degrees delay relative to excitation and from the image sets taken 5 and 35 min. after induction of anesthesia and dilation of the pupil. The phosphorescence intensity images are presented as negatives, with increasing intensity appearing as increasing darkness in the image. The images taken just 5 min. after anesthesia show significantly higher intensity from the veins and capillaries relative to the arterioles than do the images taken at 35 min. after anesthesia. This is consistent with the oxygen pressure maps, which show the oxygen pressures in the veins were much lower (15 mm Hg) at 5 min than at 35 min (50 mm Hg).

5 min, intensity image, negative   **35 min, intensity image, negative**

5 min, oxygen map   **35 min. oxygen map**

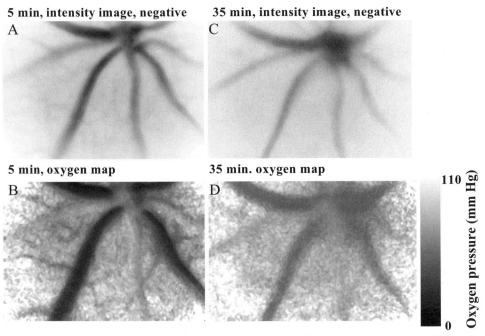

Figure 1. Dependence of the retinal oxygen pressures on the time after induction of anesthesia. The mouse was given an intraperitoneal injection of anesthetic, the pupil dilated, and retinal imaging of the phosphorescence lifetimes begun about 3 minutes later. The area of the retina that was imaged was approximately 0.92 mm high by 1.2 mm wide. The presented images and oxygen maps are from the data sets taken 5 and 35 min after induction of anesthesia.

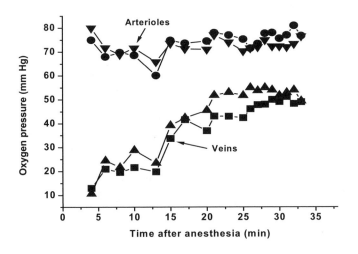

Figure 2. Time course of the oxygen pressures in the retinal veins and arterioles of the mouse retina following anesthesia. Oxygen pressure maps of the retina were repetitively measured, a total of 19 measurements over the period of 3 to 35 minutes following anesthesia. Regions of interest were selected within two different arterioles and two different veins and the average oxygen pressures for each region determined at every time point. The resulting values are plotted in the figure as a function of the time after anesthesia.

3.2. Effect of Anesthesia on the Measured Oxygen Maps

In the experiment shown in Fig. 1 and 2, a total of 19 oxygen maps, each consisting of a set of 9 phosphorescence intensity images, were taken over a period of 35 min. The venous oxygen pressures were decreased immediately following anesthesia but recovered in approximately 15 minutes to a stable level. The anesthesia response was quite variable among mice, ranging from no measurable effect (both arteriolar and venous oxygen pressures were stable from the beginning) to intermittent breathing throughout the period of anesthesia. In the latter case, the arteriolar and venous oxygen pressures were both unstable, decreasing during apnea and then quickly rising again when breathing restarted. Oxygen pressures measured in normal muscle tissue on the back leg showed similar effect of anesthesia on the oxygen levels (data not shown). Once the venous oxygen pressures increased to near normal, however, continuing measurements showed no further change. In addition, the oxygen map for the 35 min measurement (Fig. 1D) shows no evidence of local alterations in oxygen pressure, indicating that the measurements *per se* did not result in vascular injury.

3.3. Do the Oxygen Measurements Cause Vascular/Tissue Injury?

Figures 1 and 2 demonstrate that many consecutive measurements of oxygen maps can be made without significant alteration in the measured oxygen pressures or evidence of vascular injury. We have made measurements of the same region of the same eye of individual mice, repeated 3 times at 1 week intervals, 20 measurements (complete oxygen maps) per occasion. No evidence was found for vascular injury even in the 60th measurement. Further experiments are underway to determine at what point there is evidence for retinal injury.

4. ACKNOWLEDGEMENTS.

Supported in part by NS-31465, HD041484, and R43-DK064543.

5. REFERENCES

1. L'Esperance, FA and James, WA. 1983, The eye and diabetes Mellitus, in Ellenberg, M., and Rifkin, H. eds. Diabetes Mellitus: Theory and Practice, ed. 3. New Hyde Park, N. Y., Med. Examination, 727-757.
2. Linsenmeier, RA. 1986, Effects of light and darkness on oxygen distribution and consumption in the cat retina. J. Gen. Physiol. 88: 521-542.
3. Linsenmeier, RA, Braun, RD, McRipley, MA, Padnick, LB, Ahmed, J, Hatchell, DL, McLeod, DS, and Lutty, GA. 1998, Retinal hypoxia in long-term diabetic cats, Invest. Ophthalmol. & Visual Sci. 39: 1647-1657.
4. Berkowitz, BA, Kowluru, RA, Frank, RN, Kern, TS, Hohman, TC, and Prakash, M. 1999, Subnormal retinal oxygenation response precedes diabetic-like retinopathy. Invest. Ophthalmol. & Visual Sci. 40: 2100-2105.
5. Maguire AM. 1997, Management of diabetic retinopathy, J. Am. Osteopathic Assoc. 97: S6-S11.
6. Adamis, AP, Miller, JW, Bernal, MT, D'Amico, DJ, Folkman, J, Yeo, T-K, Yeo, K-T. 1994, Increased vascular endothelial growth factor levels in the vitreous of eyes with proliferative diabetic retinopathy. Am. J. Ophthalmol. 118, 445-450.

7. Pierce, EA, Avery, RL, Foley, ED, Aiello, LP, and Smith, LEH, 1995, Vascular endothelial growth factor/vascular permeability factor expression in a mouse model of retinal neovascularization. Proc Natl Acad Sci USA 92: 905-909.

8. Dorey, C.K., Aouidid, S., Reynaud, X., Dvorak, H.F. and Brown, L.F. 1996, Correlation of vascular permeability factor/vascular endothelial growth factor with extraretinal neovascularization in the rat.[comment][erratum appears in Arch Ophthalmol 1997 Feb; 115(2):192]. Arch. Ophthalmol. 114: 1210-1217.

9. Wilson, D.F., Vinogradov, S.A., Grosul, P., Kuroki, A., and Bennett, J. 2004, Imaging oxygen in the retina of the mouse eye. Adv. Exptl. Med. Biol. In press.

10. Shonat, R.D., Wilson, D.F., Riva, C.E., and Cranstoun, S.D. 1992, Effect of acute increases in intraocular pressure on intravascular optic nerve head oxygen tension in cats, Investigative Ophthalmology & Visual Science 33, 3174-3180.

11. Shonat, R.D., Wilson, D.F., Riva, C.E., and Pawlowski, M. 1992, Oxygen distribution in the retinal and choroidal vessels of the cat as measured by a new phosphorescence imaging method, Applied Optics, 33, 3711-3718.

12. Shonat, RD and Kight, AC. 2003a, Frequency domain imaging of oxygen tension in the mouse retina. Adv. Exptl. Med. Biol. 510: 243-247.

13. Shonat, RD and Kight, AC. 2003b, Oxygen tension imaging in the mouse retina. Ann. Biomed. Eng. 31: 1084-1096.

14. Vinogradov SA, Fernandez-Seara MA, Dugan BW, and Wilson DF. 2001, Frequency domain instrument for measuring phosphorescence lifetime distributions in heterogeneous samples. Rev. Sci. Inst. 72 (8): 3396-3406.

16. Vinogradov, SA and Wilson, DF. 1994, Metallotetrabenzoporphyrins. New phosphorescent probes for oxygen measurements. J. Chem. Soc., Perkin Trans. II, 103-111.

17. Dunphy, I, Vinogradov, SA, and Wilson, DF. 2002, Oxyphor R2 and G2: Phosphors for measuring oxygen by oxygen dependent quenching of phosphorescence. Analy. Biochem. 310: 191-198.

20.

RED BLOOD CELLS (RBC) DEFORMABILITY AND AGGREGABILITY: ALTERATIONS IN ALCOHOLISM

Vincenzo O. Palmieri, Giuseppe Cicco*, Francesco Minerva, Piero Portincasa, Ignazio Grattagliano, Vincenzo Memeo* and Giuseppe Palasciano[1]

1. INTRODUCTION

Alcohol is known to produce alterations in erythrocyte function leading to hemolysis association with alterations in shape and volume (macrocitosis, echinocitosis). [1,2]

The presence of macrocitosis is strictly associated with alcohol-correlated folate deficit even if modifications of the Mean Corpuscular Volume (MCV) do not correlate with the values of daily alcohol intake in alcoholic patients. [3,4,5]

Whether these effects are due to hemorheological alterations (deformability and aggregation) is not well understood.

Preliminary and isolated observations have reported some effect of *in vivo* alcohol on red blood cell (RBC) morphology and deformability evaluated by Rheodine SSD in healthy subjects consuming a large quantity of alcohol [6], but these changes have not been correlated with the selective action of alcohol per se or its metabolites (acetaldehyde, fatty acid ethyl esters) and could not be demonstrated in alcoholism (that according to WHO may be defined as a condition characterized by chronic consumption of alcohol associated with the presence of dependence and target organs damage).[7]

Recent studies have detected an *in vitro* action of ethanol, but not of acetaldehyde, on erythocytes resistance against hemolysis induced by sodium hypochlorite; the deleterious effect of ethanol consumption on erythrocyte *in vivo* may be, at least in part, the result of a direct effect of unmetabolized ethanol on erythrocyte components.[8]

No definitive data are available on *in vivo* RBC hemorheological properties in human alcoholics.

The Laser Assisted Optical Rotational Cell Analyzer (LORCA, Mechatronics, Hoon), first described by Hardeman et al.[9,10,11] is a valid technique for the measurement of various structural hemorheological parameters such as RBC deformability expressed by the elongation (EI: expressed as PA) and the RBC aggregation index (expressed as AI and T1/2).

This study therefore aimed to evaluate in patients with a history of alcoholism, RBC deformability and aggregation alterations using the LORCA, to correlate data on RBC

[1] Clinica Medica "A. Murri" University of Bari – *CEMOT Bari University of Bari

deformability with RBC morphological alterations and MCV values, to correlate data on RBC aggregability with factors able to modify this physical property, (such as the level of plasma fibrinogen), and evaluate the effects of alcohol withdrawal on these parameters.

2. MATERIALS AND METHODS

Nineteen consecutive alcoholic patients were enrolled in this study during their stay in the Clinica Medica "A. Murri" to evaluate alcohol-related damage and to plan their rehabilitation program.

The diagnosis of alcoholism was made according to the criteria suggested by the WHO[7] and confirmed by clinical questionnaires, biochemical markers (MCV, gamma GT, AST/ALT), and the presence of one or more target organs damage. In all patients liver disease was ascertained by histology. The mean daily alcohol intake at admission to the study was assessed according to a standardized questionnaire. Patients were queried about their alcohol intake during the previous 6-year period and the average daily consumption of beer, wine and spirits was quantified. Thereafter, the mean lifetime daily alcohol intake was calculated. General inclusion criteria were as follows: patient compliance, age $\geq 18 \leq$ 60yrs, history of problematic use of alcohol, daily alcohol consumption > 80g/day > 4 days/week, absence of arterial hypertension or cardiac and hematological diseases.

The hemorheological study was conduced day 1, 7, 14 and 90 of alcohol withdrawal. Preliminary data are available on days 1 and 7 of abstinence.

Alcoholic subjects that did not comply with abstinence during the observation period were excluded from the evaluation. The maintenance of abstinence was ascertained by clinical and biochemical (MCV, gamma GT, AST/ALT) parameters.

18 subjects with a negative personal history of problematic alcohol consumption and with a mean daily alcohol intake <20 g/day served as controls. They underwent the same preliminary evaluation of hepatic and hematological functions as assessed in the alcoholic patients.

The hemorheological study was conducted with the LORCA equipment of the CEMOT in the University of Bari. Deformability and aggregation of RBC were expressed as previously reported[9,10]. According to the standard procedure, 25 µl blood samples from each subject were processed the same day of the observation and diluted in 5 ml of polyvinylpyrrolidone (PVP) before running the test on the LORCA equipment.

Statistical analysis: Student's t test was used for paired (between alcoholics) and unpaired data (comparison with controls), the Mann-Withney U test in case of failure of the normality test, chi-square with Fisher correction; according to previously published data on EI values obtained by the LORCA device, we considered the EI results at pressure values of 0.3, 3 and 30 AP making it appropriate to use a direct comparison of mean values.

Results are expressed as Means (M) $\pm$ Standard Deviation (SD).

3. RESULTS

Clinical and serological characteristics of the alcoholic patients and controls are reported in Table 1. Although no differences were found between the two groups regarding sex distribution and age, the alcoholics had much higher values of AST, ALT and gamma GT than controls, only a slight increase of bilirubin and alkaline phosphatase, and no difference in serum albumin levels. These data and the normal coagulation values in the alcoholic group (data not shown) are in accordance with histological hepatic findings that demonstrate variable severity degrees of chronic alcoholic steatohepatitis and no case of cirrhosis.

Table 1. Comparison between clinical and serological parameters of alcoholic patients and normal controls

	Alcoholics (n=19)	Controls (n=18)	p-value
Age	40.6 ± 6.3	35.7 ± 5.4	NS
M/F	14/5	11/7	NS
BMI	24.2 ± 16.6	22.8 ± 1.1	< 0.01
Smoking habit	6/19	-	-
ALT (U/l)	88 ± 121	22 ± 6	< 0.006
AST (U/l)	71.8 ± 47.6	32.16 ± 5	<0.001
GGT (U/l)	306 ± 147	41.3 ± 117.5	< 0.001
Total bilirubin [mg/dl]	1.15 ± 0.19	0.88 ± 0.22	<0.001
Alkaline phosphatase [mg/dl]	99.3 ± 29.72	77.2 ± 11.34	< 0.02
Albumin [mg/dl]	3.72 ± 0.72	4.03 ± 0.43	NS
Fibrinogen [mg/dl]	376.36 ± 36	271.1 ± 44.3	< 0.001

Mean alcohol daily intake among the alcoholics was 141 ± 44 g/day (range 80-250 g/day). Comparison of alcohol consumption between alcoholics with MCV > 100 fl (macrocitosis) and alcoholics with normal MCV values showed no significant difference (respectively 136 ± 36 vs 141 ± 56 g/day), confirming previous observations by others.

Fibrinogen levels are also much higher in alcoholics than in controls (376 ± 36 vs 271 ± 44 mg/dl, p <0.001); these data are interesting in the evaluation of the aggregability results that are strictly related to the fibrinogen concentration.

Figure 1 and Table 2 show the results on the deformability of RBC expressed by the modifications of the EI for each level of shear stress. Corresponding with the intermediate values of shear stress (1.69, 3 and 5.33 mPA) we did not find any difference between alcoholics and controls. At lower levels of shear stress (0.3, 0.53 and 0.95 mPA), the EI values were significantly higher in the alcoholics than in controls; at higher levels of shear stress (9.49, 16.87 and 30 mPA), the deformability of RBC of the alcoholics was lower than in controls as demonstrated by the lower values of the EI.

In order to evaluate the direct effect of morphological alterations of alcoholic's RBC on their own hemorheological properties, we stratified the values of the EI on the basis of the MCV values in two groups: alcoholics with (MCV $\geq$100 fl) and alcoholics without (MCV <100 fl) macrocitosis (Table 3). Analysis of the data shows that the reduction of the RBC deformability is much more evident for the alcoholics with an elevated MCV (namely at higher values of shear stress), while the RBC of alcoholics with MCV values within the normal range are more deformed at the lowest values of pressure.

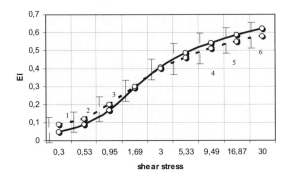

Figure 1. Elongation Index (EI) values (M±SD) at increasing doses of shear stress (bold line: controls n=18, dashed line: alcoholics n=19) p1 <0.02; 2 <0.001; 3<0.04; 4<0.03; 5<0.008; 6<0.003

The results on RBC aggregability as expressed by the Aggregation Index and T ½ are shown in Table 4. It is evident that the Aggregation Index is significantly higher (and T ½ significantly lower) in alcoholics than in controls. In contrast to the profile of RBC deformability, the alterations in aggregability parameters in alcoholics do not depend on MCV values and therefore do not differ significantly in the presence of macrocitosis (Table 5).

Finally, after one week of alcohol abstinence no difference was found in the E.I. or the aggregation index. Similarly, the MCV values did not change significantly in the same period (data not shown).

Table 2. Elongation Index values at increasing doses of shear stress.

SUBJECTS	Shear Stress (PA)	Elongation Index (M± SD)	MAX	MIN	p-VALUE
Alcoholics	0.3	0.0675 ± 0.018	0.12	0.035	< 0.02
Controls		0.054 ± 0.013	0.086	0.036	
Alcoholics	0.53	0.098 ± 0.021	0.152	0.078	< 0.001
Controls		0.091 ± 0.019	0.127	0.061	
Alcoholics	0.95	0.191 ± 0.019	0.215	0.144	< 0.04
Controls		0.176 ± 0.23	0.214	0.116	
Alcoholics	1.69	0.297 ± 0.025	0.329	0.234	NS
Controls		0.292 ± 0.017	0.324	0.266	
Alcoholics	3	0.398 ± 0.027	0.43	0.332	NS
Controls		0.406 ± 0.01	0.424	0.39	
Alcoholics	5.33	0.475 ± 0.025	0.503	0.402	NS
Controls		0.485 ± 0.015	0.501	0.473	
Alcoholics	9.49	0.528 ± 0.025	0.551	0.444	< 0.03
Controls		0.543 ± 0.007	0.555	0.528	
Alcoholics	16.87	0.570 ± 0.028	0.6	0.467	< 0.0008
Controls		0.587 ± 0.008	0.599	0.575	
Alcoholics	30	0.603 ± 0.031	0.631	0.483	< 0.003
Controls		0.619 ± 0.006	0.632	0.608	

Table 3. Elongation Index values (M±SD) at increasing doses of shear stress in alcoholics with pathological (≥100 fl) or normal (<100 fl) values of Mean Corpuscular Volume (MCV)

Shear stress (PA)	Controls (n=18)	Alcoholics with MCV ≥ 100 fl (n = 8) (A)	Alcoholics with MCV ≥ 100 fl (n = 11) (B)
0.3	0.054± 0.01 *[1]	0.057± 0.01 *[2]	0.075± 0.01
0.53	0.089± 0.02 *[3]	0.11± 0.01	0.12± 0.01
0.95	0.17± 0.02 *[4]	0.18± 0.02 *[5]	0.20± 0.01
1.69	0.29± 0.01	0.28± 0.02	0.30± 0.02
3	0.40± 0.01	0.39± 0.02	0.40± 0.02
5.33	0.48± 0.01 *[6]	0.46± 0.03	0.48± 0.02
9.49	0.54± 0.007 *[7]	0.51± 0.03	0.53± 0.01
16.87	0.58± 0.008 *[8]	0.55± 0.03	0.58± 0.01
30	0.62± 0.006 *[9]	0.58± 0.04	0.61± 0.01

*[1] p<.003 vs B, *[2] p<.03 vsB, *[3] p<.007 vs B and A, *[4] p<.006 vs B, *[5] p <.03 vs B, *[6] p<.01 vs A, *[7] p<.006 vs A, *[8] p<.003 vs A, *[9] p<.001 vs A

Table 4. Aggregability indices (M± SD) in alcoholics and controls

RBC Aggregability	Alcoholics (n=16)	Controls (n= 18)	P
AI	66.34 + 7.9	51.48 + 4.8	< 0.001
T½	1.8 + 0.62	3.1 + 0.5	< 0.001

Table 5. Aggregability indices (M± SD) in alcoholics and controls in relation to MCV values in alcoholics.

RBC Aggregability	Controls (n=18)	Alcoholics with MCV > 100 fl	Alcoholics with MCV < 100 fl
AI	66.34 + 7.9	51.48 + 4.8	< 0.001
T½	1.8 + 0.62	3.1 + 0.5	< 0.001

4. DISCUSSION

This study gives new insights into the evaluation of RBC alterations induced by alcohol. Namely, we show for the first time that the hemorheological properties (both deformability and aggregability) of RBCs of human alcoholics, studied *in vivo* by the LORCA equipment, are changed compared with normal subjects.

Our population is composed of very heavy drinkers that fulfil the WHO criteria for the diagnosis of alcoholism. Potential effects of hepatic advanced fibrosis on the evaluation of RBC alterations (as suggested by Maruyama[12]) are excluded by the histological confirmation of the absence of cirrhosis in all subjects enrolled in this study, and therefore the observed alterations are mainly (if not entirely) due to the action of chronic alcohol consumption.

We show that the deformability of RBCs in alcoholism differs depending on the increase of MCV and is markedly reduced at higher values of macrocitosis. This effect is particularly evident at higher values of shear stress, while at lower values of pressure, the deformability of RBC is higher than normal. On the basis of these results, we believe that the condition of macrocitosis observed in 8/19 alcoholic subjects leads to a reduction of

RBC deformability for high shear stress values, while the deformability is increased in alcoholics with normal MCV at low levels of shear stress. Alcohol may induce some alterations of RBC membranes that may modify the deformability for low levels of shear stress in the presence of normal MCV values. These data are consistent with an earlier study showing a positive effect of low and so-called "physiological" doses of ethanol *in vitro* on the filterability of erythrocytes not subject to mechanical stress.[13] On the other hand, previous human studies on the chronic effects of alcohol ingestion have focused on the effects on the erythrocyte membrane, with conflicting results. Studies have shown an increase[14] a decrease[15,16] and no change[17,18] on baseline membrane fluidity in alcohol abusers relative to controls. Furthermore, our results could be due to alterations of membrane properties (lipid composition?) of RBC during alcoholism that still do not modify cellular volume or shape.[19] On the other hand, the reduced deformability of RBCs in alcoholics with macrocitosis may be attributed to modification of the surface area/volume ratio. It is well known that one of the most important factors related to RBC deformability is represented by this ratio. The reduction of this value may produce a decrease of RBC deformability.[20] It is possible that in alcoholic subjects with macrocitosis, the increase of the surface of the RBC membrane implies a reduction of the aforementioned ratio and explains the altered RBC deformability.

In alcoholism, the RBCs are characterized by an elevated and rapid aggregability. We show that this property does not depend on MCV values, while we hypothesize a possible dependence on the high value of serum levels of fibrinogen, as already shown in different diseases characterized by an increase of serum fibrinogen (e.g. arterial hypertension[21]). In these conditions, RBC present alterations of surface electric charges that may be an important factor in the aggregability increase demonstrated in our experiments. Furthermore, we cannot exclude the importance of the oxidation of circulating proteins in alcoholics and in the behaviour of RBC aggregability, since it has been shown that proteins are oxidatively modified in plasma and in erythrocytes of active alcoholics.[22]

Further studies are needed to evaluate the effect of alcohol withdrawal since our observation is complete only in relation to the first seven days of abstinence. Nevertheless, other authors could not observe any modification of RBC membrane fluidity (evaluated by fluorescence polarization) in alcoholics after two weeks of withdrawal[18] while they could show a regression of platelet membrane fluidity properties during the same study period. It is therefore possible that the lack of modification of the physical properties of RBC revealed in the first week of abstinence of our patients is an expression of structural damage (e.g. morphological, protein or electric charge surface composition), that is not completely reversible during the first phases of abstinence. This hypothesis is partially confirmed by the persistence of macrocitosis observed in our group of patients during that period. Therefore, these results have to be confirmed by a longer observation period (at least 90 days).

5. REFERENCES

1. R. Cooper, Effects of alcohol and liver disease on the blood, *Delaware Med. J.* **45**: 297-298 (1973)
2. C.S. Lieber, Hepatic and metabolic effects of ethanol: pathogenesis and prevention, *Ann. Intern. Med.* **26**: 325-330 (1994).
3. H. Kanli, D.A. Terreros, Acute ethanol effects on cell volume regulation, *Ann. Clin. Lab. Sci.* **20**:205-13 (1990).
4. O.V. Tyulina, M.J. Huentelman, V.D. Prokopieva, A.A. Boldyrev, and P. Johnson, Does ethanol affect erythrocyte hemolysis? *Biochimica et Biophysica Acta* **1535**:69-77 (2000).

5 C.H. Halsted, J.A. Villanueva, A.M. Devlin, C.J. Chandler, , Metabolic interactions of alcohol and folate, *J. Nutr.* **132**: 2367S-2372S (2002).
6. B.A. Chmiel, Z.B. Olszowy, B.B. Turczynski, S.A. Kusmierski, Effect of controlled ethanol intake on arterial blood pressure, heart rate and red blood cells deformability, *Clin. Hemorheol. Microcirc.* **21**:325-328 (1999).
7. World Health Organization, The ICD-10 Classification of Mental and Behavioural Disorders, *World Health Organization, Geneva* (1992).
8. O.V. Tyulina, V.D. Prokopieva, R.D. Dodd, J.R. Hawkins, S.W. Clay, D.O. Wilson, A.A. Boldyrev and P. Johnson, In Vitro effects of ethanol, acetaldehyde and fatty acid ethyl esters on human erythrocytes, *Alcohol and Alcoholism* **37**: 179-186 (2002).
9. M.R. Hardeman, J.G.G. Dobbe, C. Ince, , The Laser-assisted Optical Rotational Cell Analyzer (LORCA) as red blood cell aggregometer, *Clin. Hemorrheol* **25**: 1-11(2001).
10. M.R. Hardeman, P.T. Goedhart, J.G.G. Dobbe, K.P. Lettinga, Laser-assisted Optical Rotational Cell Analyzer (LORCA); A New instrument for measurement of various structural hemorrheological parameters, *Clin. Hemorrheol* **14**: 605-618 (1994a).
11. M.R. Hardeman, P.T. Goedhart, N.H. Schut, Laser-assisted Optical Rotational Cell Analyzer (LORCA); I.Red blood cell deformability; elongation index versus cell transit time, *Clin. Hemorrheol* **14**: 619-630 (1994b).
12. S. Maruyama, C. Hirayama, S. Yamamoto, M. Koda, A. Udagawa, Y. Kadowaki, M. Inoue, A. Sagayama, K. Umeki, *J. Lab. Clin. Med.* **138**:332.337 (2001).
13. T. Oonishi, K. Sakashita, Ethanol improves filterability of human red blood cells through modulation of intracellular signalling pathways, *Alcohol. Clin. Exp. Res.* **24**:352-356 (2000).
14. S. Hrelia, G. Lercker, P.L. Biagi, A. Bordoni, F. Stefanini, P. Zunarelli, C.A. Rossi, , Effect of ethanol intake on human erythrocyte membrane fluidity and lipid composition, *Biochem.Int.* **12**: 741-750 (1986).
15. F. Beaugé, J. Gallay, H. Stibler, S. Borg, Alcohol abuse increases the lipid structural order in human erythrocyte membranes, *Biochem. Pharmacol.* **37**:3823-3828 (1988).
16. H. Stibler, F. Beaugé, A. Leguicher, S. Borg, Biophysical and biochemical alterations in erythrocyte membranes from chronic alcoholics, *Scand. J. Clin. Lab. Invest.* **51**:309-319 (1991).
17. F. Beaugé, E. Niel, E. Hispard, R. Perrotin, V. Thepot, M. Boynard, B. Nalpas, , Red blood cell deformability and alcohol dependence in humans, *Alcohol and Alcoholism* **29**:59-63 (1994).
18. P. Thompson, Platelet and erythrocyte membrane fluidity changes in alcohol-dependent patients undergoing acute withdrawal, *Alcohol and Alcoholism* **34**:349-354 (1999).
19. M. Parmahamsa, K. Rameswara Reddy, N. Varadacharyulu, Changes in composition and properties of erythrocyte membrane in chronic alcoholics, *Alcohol and Alcoholism* **39**:110-112 (2004).
20. S. Chien, , Red cell deformability and its relevance to blood flow, *Ann. Rev. Physiol.* **49**:177-192 (1987).
21. G. Cicco, M.C. Carbonara, G.D. Stingi, A. Pirrelli, , Cytosolic calcium and hemorheological patterns during arterial hypertension, *Clin Hemorheol Microcirc* **24**:25-31 (2001).
22. Grattagliano, G. Vendemiale, C. Sabbà, P. Buonamico, E. Altomare, Oxidation of circulating proteins in alcoholics: role of acetaldehyde and xanthine oxidase, *J. Hepatol.* **25**:28-36 (1996).

21.

EFFECTS OF MAJOR ABDOMINAL SURGERY ON RED BLOOD CELL DEFORMABILITY

Luigi Greco[1], Antonella Gentile[1], Piercarmine Panzera[2], Giorgio Catalano[2], Giuseppe Cicco [2], Vincenzo Memeo[2] *

1. INTRODUCTION

Erythrocyte deformability is defined as the ability of red blood cells (RBC) to change their shape while travelling through the capillary bed of the general circulation[1]. Such deformability, influencing the blood viscosity, is the important factor in the microcirculation, and in the delivery of oxygen to the tissues[2].

Erythrocyte deformability depends mainly on three interrelated factors: the surface/volume ratio, the viscoelastic properties of the membrane, and the viscosity of the intracellular hemoglobin solution[3]. In fact, a decrease in deformability increases the erythrocyte transit time and might reduce peripheral perfusion[4].

A reduction in erythrocyte deformability has been detected in several pathological conditions such as sepsis, chronic liver diseases and drugs therapy[5, 6, 7].

Aim of this study is to assess the modification of erythrocyte deformability, measured by LORCA essays, in patients that underwent radical surgery for major abdominal neoplasms.

2. MATERIALS AND METHODS

2.1 Patients

The study included 14 patients: 7 men and 7 women, with a mean age of 57 years. Six had gastric cancer, 5 rectal cancer and 3 colon cancer.

---

* [1] Department of General Surgery and Liver Transplantation, Faculty of Medicine , University of Bari
  [2] CEMOT Centre of Research in Haemorheology, Microcirculation and Oxygen Transport
University of Bari

2.2 Surgery

All patients underwent a radical surgery. Ten patients were transfused less than 500ml of concentrated red blood cells, 4 patients were not transfused at all. All had a regular recovery without any surgical complication.

2.3. Preparation of Blood Samples

Venous blood samples were obtained by venepuncture from an antecubital vein and anticoagulated with EDTA. The sampling times were the pre-operative day (T0) and the first (T1), the 5th (T5) and the 12th post-operative days. 200 µl of each blood sample was diluted in 5 ml of a polyvinylpyrovidone solution. The experiments were done at 37°C.

2.4. Determination of Erythrocyte Deformability

RBC deformability was quantified using a laser-assisted optical rotation red cell analyser (LORCA RR Mechatronics, Hoorn, The Netherlands)[8-9]. This instrument consists of a laser light, a rotating thermostatted cup and a video camera connected to a dedicated ellipse-fit computer program. It measures the diffraction pattern of the RBC under various shear stresses in the range of 0.3 to 30 Pa.

The Elongation Index (EI) was the parameter used to express RBC deformability at several shear stresses. An increased EI indicates greater cell deformability.

2.5. Statistical Analysis

Results are expressed as mean ± standard error (S.E.). Statistical comparison between groups was done by Student's paired *t* test. P values <0.05 were taken as statistically significant.

3. RESULTS

The mean preoperative EI was 0.387± 0.006 at a shear stress of 3 Pa, 0.473± 0.005 at a shear stress of 5.33 Pa, 0.568± 0.008 at a shear rate of 16 Pa and 0.595± 0.01 at a shear rate of 30 Pa. No statistically significant change of erythrocyte deformability was found on the 1st and 5th post-operative days. On the 12th post-operative day there was a small increase in the EI but only at higher shear stress: 0.586± 0.002 (P<0.05) at a shear stress of 16 Pa, and 0.6178± 0.002 (P<0.05) at a shear stress of 30 Pa.

4. DISCUSSION

In the last 10 years interest in the deformability of RBCs in different pathological diseases has increased dramatically, mainly due to the development of more appropriate and sensitive diagnostic techniques[10].

Photodynamic treatment is at the moment the best method that may be applicable in this field[11].

When photosensitizer solutes are illuminated with light of the appropriate wavelength, reactive oxygen species are formed that can inactivate nucleic acids, with harmful effects on proteins or lipids.

Therefore RBCs possess natural defence mechanisms that may resist attack from reactive oxygen species. It is important to use a sensitizer such as red light that is not absorbed by Hb.

The laser-assisted optical rotational cell analyzer (LORCA), is a reliable technique for measuring RBC deformability, studying the effects of in vitro manipulations of cells suspensions[12, 13].

After photodynamic treatment, the RBC suspensions are diluted 200 times in 0.14 ml per liter polyvinylpyrrollidone. One ml of this suspension is transferred into the LORCA measuring system and subjected (fully automatically) to varying shear stresses by increasing the rotation speed.

Deformation is expressed by the elongation index, derived from the resulting ellipsoid diffraction pattern. The deformation curve is obtained by plotting the calculated values for the elongation index versus the corresponding shear stress.

An absolute prerequisite for RBCs, which have a mean diameter of 7 to 9 microns, to pass the microvascular bed, in which the smallest capillaries have a mean diameter of 4 microns, is their capacity to deform. It is evident that a reduction in their deformability capacity causes an important alteration of the hemorrheologic setting.

Decrease of RBC deformability has been associated with pathogenesis of impaired microcirculatory flow in different diseases[14].

The erythrocytes of many patients with chronic liver diseases seem to be abnormal in shape and function and it is known that this depends on altered lipid composition of the cellular membrane which decreases their deformability[15].

The membrane lipid composition is affected by abnormal metabolism in liver disease, probably due to an altered synthetic system, so the cellular membrane lipid composition depends overall on the surrounding plasma.

Horii et al showed changes in RBC deformability after hepatic resection for liver tumours in cirrhotic patients. They also demonstrated correlations between the scale of the operation and decrease of RBC deformability that is related with the development of postoperative complications[16].

The mechanism of this effect has not yet been identified, but probably decreased erythrocyte deformability may disturb the microcirculation and impair liver function.

Langenfeld et al.[10] showed that the measurement of RBC deformability gives information useful for the early detection of infection in patients with multiple injuries and erythrocyte rigidification has been demonstrated during cyclosporin therapy[17]. The effect of surgery on RBC deformability has also been investigated.

Scholz et al. found that deformability of erythocytes decreased in operated patients who received more than 7 units of transfused blood[18].

A postoperative decrease in deformability of erythrocytes has been reported in patients after arterial surgery or varicose vein stripping and also in patients undergoing open heart surgery[19, 20, 21].

We studied the effects of major abdominal operations on RBCs using a laser assisted rotational cell analyzer, currently the most sensitive technique available.

All patients underwent radical surgery for neoplastic disease of the stomach or colon, with minimal blood transfusion, never more than two units. The postoperative course was uneventful in all cases.

Our results show minimal modification of RBC deformability after major abdominal surgery without complications, with a slight increase in the elongation index, not influenced by antithrombotic prophylaxis with calciparine.

We found a surprising, significant improvement in RBC deformability on the 12th post-operative day .

It is difficult to explain this result and more investigations are needed.

We hypothesize that radical removal of tumours can reduce the circulating factors having an effect on RBC deformability, as has been suggested in the past [22].

5. REFERENCES

1. P.C. Mokked,M. Kedari, C.P. Henny et al. The clinical importance of erythrocytes deformability; a hemorrheological parameter.Ann. Hematol. 1992: 64: 113-22.
2. T. Shiga, N. Maeda, K Kon. Erythrocyte rheology. Crit.Rev.Oncol.Hematol. 1990: 10:9.
3. H. J. Voerman, A.B.J. Groeneveld. Blood viscosity and circulatory shock. Inten.Care Med. 1989: 15:72.
4. S.Chien Determinants of blood viscosity and red cell deformability. Scan.J.Lab.invest. 1981, 41(suppl.156) : 7.
5. M. Garnier , M Hanss., A Paraf. Erythrocytes filterability reduction and membrane lipids in liver cirrhosis. Clin.Hemorheol: 1895: 3:45.
6. D. Bareford, P. C. W. Stone, N.M Caldwell, J. Stuart. Erythrocyte morphology as a determinant of abnormal erythrocyte deformability in liver disease. Clin. Hemorheol. 1985: 5:473.
7. G.W. Machiedo.,R.J. Powell, B. F. Rush, N.I. Swislocki, G. Dikdan The incidence of decreased red blood cell deformability in sepsis and the association with oxygen free radical damage and multiple-system organ failure. Arch.Surg. 1989: 124: 1386.
8. M. R. Hardeman, P. T. Goedhart, K.P. Lettinga. Laser-assisted optical rotational cell analyser (L.O.R.C.A.). A new instrument for measurement of various structural hemorheological parameters. Clin Hemorheol 1994;14:605-618.
9. K. Osterloh, P Gaehtgens Pries AR. Determination of microvascular flow pattern formation in vivo. Am J Physiol Heart Circ Physiol 2000;278:H1142-H1152.
10. J.E. Langenfeld, D.H. Lingstone, G.W. Machiedo. Red blood deformability is an early indicator of infection. Surgery 1991: 110: 398
11. E. Ben-Hur, A.C. Moor, H. Margolis Nunno. The photodecontamination of cellular blood components: Mechanisms and use of photosensitization in transfusion. Transf. Med.Rev. 1996. 10:15.
12. M.R. Hardeman,G.A.J. Besselink, I. Ebbing, C. Ince, A.J. Verhoeven. Laser associated optical rotation cell analyzer measurements reveal early changes in human RBC Deformability induced by photodynamic treatment. Transfusion 2003: 43: 1533.
13. M.R. Hardeman, C. Ince. Clinical potential of in vitro measured red cell deformability; a myth? Clin. Hemorheol. Microcirc. 1999: 21: 277.
14. O.K. Baskurt, A. Temiz, H.J. Meiselman. Effect of superoxide anions on red blood cell rheology properties. Free radical Biology & Medicine: 1998: 24:102.
15. J.S. Owen, K.R. Bruckdorfer, R.C. Day Mc Intyre N. Decreased erythrocyte membrane fluidity and altered lipid composition in human liver disease. J.Lipid Res. 1982: 23:124.
16. K. Horii, S. Kubo, K. Hirohashi, H. Kinoshita. Changes in erythocyte deformability after liver resection for hepatocellular carcinoma with chronic liver disease. W.J.Surg. 1999: 23: 85.
17. M.R. Hardeman, M.M. Meinardi, C. Ince, J. Vreeken. Red blood cell rigidification during cyclosporin therapy: a possible early warning signal for adverse reactions. Scand. J.Clin.Lab.invest. 1998: 58: 617.
18. P.M. Scholz,, J.M. Kinney, S. Chien. Effects of major abdominal operations on human blood rheology. Surgery 1975: 77: 351.
19. A.J. Dodds, P.N. Matthews, M.J. Bailey, P.T. Flute, J.A. Dormandy. Changes in red cell deformability following surgery. Thromb. Res. 1980: 18:561.
20. S. Ekestrom, B.L. Koul, T. Sonnenfeld. Decreased red cell deformability following open heart surgery. Scand. J. Thorac. Cardiovasc. Surg. 1983: 17:41.
21. R. Muller, P. Musikic. Hemorheology in surgery: a review. Angiology 1987: 35:581.
22. G.W. Tietjen, S. Chien, P. Scholz, F.E. Gump, J.M. Kinney. Changes in blood viscosity and plasma proteins in carcinoma. J.Surg. Oncol. 1977 : 53: 9.

22.

INFLUENCE OF WHOLE-BODY VIBRATION STATIC EXERCISE ON QUADRICEPS OXYGENATION

Vittorio Calvisi,[1,2] Massimo Angelozzi,[2] Antonio Franco,[3] Leonardo Mottola,[4] Stefano Crisostomi,[3] Cristiana Corsica,[2] Marco Ferrari,[2,4] and Valentina Quaresima[2,4] *

1. INTRODUCTION

Whole-body vibration (WBV) is a neuromuscular training method recently designed to improve muscle strength and flexibility.[1-9] More recently, WBV has been proposed to be a suitable training method as efficient as conventional resistance training to improve knee-extension strength and speed of movement and counter movement jump performance in older women.[10] The acute effects of vibration seem to be connected to the duration of the stimulation, the characteristics of the subjects (well trained vs. untrained), and the magnitude of the vibration stimulus (amplitude, frequency and acceleration). When the human body undergoes vibratory stimuli, muscle activity is necessary for damping the vibratory waves. It is assumed that vibrations evoke muscle contraction, probably via the monosynaptic stretch reflex. Although the electromyography (EMG) activity of the *vastus lateralis* (VL) muscle during WBV has been investigated,[4] there are no studies about the effects of WBV on the oxygenation (oxidative metabolism) of leg skeletal muscles.

This study aimed at investigating the oxygenation response (measured as tissue oxygenation index (TOI)) in *rectus femoris* (RF) and VL muscle groups during different frequencies of WBV.

* [1]Dipartimento di Scienze Chirurgiche, [2]Centro Interdipartimentale di Scienza dello Sport, [3]Scuola di Specializzazione in Medicina dello Sport, [4]Dipartimento di Scienze e Tecnologie Biomediche, Università di L'Aquila, 67100 L'Aquila, Italy; e-mail: vale@univaq.it

2. METHODS

Seven volunteers (age: 23 ± 2 years; body mass: 79 ± 9 kg) participated in this study. Subjects were physically active although none were engaged in daily, intensive or specific training programs. All subjects gave their informed consent prior participation after a full oral and written explanation of the experiments. Subjects were asked to stand in half-squat (HS) position (knee angle $110°$) on a vibration platform (NEMES, OMP, Italy) in the following conditions: no vibrations, and randomly 30, 40, and 50 Hz WBV. Each condition lasted 110 s, and the interval between sets was 45 min. Muscle oxygenation was monitored by a 2-channel NIRO-300 oximeter (Hamamatsu Photonics, Japan). The emission and detection probes were kept at a constant geometry and distance (4.5 cm apart) by a rigid rubber probe holder. Muscle O_2 saturation in RF and VL was measured as TOI (%). TOI reflects the balance between O_2 supply and O_2 consumption in the examined muscle volume. TOI was also measured during the standing (S) position that preceded the HS position at the beginning of each set. Concomitantly total hemoglobin volume changes (ΔtHb, $\mu M*cm$) were monitored. The sampling rate of NIRS data was 6 Hz. Adipose tissue thickness underlying the monitored VL and RF area was measured with a skinfold caliper. Adipose tissue thickness (ATT) was 4.2 ± 0.9 mm and 4.4 ± 1.8 mm for VL and RF, respectively. These similar ATT values allowed the correct comparison of NIRS data between the two muscle groups.

Data are reported as mean $\pm$ standard deviation of TOI and tHb changes (average over the last 5 s in every subject) for each condition. TOI and changes in tHb were compared between conditions by repeated measures analysis of variance. Significance level was set at $P<0.05$.

3. RESULTS

Oxygenation responses observed in VL and RF of one representative subject for each experimental condition are reported in Fig. 1. A significant TOI decrease was observed in RF muscle after about 40 s of WBV at 30 Hz. Concomitantly, tHb was almost stable. TOI slightly decreased in VL over the last 50 s of WBV exposure, and tHb gradually increased. TOI did not change in RF and VL in the remaining conditions, whilst tHb tended to increase. Furthermore, the pattern and the amplitude of tHb raise were different among WBV frequency conditions. Considering the averaged response over the 7 subjects (Fig. 2), WBV did not affect TOI value in VL muscle, while induced a consistent decrease of TOI in RF muscle only at a frequency of 30 Hz compared to TOI measured during baseline condition (standing). Comparing TOI of half-squat condition with TOI during WBV condition provoking the highest TOI decrease, a significant difference was found in RF (from 59.3 ± 5.3 to $53.0\pm6.4\%$, $P=0.04$) and VL (57.6 ± 2.5 to $50.3\pm7.9\%$, $P=0.03$).

4. DISCUSSION

Initially, WBV training was used in elite athletes to improve speed-strength performance. More recently, it is becoming extremely popular in European health and fitness clubs as an alternative training method. However, a consistent scientific support about the benefits of WBV on fitness and health is still missing. Much work has been done by Bosco et al.,[1,2] and Cardinale et al.[3-5] An increase in force-velocity, force-power and vertical-jump performance immediately after one WBV session was found.[1,2]

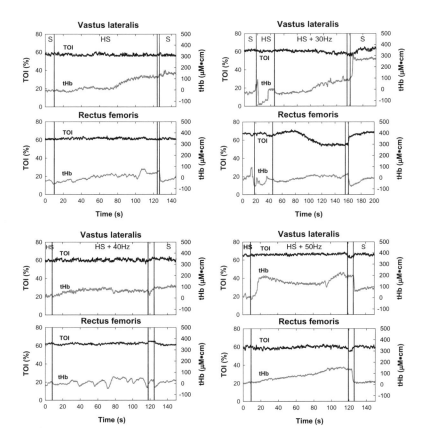

Figure 1. Time course of *vastus lateralis* and *rectus femoris* TOI and tHb changes during static HS condition only or with 30 Hz whole-body vibration (upper panels), and with 40 or 50 Hz whole-body vibration (lower panels). S: standing; HS: half-squat position (knee angle 110°).

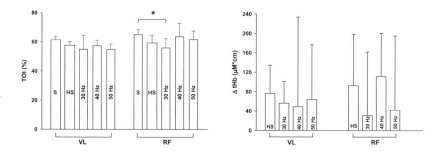

Figure 2. TOI and tHb of the *vastus lateralis* (VL) and the *rectus femoris* (RF) muscles measured at the end of 30 s standing (S), 110 s of half-squat (HS), and at the end of 110 s of the three whole-body vibrations (30, 40, and 50 Hz). *: indicates a significant difference between TOI at the end of 110 s of 30 Hz whole-body vibration and TOI measured in standing position.

The effects of WBV training on muscle performance over a longer period were also investigated.[3] It was suggested that WBV training might result in neuromuscular adaptations similar to the effects produced by explosive strength training, and that the long-term effects of WBV might depend on duration of WBV-training programs.

Recently, Delecluse et al.[6] reported that WBV (and the resulting reflexive muscle contraction) has the potential to induce strength gain in knee-extensor of previously untrained females to the same extent as resistance training at moderate intensity. Furthermore, they demonstrated that strength increases after WBV training are not attributable to a placebo effect. The EMG responses of VL muscle to different WBV were previously investigated.[4] The highest EMG root mean square was found at 30 Hz, suggesting this frequency as one eliciting the highest reflex response in VL muscle during WBV in static half-squat position.

Our study reports, for the first time, the oxygenation responses in VL and RF muscle groups during different (30, 40, 50 Hz) WBV in static half-squat position. A significant decrease of TOI was found in RF muscle during 30 Hz WBV. This suggests that the metabolic activity (as a consequence of muscle activity) in RF was higher than in VL. The oxidative metabolic activity of RF was not satisfied by the O_2 blood inflow, and the required O_2 was extracted by the local oxy-Hb with a consequent TOI decrease. Considering that, tHb is related to changes in the total volume of Hb in the muscle region of interest,[11] the observed tHb rise (Fig. 1) could reflect a local blood flow increase (vasodilatation in response to WBV). The fluctuations observed in tHb tracing of RF (Fig. 1, 50 Hz WBV panel) could be attributable to the posture changes resulting from the difficulty to maintain HS position at the highest WBV frequency used.

According to the observed inter-subjects variability of VL or RF oxidative metabolic response (different ΔTOI) to each WBV frequency, WBV training program should be individualized. However, other studies should be performed to better understand the physiological and biochemical mechanisms underlying the individual responses to WBV. This would allow the establishment of criteria to select the most effective WBV frequency and duration in planning WBV protocols for training or rehabilitation purposes.

5. REFERENCES

1. C. Bosco, R. Colli, E. Introini, M. Cardinale, M. Iacovelli, J. Tihanyi, S. P. Von Duvillard, and A. Viru, Adaptive response of human skeletal muscle to vibration exposure, *Clin. Physiol.* **19**(2), 183-187 (1999).
2. C. Bosco, M. Iacovelli, O. Tsarpela, M. Cardinale, M. Bonifazi, J. Tihanyi, M. Viru, A. De Lorenzo, and A. Viru, Hormonal responses to whole body vibrations in man, *Eur. J. Appl. Physiol.* **81**(6), 449-454 (2000).
3. M. Cardinale and M. H. Pope, The effects of whole body vibration on humans: dangerous or advantageous?, *Acta Physiol. Hung.* **90**(3), 195-206 (2003).
4. M. Cardinale and J. Lim, Electromyography activity of vastus lateralis muscle during whole-body vibrations of different frequencies, *J. Strength Cond. Res.* **17**(3), 621-624 (2003).
5. M. Cardinale and C. Bosco, The use of vibration as an exercise intervention, *Exerc. Sport Sci. Rev.* **31**(1), 3-7, (2003).
6. C. Delecluse, R. Machteld, and S. Verschueren, Strength increase after whole-body vibration compared with resistance training, *Med. Sci. Sports Exerc.* **35**(6), 1033-1041 (2003).
7. V. B. Issurin, D. G. Liebermann, and G. Tenenbaum, Effect of vibratory stimulation training on maximal force and flexibility, *J. Sport Sci.* **12**(6), 261-566 (1994).
8. J. Rittweger, G. Beller, and D. Felsenberg, Acute physiological effects of exhaustive whole-body vibration exercise in man, *Clinical Physiology* **20**(2), 134-142 (2000).
9. S. Torvinen, P. Kannus, H. Sievanen, T. A. Jarvinen, M. Pasanen, S. Kontulainen, T. L. Jarvinen, Jarvinen M., P. Oja, and I. Vuori, Effect of four-month vertical whole body vibration on performance and balance, *Med. Sci. Sports Exerc.* **34**(9), 1523-1528 (2002).
10. M. Roelants, C. Delecluse, and S. M. Verschueren, Whole-body-vibration training increases knee-extension strength and speed of movement in older women, *J. Am. Geriatr. Soc.* **52**(6):901-908, (2004).
11. M. Ferrari, L. Mottola, and V. Quaresima, Principles, techniques, and limitations of near infrared spectroscopy, *Can. J. Appl. Physiol.* **29**(4):463-487 (2004).

23.

INDOCYANINE GREEN LASER RETINAL OXIMETRY: PRELIMINARY REPORT

Nicola Cardascia, Raffaele Tommasi, Michele Vetrugno, Giancarlo Sborgia, Pietro Mario Lugarà, Carlo Sborgia[*]

1. INTRODUCTION

It is known that any alteration in blood circulation such as seen in diabetic retinopathy, hypertension, sickle cell disease, and vascular occlusive diseases results in functional impairment and extensive retinal tissue damage.[1] However little is known about retinal oxygen distribution or consumption and how it changes in response to disease. Measuring these changes non-invasively may improve the ability to diagnose, or monitor the progression of, such eye diseases. Oxygen is monitored by using its effect on blood hemoglobin color, its concentration in gas, its pressure in blood or tissue and using infrared light through the scalp and skull. In the eye non-invasive optical methods have already been reported.[2][3] These techniques depend on quantifying differences between hemoglobin (Hb) and oxyhemoglobin (HbO) light absorption.

2. PURPOSE

Most of the methods used to measure oxygen concentration are based on the relation between light transmission and oxygen saturation. We would suggest a new method to determine oxygen saturation in retina and choroid, not affected by the anatomical and physical characteristics of the eye.

3. METHODS

Our investigation is twofold: (1) to determine an angiographic dye that can modify its fluorescence according to blood oxygen concentration, and (2) to investigate the variation of fluorescence in different oxygen saturation.

---

[*] Nicola Cardascia, Dipartimento di Oftalmologia, Università di Bari, 70124 Bari, Italy. Raffaele Tommasi, Dipartimento di Biochimica Medica, Biologia Medica e Fisica Medica, Università di Bari, 70100 Bari, Italy. Michele Vetrugno, Dipartimento di Oftalmologia, Università di Bari, 70124 Bari, Italy. Pietro Mario Lugarà, Dipartimento interateneo di Fisica and Unità INFM, Università di Bari, 70126 Bari, Italy. Carlo Sborgia, Dipartimento di Oftalmologia, Università di Bari, 70124 Bari, Italy.

3.1 The Fluorescent Dye

In the eye, non-invasive methods are based on analysis of retinal reflectivity using appropriate light sources, chosen because of their spectral properties and availability.[4] Considering retinal reflectivity may influence spectral determination of oxygen saturation,[5] we prefer to consider the fluorescence of an angiographic dye. The selection of the dye is based on precise characteristics: (1) widely diffuse in clinical practice, (2) few adverse effects, (3) fluorescence not affected by pigment epithelium and other low reflective media.

We investigate, then, three fluorescent molecules: Sodium Fluorescein, Verteporfin and Indocyanine Green (ICG).

3.1.1 Sodium Fluorescein

The technique of Fluorescein Angiography was first demonstrated in the human eye by two medical students, Novotny and Alvis in 1961.[6] This paved the way for contrast studies of the ocular circulation which has become the gold standard of imaging for ocular circulation for the diagnosis of vascular disease. Moreover, retinal oxygen saturation was determined by means of fluorescein to obtain a map of relative oxygen saturation in retinal structures and the optic nerve head in nonhuman primate eyes.[7] Fluorescein dye leaks from vessels in several pathologies involving vascular disorders.[8][9]

Even through fluorescein has a long history of use in clinical practice[10] and adverse effects are well known,[11] we exclude it principally because the dye can impair vessel fluorescence leaking from vessel walls, affecting oxygen determination.

3.1.2 Verteporfin

Benzoporphyrin derivative, or verteporfin, is a modified porphyrin that has an absorption maximum near 690 nm and is phototoxic in vivo.[12] All porphyrins have fluorescent spectra in the red band and show oxygen fluorescence quenching.[13] Even though porphyrins have broad fluorescent spectra, their photo-toxicity induced us to exclude them to avoid irremediable damage of retinal and vascular structures.[14]

3.1.3 Indocyanine Green

Considering the strict inclusion criteria, ICG represents the ideal dye for our purpose. The inclusion characters of ICG are hereby discussed:

3.1.3a. Clinical practice. ICG is a dye employed in the photographic industry. It was first used in ophthalmology by Flower and Hochheimer in the early 1970s to image the choroidal circulation.[15] Although both experimental and clinical investigations with ICG continued, it was not until the early 1990s that it became an established method of investigation.[16] This was because of the increasing interest in the contribution of the choroid to retinal diseases and improvements in technology.

3.1.3b. Adverse effects. Adverse reactions to ICG are rarer than those with intravenous fluorescein angiography. Mild reactions such as nausea, vomiting, sneezing, and transient itching occur in 0.15% of cases.[17] More severe reactions such as urticaria, syncope, fainting, and pyrexia may also occur. Severe reactions such as hypotensive shock[18] and anaphylactic shock[19] have been reported. Crossover allergy to iodine can occur in patients with seafood allergies. Thus, seafood allergy is a contraindication to ICG angiography.

Because ICG is metabolized primarily by the liver, it should be avoided in patients with hepatic disease. Those undergoing hemodialysis are also at increased risk of complications from ICG.

3.1.3c. Fluorescence. ICG dye absorbs and emits light in the near infrared part of the electromagnetic spectrum, enhancing transmission through pigment, turbid exudation, and blood for more highly resolved images of the choroidal circulation and its associated disease. The large ICG molecule is almost completely protein-bound in blood, rendering it relatively impermeable to the choriocapillaris.[20] It was seen that with infrared wavelengths the background reflectance is affected by the reflectance of choroidal vessels.[21] The ICG prevents abnormal choroidal reflectivity filling the vessels, superimposing the reflectivity of the vessels walls and pigment epithelium.[22]

3.2 The Variation of Fluorescence of ICG in Different Oxygen Saturation

To clarify the relation between ICG and oxygen saturation we have to consider the interaction of the fluorescent dye with plasmatic compounds and the variation of fluorescence induced by oxygen concentration.

3.2.1. ICG and plasma compound.

ICG molecule is composed of two polycyclic parts (benzoindotricarbocyanin), which are quite lipophilic and are linked by a carbon chain. A sulfate group is bound to each polycyclic part, leading to some water solubility. This complex molecular structure leads to amphiphilic properties, that is, both hydrophilic and lipophilic properties.[23] Like many carbocyanine dyes, ICG tends to form aggregates.[24] The interaction of the polymerization of ICG molecules and the aggregation to the plasmatic proteins or lipoproteins, explains the stability of the absorption spectrum after the dilution that occurs during intravenous injection.[25] The spectroscopic parameters of ICG indicate that the dye is a fast saturable absorber and laser dye in near-IR spectral region.[26] Anyway, after intravenous injection, some variations of the fluorescence emission peak occur.[27][28]

3.2.2. Hemoglobin and oxygen saturation

Although hemoglobin is not a one-component system, it has been demonstrated that the Lambert-Beer law holds for solutions of hemoglobin. At minimum, there are two optically active ingredients: oxyhemoglobin and deoxyhemoglobin. Each has a characteristic extinction coefficient at different wavelengths.[29][30] Oxygen saturation is determined by the percentage of oxygenated hemoglobin within total hemoglobin.

Taking into consideration these issues, oxygen saturation can be measured by the quenching of fluorescence of an injected dye.[31] It is well demonstrated that fluorescence is affected by molecular oxygen.[32] The decay of the fluorescence is not influenced by: oxygen molar concentration, dye concentration, and aberrations induced by anatomical structures.[33] Fluorescent lifetime quenching, therefore, is strictly dependent on oxygen saturation of hemoglobin surrounding the dye.[34]

3.3 Experimental Set-up

3.3.1. ICG solution

ICG-NaI (Arkon Inc.) was prepared as a single 4.83×10^{-4}M solution[35] in methanol (HPLC grade), a solvent in which ICG is known to be stable.[36] Inspection cuvette was covered by a plastic tip with drilled holes to introduce gases. To obtain an ICG solution desaturated by molecular oxygen, N_2 was insufflated for 45 min. in the solution. Otherwise a full-oxygen saturated solution was obtained by insufflating pure O_2 in the solution for the same amount of time.

3.3.2. Laser system and spectroscopic set-up

ICG fluorescence was induced by focusing the third-harmonic ($\lambda=355$nm) of a nanosecond ($\tau\approx10$ ns) Nd:YAG laser (Quantel mod. YG580), emitting at a 10 Hz repetition rate onto the inspection cuvette. The fluorescence was collected using a set of lenses and analyzed by a ARC monochromator (mod. SpectraPro 275) followed by a photomultiplier tube with extended near-infrared response (Hamamatsu R943-02). The signal was sent to a boxcar averager (EG&G mod.162), combined with a gate integrator (mod. 164 Processor Module-Gated Integrator) to shift the detecting gate at a settable delay of time after photoexcitation. The boxcar trigger was driven by the laser pulses. The boxcar output was visualized using an oscilloscope and stored in a personal computer using Spectra Card (ACR Instruments) that also controlled the monochromator by means of a dedicated software package. The gate width was set to ≈25 ns with a minimum time delay of ≈80 ns.

4. RESULTS

Several factors may induce errors in the determination of fluorescence. The effects of laser power fluctuations were reduced by extending the acquisition time of Spectra card (3sec). Set-up alignment and cuvette cleaning were performed frequently.

Figure 1, left graph, shows the spectrum of desaturated ICG-NaI solution, which represents the sample fluorescence at a time delay of 325 nsec after photoexcitation. A bell shaped spectrum from 800 nm to 925 nm, peaked at 870 nm is recorded with a spectral resolution of 1 nm.

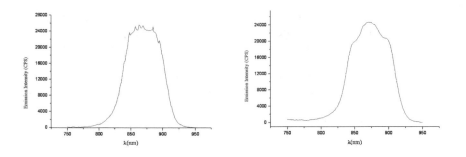

Figure 1. Left: Spectrum of oxygen desaturated ICG-NaI solution after 325 ns from excitation. Right: Spectrum of oxygen saturated ICG-NaI solution after 325 ns from excitation.

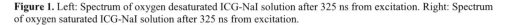

Figure 1, right graph, represents the spectrum of an oxygen-saturated ICG-NaI sample. A peak at 870 nm and two shoulders at ~850nm e ~895nm are clearly visible. The observed spectral features suggest to analyze fluorescence intensity in the spectral range 800-925nm as a function of the time delay. Such scans were performed at different delays, ranging from 0.325 to 50.075μs. We always found a higher fluorescence intensity in oxygen saturated ICG-NaI solution than in desaturated one. Those results were confirmed with the theoretical fits of the energy state at laser excitation and at oxygen interaction.

5. DISCUSSION

We have made an important point in the determination of retinal oxygen saturation. We believe we have freed the measurement of plasma oxygen concentration from the reflectivity of retinal structures and from the opacity of optical means. We demonstrate that ICG can vary its fluorescence according to plasma oxygen concentration. We indicate the wavelength and the delayed window to detect the ICG spectra. Our data could be used to set digital enhanced scanning laser ophthalmology in the measurement of retinal oximetry.

6. REFERENCES

1. Delori FC, Noninvasive technique for oximetry of blood in retinal vessels. Applied Optics 1988; 27: 1113–1125.
2. Beach JM, Schwenzer KJ, Srinivas S, Kim D, Tiedman JS, Oximetry of retinal vessels by dual wavelength imaging: calibration and influence of pigmentation. Journal of Applied Physiology 1999; 86(2): 748–758 (Abstract).
3. Cohen AJ, Laing RA, Multiple scattering analysis of retinal blood oximetry. IEEE Transactions on Biomedical Engineering 1976; bme-23: 391–400.
4. Smith M, Optimum wavelength combinations for retinal vessel oximetry. Applied Optics 1999; 38: 258.
5. Ashman RA, Reinholz F, Eikelboom RH. Oximetry with a multiple wavelength SLO. International Ophthalmology 23: 343–346, 2001.
6. Novotny HR, Alvis DL. A method of photographing fluorescence in circulating blood in the human retina. Circulation. 1961 Jul;24:82-6.
7. Khoobehi B, Beach JM, Kawano H. Hyperspectral imaging for measurement of oxygen saturation in the optic nerve head. Invest Ophthalmol Vis Sci.2004;45(5):1464-72
8. Singh JK, Dhawahir FE, Hamid AF, Chell PB. The use of dye in ophthalmology. J Audiov Media Med. 2004 Jun;27(2):62-7
9. Sander B, Larsen M, Moldow B, Lund-Andersen H: Diabetic macular edema: passive and active transport of fluorescein through the blood-retina barrier. Invest Ophthalmol Vis Sci 42:433–8, 2001
10. Merin LM. Fluorescein angiography printouts. Ann Ophthalmol. 1980 Apr;12(4):441-3
11. Trindade-Porto C, Alonso-Llamazares A, Robledo T, Chamorro M, Dominguez J, Plaza A, Martinez-Cocera C. Fluorescein-induced adverse reaction. Allergy. 1999;54(11):1230.
12. A. Richter et al., Photosensitizing potency of structural analogues of benzoporphyrin derivative (BPD) in a mouse tumor model, Br. J. Cancer 63 (1991) 87.
13. Aveline B, Hasant T, Redmond RW. Photophysical and photosensitizing properties of Benzoporphyrin derivate monoacid ring A (BPD-MA). Photochemistry and Photobiology, 1994: 59 328-335
14. D. Husain et al., Effects of photodynamic therapy using verteporfin on experimental choroidal neovascularization and normal retina and choroid up to 7 weeks after treatment. Invest. Ophthalmol. Vis. Sci. 40 (1999) 2322–2331.
15. Flower RW, Hochheimer BF. Clinical infrared absorption angiography of the choroid [letter]. Am J Ophthalmol 1972; 73:458.
16. Bischoff PM, Flower RW. Ten years experience with choroidal angiography using indocyanine green dye: a new routine examination or an epilogue? Doc Ophthalmol 1985;60:235–91.

17. Hope-Ross M, Yannuzzi LA, Gragoudas ES, et al. Adverse reactions due to indocyanine green. Ophthalmology 1994;101:52 9–33.
18. Bonte CA, Ceuppens J, Leys AM. Hypotensive shock as a complication of indocyanine green injection. Retina 1998;18:476–7.
19. Olsen TW, Lim JI, Capone A Jr, et al. Anaphylactic shock following indocyanine green angiography. Arch Ophthalmol 1996;114:97.
20. Fox IJ, Wood EH. Indocyanine green: physical and physiological properties. Mayo Clin Proc 1960;35:732.
21. Thamm E, Schweitzer D, Hammer M, A data reduction scheme for improving the accuracy of oxygen saturation calculations from spectroscopic in vivo measurements. Phys Med Biol. 1998 Jun;43(6):1401-11
22. Desmettre T, Devoisselle JM, Mordon S. Fluorescence Properties and Metabolic Features of Indocyanine Green (ICG) as Related to Angiography. Surv Ophthalmol 2000;45:15–27
23. Devoisselle JM, Mordon S, Soulie S, Desmettre T, Maillols H: Fluorescence properties of indocyanin green/part 1: in vitro study with micelles and liposomes, in Lakowicz JR, Thompson JB (eds): Advances in Fluorescence Sensing Technology III; 1997. Bellingham, CA, USA, SPIE, 1997, pp. 530–37
24. West W, Pearce S: The dimeric state of cyanine dyes. J Phys Chem 69:1894–903, 1965
25. Landsman ML, Kwant G, Mook GA, Zijlstra WG: Light-absorbing properties, stability, and spectral stabilization of Indocyanine green. J Appl Physiol 40:575–83, 1976
26. Philip R, Penzkofer A, Baumler W, Szeimies RM, Abels C. Absorption and fluorescence spectroscopic investigation of indocyanine green. J Photochem and Photobiol A: Chem 1996; 137-148
27. Mordon S, Devoisselle JM, Soulie-Begu S, Desmettre T: Indocyanine green: physicochemical factors affecting its fluorescence in vivo. Microvasc Res 55:146–52, 1998
28. Desmettre T, Devoisselle JM, Soulie-Begu S, Mordon S: Propriétés de fluorescence et particularitiés métaboliques du vert d'indocyanine (ICG). J Fr Ophtalmol. 1999 Nov;22(9):1003-16
29. van Assendelft OW. Spectrophotometvy of Haemogtobin Derivatives. Assen. The Netherlands: Royal Vangorcum Ltd 1970.
30. Horecker BL. The absorption spectra of hemoglobin and its derivatives in the visible and near infra-red regions. JBiol Chem. 1943:148:173.
31. Wilson D. F. e al., "A versatile and sensitive method for measuring oxygen", Adv. Exp. Med. Biol. 215, 71 (1987).
32. Gilbert A., Barrott J., "Essentials of Molecular Photochemistry", Blackwell Science (1990) ISBN: 0632024283.
33. Wilson D. F., Rumsey W. L., Green T. J., Vanderkooi J. M., "The oxygen dependence of mitochondrial oxidative phosphorylation measured by a new optical method for measuring oxygen concentration", J. Biol. Chem. 263, 2712 (1988).
34. Shonat R. D., Wilson D. F., Riva C. E., Pawlowski M., "Oxygen distribution in the retinal and choroidal vessels of the cat as measured by a new phosphorescence imaging method", Applied Optics 31(19), 3711 (1992).
35. Yu I-Ju, Hsies W.-F., "Lasing spectral blue shifts of fluorescent saturable absorbing dye in microdroplets", Chinese Journal of Physics, 36(3) (1998).
36. Bjornsson OG, Murphy R, Chadwick VS, Bjornsson S. Physiochemical studies on indocyanine green: molar lineic absorbance, pH tolerance, activation energy and rate of decay in various solvents.J Clin Chem Clin Biochem. 1983;21(7):453-8

24.

PREDICTION OF SURGICAL SITE INFECTIONS USING SPECTROPHOTOMETRY: PRELIMINARY RESULTS

Charlotte L. Ives[*], David K. Harrison[*] and Gerard Stansby[†]

1. INTRODUCTION

Wound infections occur in approximately 5% of operations[1]. These infections cost the U.K. National Health Service up to £65 million per year and lower patient quality of life[2].

If wound infections could be predicted there would be potential for prevention and a reduction in cost and morbidity would ensue. Work has previously been carried out to predict when these infections will occur, and attempts have been made to create scoring systems for patients undergoing operations. The most simple of these is to classify the type of operation into one of the four following groups:

- Clean surgery (no inflammation present, respiratory, alimentary and genitourinary system are not breached) carries a 1-2% risk.
- Clean-contaminated surgery (above systems are entered, but there is not significant spillage of contents) – risk of <10%.
- Contaminated surgery (where inflammation is present, and there is spillage of contents) carries a risk of 15-20%.
- Dirty surgery is classified as that carried out where there is an abscess present or there has been spillage of contents for more than 4 hours. This carries a risk of around 40% of developing a wound infection.

Although in a broad sense this can be applied, some series report higher infection rates than should be encountered for the surgery type[3]. More sophisticated scoring systems are available which can be used to predict infections with higher accuracy. The best known is the SENIC score (Study on the Effect of Nosocomial Infection Control)

[*] University Hospital of North Durham, North Road, Durham, DH1 5TW UK
[†] Northern Vascular Centre, Freeman Hospital, Freeman Road, Newcastle-upon-Tyne, NE7 7DN UK

which was evaluated in the US between 1975 and 1985[4]. This scoring system gave a point for each of the following:

- Abdominal operation
- Notable wound contamination
- Operation lasting for longer than 2 hours
- 3 or more diagnoses on discharge from hospital

The wound infection risk was from 1% for 0 points to 27% for 4 points. Comparing this again to the work by Israelsson[3] the infection rate for patients in the high risk group may underestimate the actual surgical site infections (SSI). However, it has been shown that the use of scoring systems per se does not reduce wound infection rates[5]. The NNIS (National Nosocomial Infection Score) was developed after this[6], and is similar to the SENIC score. However it instead measures the ASA (American Society of Anaesthetists) grade of the patient, the length of operation and the type of operation. Again it has not been found to be useful for all types of operation.

The problem with these scoring systems is that they can be time consuming, are often subjective and are not accurate. The ideal tool for prediction of SSI would be one that is easy to use, is objective, reproducible and gives the clinician an accurate risk of wound infection in a time scale where there is opportunity for intervention. Measuring oxygen saturation in the tissues may fit these criteria. It is easy to measure in a non-invasive way and is objective.

Determining oxygen levels in the tissues is logically a sound marker for risk of wound infection. Factors which increase the risk of wound infection are related to the oxygenation (or perfusion) of the tissues. For example a longer operation will lead to wound ischemia or a patient with other morbidities, such as chest disease, may be hypoxic. Furthermore it is well known that oxygen is the most important substrate needed for wound healing. Among some of its actions it provides a nutrient for the dividing cells, it encourages collagen accumulation[7] and acts as a substrate in the oxidation of bacteria by macrophages[8].

Hopf et al.[9] measured wound pO_2 (partial pressure of oxygen) with subcutaneous probes in specially created wounds in the upper arm. They recorded data on 130 patients undergoing general elective surgery and found pO_2 was more accurate in predicting wound infections than the SENIC score of the patient. They also found that the pO_2 could be manipulated clinically, and so shows a potential for prevention of wound infections. However their technique has the drawbacks of being invasive as a wound is created in the upper arm and the probe needs time to calibrate. It would also be beneficial to find a method that could measure oxygen at the actual site of surgical insult. Oxygen saturation (SO_2) can be measured with spectrophotometry. This method uses probes that are placed onto the skin at the surgical site and no time is needed for calibration when in use. It has been shown that oxygen saturation can predict wound healing in amputation stumps[10] and can assess degree of peripheral vascular disease[11-14].

Wavelengths of light in the visible range are used to measure skin oxygen saturation (SSO_2) as they only penetrate the first few millimetres of tissue, whereas the absorptive properties of tissue are such that wavelengths in the near-infrared range can penetrate a few centimetres and measure muscle oxygen saturation (MSO_2).

2. AIM

The main aim of this investigation is to establish whether oxygen saturation of tissues perioperatively predicts wound infections after major surgery.

3. MATERIALS

A Whitland Research RM200 was employed as the visible lightguide spectro-photometer (LGS).

For a near-infrared spectrophotometer (NIS) an Inspectra tissue spectrophotometer Model 325 (kindly loaned by Hutchinson Technology Inc, Hutchinson, USA) was used. Please refer to Ives et al. and Myers et al. (this volume) for details on wavelengths used.

4. METHODS

After gaining ethical approval from the Local Research Ethics Committee patients undergoing major elective surgery were invited to participate. Written informed consent was obtained by the main researcher from each patient prior to procedure.

Age, sex, weight and height (to calculate body mass index, BMI) and triceps skinfold thickness (TST, using Haltain callipers) were recorded. Previous surgery and factors that could affect wound healing (for example vascular disease, anaemia and diabetes) were also noted. SSO_2 and MSO_2 were then measured on the arm and abdomen pre-operatively and subsequently at 6, 12, 24 and 48 hours post-operatively. At all stages temperature and arterial oxygen saturation (measured with a pulse oximeter) were measured.

An independent assessor monitored for signs of infection at one week and one month (30 days) using a proforma based upon the Control of Diseases Centre guidelines for diagnosis of wound infection[15].

5. RESULTS

In the preliminary phase of the trial 17 patients (10 male and 7 female) have been recruited. The patient characteristics include an age range of 59.5 +/-20.9 years (mean +/- standard deviation), BMI 27.2 +/ 5.6 and TST 16.6 +/- 7.3.

Of the 17 patients in the study eight operations were for cancer, eight were for other gastrointestinal problems and one was for peripheral vascular disease. Of the operations for cancer five wounds became infected and three healed without problems. In those for non-cancerous conditions two wounds became infected and six wounds were not infected. The one operation for vascular disease did not develop an infection. This would imply that SSI are more likely in patients with cancer, however this is not a large enough sample to show significance.

Two groups were classified retrospectively; patients with uneventful wound healing (Group A), and those who had wound infections within 30 days post-surgery (Group B). Group A comprised 10 patients; 4 male and 6 female. Group B comprised 7 patients: 6 male and 1 female. Please see Table 1 for group characteristics.

Table 1. Characteristics of patient groups

Characteristic	Group A	Group B
Age	56.3 years +/- 20.4[a]	64.1 years+/- 22.4
Gender	2:3 M:F ratio	6:1 M:F ratio
BMI	27.1 +/- 6.9	27.4 +/- 4.1
TST	18 +/- 8.0	14.5 +/- 6.3

[a] All values given in this paper are mean +/- standard deviation

There were no significant differences in age, BMI or TST between these two groups. Nor were there any significant differences in arterial oxygen saturation, temperature, skin oxygen saturation or arm muscle oxygen saturation between Groups A and B at any stage. Skin and muscle SO_2 of arm and abdomen were not related to each other within either group.

At 12 hours mean abdomen MSO_2 in Group A was nearly double that in Group B (A=62.9%+/- 20.2, B=35.1% +/- 13.9) and this was significantly different (p=0.02). Mean abdomen MSO_2 value in Group A was also higher than in Group B at 6 hours (A=66.6%+/-17, B=42.7% +/- 25.9) but this was not significant at a 5% level (p=0.075). A paired T-test showed that in Group A abdomen MSO_2 was significantly higher at 6 and 12 hours compared to the pre-operative values. Preoperative=55.2%+/-14.8, 6hr =66.6%+/-17.0 (p =0.01), 12hr=62.9% +/-20.2 (p=0.05); this rise was not observed in Group B.

6. DISCUSSION

This study, and previous work, has shown that skin SO_2 varies depending on site[16]. This suggests that skin blood flow is controlled locally, e.g. affected by temperature or emotion, and so measurements on the arm are likely to be of little use in predicting what is happening on the abdomen. Work on the assessment of oxygenation of amputation levels by Harrison et al.[10] looks at the degree of tissue hypoxia (number of readings below a particular level) as well as the mean skin SO_2 at the site to be operated upon. By analysing our data in a similar way looking at the abdomen SSO_2 values (see figures 1-5 below) it is seen that those who developed an infection have mostly high degrees of tissue hypoxia. Values less then 30% saturation were considered 'hypoxic'. Some still developed an infection with good readings and with the small numbers of patients tested in this study it is difficult to put these patients into a subset (for example may have very superficial infections, or are smokers) to account for the infection. In any case the presence of inflammation or infection would cause erythema and hence a rise in skin SO_2 and so differences between those developing an infection or those who heal uneventfully would be obscured. The amputation level assessments exclude patients with known infections for this reason.

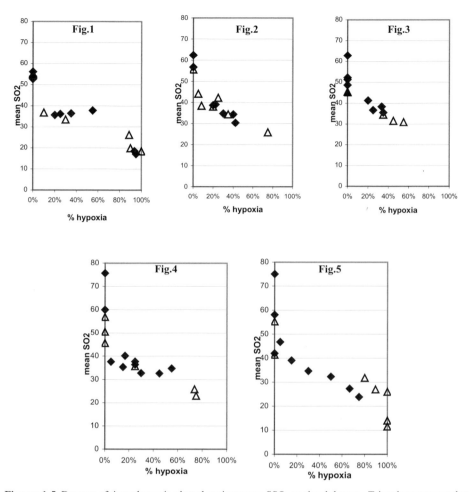

Figures 1-5. Degrees of tissue hypoxia plotted against mean SSO_2 on the abdomen. Triangles represent those patients who subsequently developed SSI. The figures show the readings (1) pre-operatively, (2) at 6 hours, (3) 12 hours, (4) 24 hours and (5) 48 hours respectively

A trial on healthy volunteers found a correlation between arm and abdomen MSO_2 (see Ives et al., this volume); however, this finding is not supported in the patient trial to date.

The previous work with healthy volunteers also showed that age did not affect MSO_2 or SSO_2; this was supported in the patient trial. Old age is a factor in poor wound healing, although no correlation was found between age and infections in this study to date. We hypothesise that this was not seen in the trial because those selected for operation are medically fit. An alternative hypothesis is that old age does not affect tissue oxygenation but perhaps healing is impaired due to other factors such as weaker connective tissues and weaker immune systems.

The preliminary results indicate that a post-operative increase in tissue oxygen saturation at the muscle level within the first 12 hours may be necessary to protect against surgical site infection. This is supported by work cited in Hopf et al.[9] which found that in the first six hours after bacterial inoculation increasing oxygen reduced wound infection. Those patients with pre-existing vascular disease, post-operative hypotension or anaemia may not be able to perfuse the tissues adequately despite having adequate arterial oxygen saturation as measured by pulse oximeter. By measuring tissue SO_2 within 6-12 hours post-operatively patients at risk of wound infection will be identified early and there will be a window of opportunity for intervention. Fluid infusions have been shown to improve tissue pO_2[17]. Methods to improve tissue oxygen saturation post-operatively need to be analysed in further trials.

This study shows encouraging results that we have developed a better method to assess risk of SSI after surgery than risk scoring, and there may be an opportunity to decrease infection rates.

7. REFERENCES

1. CDSC. Surveillance of surgical site infections. *Commun Dis Rep CDR Wkly* 2000; 10:21
2. R. Plowman, N. Graves, M. Griffin et al. The socio-economic burden of hospital acquired infection. *Public Health Laboratory Service* 2000
3. L.A .Israelsson. The surgeon as a risk factor for complications of midline incisions. *Eur J Surg* 1998; **164**(5):353-9.
4. R.W. Haley, D.H. Culver, W.M. Morgan, et al. Identifying patients at high risk of surgical wound infection. A simple multivariate index of patient susceptibility and wound contamination. *Am J Epidemiol* 1985; **121**(2):206-15.
5. F. Gottrup. Prevention of surgical wound infections (editorial). *N Engl J Med* 2000; **342**(3):202-204.
6. D.H. Culver, T.C. Horan et al. Surgical wound infection rate by wound class operative procedure and patient risk index. National Nosocomial Infectious Surveillance System. *Am J Med* 1991; 9 (suppl 3B):152S-157S.
7. T.K. Hunt, M.P. Pai. The effect of varying ambient oxygen tensions on wound metabolism and collagen synthesis. *Surg Gynecol Obstet* 1972;135:561-7
8. D.B. Allen, J.J. Maguire, M. Mahdavian et al. Wound tissue oxygen tension and acidosis limit neutrophils bacterial killing mechanisms. *Arch Surg* 1997;132:997-1004
9. H.W. Hopf, T.K .Hunt, J.M. West, et al. Wound tissue oxygen tension predicts the risk of wound infection in surgical patients. *Arch Surg* 1997;132:991-6
10. D.K. Harrison, P.T. McCollum, D.J. Newton, et al. Amputation level assessment using lightguide spectrophotometry. *Prosthet Orthot Int* 1995;19:139-47
11. D. Choudhury, B. Michener, P. Fennelly, et al. Near-Infrared spectroscopy in the Early Detection of Peripheral vascular disease. *J Vasc Tech* 1999; **23**(3): 109-113
12. T. Komiyama, H. Shigematsu, H.Yasuhara, et al. Near-infrared spectroscopy grades the severity of intermittent claudication in diabetics more accurately than ankle pressure measurement. *BJS* 2000; 87: 459-499
13. U. Wolf, M. Wolf, J. Choi, et al. Localized irregularities in hemoglobin [sic] flow and oxygenation in calf muscle in patients with peripheral vascular disease detected with near-infrared spectrophotometry. *J Vasc Surg* 2003; 37: 1017-1026
14. D.K. Harrison, C. Voss, H.S .Vollmar, et al. Response of muscle oxygen saturation to exercise, measured with near infrared spectrophotometry in patients with peripheral vascular disease. *Oxygen Transport to Tissue XX.* Plenum Press; 1998: 45-52
15. T.C .Horan, R.P. Gaynes, W.J. Martone, et al. CDC definitions of nosocomial surgical site infections, 1992: a modification of CDC definitions of surgical wound infections. *Am J Infect Control* 1992; 20: 271-274
16. L. Caspary, J. Thum, A.Creutzig, et al. Quantitative reflection spectrophotometry: spatial and temporal variation of Hb oxygenation in human skin. *Int J Microcirc Clin* Exp 1995; **15**(3):131-6.
17. N. Chang, W.H. Goodson, F. Gottrup et al. Direct measurement of wound and tissue oxygen tension in postoperative patients. *Ann Surg* 1983; **197** (4): 470-8

25

SPATIALLY RESOLVED BLOOD OXYGENATION MEASUREMENTS USING TIME-RESOLVED PHOTOACOUSTIC SPECTROSCOPY

Laufer JG, Elwell CE, Delpy DT, Beard PC[*]

1. INTRODUCTION

Photoacoustic spectroscopy relies on the generation of acoustic waves as a result of the absorption of short pulses of light in tissue. The absorption of the optical energy produces rapid heating and a consequent pressure rise in the illuminated volume, which generates acoustic waves that propagate away from their origins and are detected at the tissue surface. By measuring the time-of-arrival of the acoustic waves, the spatial distribution of photoacoustic sources can be determined. When optical excitation in the near-infrared wavelength range is used, the waves originate predominately from blood vessels due to the relatively strong absorption by haemoglobin. The amplitude of the photoacoustic signal is determined by the local concentration of haemoglobin, its oxygenation, and by the optical absorption and scattering in the surrounding tissue. Since blood exhibits wavelength-dependent changes in absorption as a result of varying concentrations of oxy- (HbO_2) and deoxyhaemoglobin (Hb)[1], blood oxygen saturation (SO_2) can be determined by making multi-wavelength measurements of the amplitude of the photoacoustic signals originating from a blood vessel.

By using an array of acoustic transducers, the principle of photoacoustic spectroscopy could be incorporated into photoacoustic imaging. Photoacoustic imaging has already been used for mapping the brain of small mammals *in vivo*[2] and blood vessel phantoms[3]. Combining photoacoustic spectroscopy with imaging would not only allow the reconstruction of a structural image with high spatial resolution (<100μm) but also the mapping of SO_2 in the microvasculature and hence the collection of functional information[4]. This technique may be particularly suitable for the study of the development of microvasculature in tumours, skin grafts and inflamed or healing tissue. However, photoacoustic spectroscopy for the quantitative measurement of SO_2 still requires experimental validation and an assessment of its accuracy.

[*] Jan G. Laufer, Department of Medical Physics & Bioengineering, University College London,
11-20 Capper Street, London WC1E 6JA, email: jlaufer@medphys.ucl.ac.uk,
Web: www.medphys.ucl.ac.uk/research/mle/index.htm

The aim of this study was to demonstrate for the first time the ability of photoacoustic spectroscopy to make quantitative, non-invasive, spatially resolved measurements of SO_2 and to assess the accuracy of the technique by comparing the photoacoustically determined values to independent SO_2 measurements made with a CO-oximeter.

2. BACKGROUND & METHODS

Short laser pulses were used to generate photoacoustic signals in a tissue phantom, which contained three blood filled capillaries in a scattering medium with optical properties similar to extravascular tissue (Figure 1). The spatial distribution of absorbed optical energy produced by the excitation laser pulses is directly related to the spatial distribution of the optical properties of the Intralipid suspension and the blood-filled capillaries. This is illustrated in Figure 1a, which shows the distribution of absorbed optical energy along the line of sight of a single-element ultrasound detector as indicated by the dashed arrow. Starting from the surface, the absorbed energy shows a steady decrease with depth, which is determined by the optical properties of the Intralipid. Peaks of absorbed optical energy due to the presence of haemoglobin occur wherever a capillary is situated. The conversion of optical energy to heat and the subsequent thermoelastic expansion produces a pressure source, which emits an acoustic pulse that is measured by the ultrasound transducer. This principle is illustrated in Figure 1b, which shows the three distinct photoacoustic signals generated in the capillaries. Under conditions of stress confinement, i.e. where the optical excitation pulse is shorter than the time it takes the stress wave to travel across the source region, the spatial pressure distribution is directly proportional to the distribution of absorbed energy.

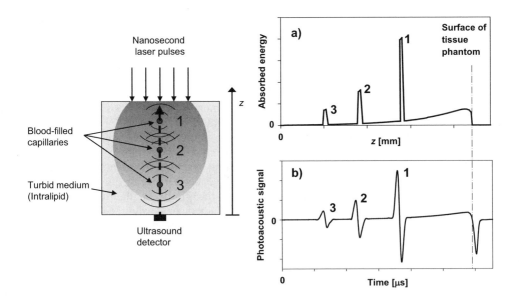

Figure 1. The tissue phantom consisted of an Intralipid bath and three capillaries through which blood was circulated. Figure 1a shows the depth profile of absorbed optical energy along the line-of-sight of the ultrasound detector.   Figure 1b shows a schematic of a typical photoacoustic signal detected in the phantom.

This makes the technique particularly powerful since the generated signal carries information about the optical coefficients at any given location within the illuminated volume from depths of several cm with a resolution of 50-100μm. By capturing signals at multiple excitation wavelengths, spatially resolved measurements of the spectral dependence of the absorption by specific chromophores, such as haemoglobin, can therefore be made in order to obtain physiological information, such as blood oxygen saturation. In the case shown in Figure 1, measurements of the amplitude of the photoacoustic waves produced in capillary 1, 2, and 3 were made as a function of wavelength.

2.1.   Experimental Set-up

Photoacoustic signals were generated in a tissue phantom using nanosecond laser pulses. The phantom consisted of three blood-filled capillaries (460μm internal diameter, 75μm wall thickness, PMMA) which were placed at depths of 3.0, 5.1 and 6.8mm in a 2.5% Intralipid suspension with an absorption coefficient of 0.05mm^{-1} and a reduced scattering coefficient of 0.5mm^{-1}. These values are similar to those found in tissues such as the breast[5] and skin[6] at 1064nm. Figure 2 shows the experimental set-up. A wavelength tuneable optical parametric oscillator (OPO) laser system provided 7ns excitation pulses between 740 and 1100nm. The output of the OPO was coupled into an optical fibre in order to homogenise the beam. The fibre output was directed onto the tissue phantom. The beam diameter incident on the phantom was approximately 6mm and the fluence was between 60 and 100 mJ cm^{-2}. The generated photoacoustic waves were detected by a single ultrasound transducer. Its sensing mechanism is based upon the detection of acoustically induced changes in the optical thickness of a 75 μm thick Fabry Perot polymer film interferometer and provides a broadband (15 MHz) detection sensitivity of 1.0 kPa[7] (noise equivalent pressure).

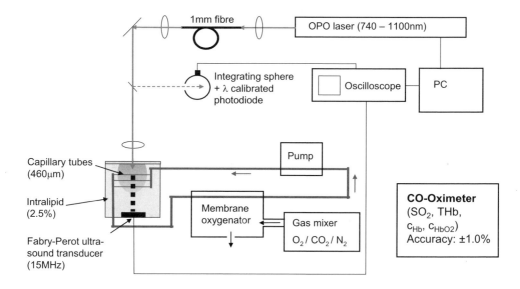

Figure 2 Experimental set-up for spatially resolved measurements of blood SO$_2$ in a tissue phantom

A small portion of the incident excitation light was directed to an integrating sphere and measured with a wavelength-calibrated photodiode in order to normalise the photoacoustic waveforms with respect to the incident pulse energy. The photodiode signal and the photoacoustic waveform were captured and averaged over 60 acquisitions using a digital oscilloscope. Blood was continuously pumped through the capillaries and a membrane oxygenator. An air mixer provided constant flow rates of nitrogen, oxygen, and carbon dioxide. SO_2 was controlled by varying the ratio of the gases flowing through the oxygenator. At each level of SO_2, a photoacoustic spectrum was measured. During the wavelength scan, blood samples were withdrawn from the circuit to be analysed by a co-oximeter (Instrumentation Labs), which measured SO_2 and the total haemoglobin concentration with high accuracy ($\pm 1\%$). These values were later compared to the photoacoustically determined SO_2.

2.2.   Blood Preparation

Blood samples were prepared from expired blood donations (older than 30 days). 2000 units of heparin were added to approximately 200ml of concentrated, leucocyte depleted blood. The samples were centrifuged and the plasma was replaced with a phosphate buffered saline solution (pH 7.4, Sigma Aldrich). This process was repeated four times. The final haemoglobin concentrations in the samples ranged from 120 to 150 g l^{-1}.

2.3   Determination of Blood SO_2 Using a Finite Element Model

Figure 3 shows amplitude spectra which were measured at different blood SO_2 on the vessel closest to the surface (capillary 1 in Figure 1) over the wavelength range 740-1100nm.

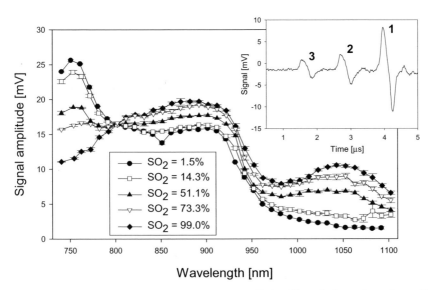

Figure 3 Inset shows the photoacoustic signals from the three capillaries. The main figure shows the amplitude of the signal from capillary 1 as a function of wavelength for five different blood oxygen saturations.

Between 740nm and 900nm, the spectra show the expected changes in haemoglobin absorption with SO_2. However, beyond 900nm, the detected spectra are corrupted by the wavelength-dependent optical absorption and scattering in Intralipid - the influence of absorption is especially noticeable at 975nm due to the water absorption peak. This has a secondary effect in that it causes the spatial distribution of the internal light fluence to become wavelength-dependent, which also affects the measured photoacoustic spectrum. In order to take account of these effects, a 2D finite element model (FEM) of light transport[8] based on the Delta-Eddington diffusion approximation[9] was used to calculate the wavelength-dependent absorbed energy distribution in the phantom. The model consisted of a two-dimensional grid, which represented a cross-section of the phantom perpendicular to the capillaries. The location of the capillaries in the model was based on the measured time-of-arrival of the photoacoustic waves from each capillary. The beam diameter, the known wavelength-dependent absorption and scattering coefficients of Intralipid and the specific absorption spectra of Hb and HbO_2 were input as fixed parameters to the model, while the concentrations of Hb and HbO_2, which determine the absorption coefficient in the capillaries, could be varied. The model was used to calculate a theoretical spectrum for each capillary, which was then fitted to the corresponding measured spectrum by varying the relative concentrations of Hb and HbO_2. From these concentrations, SO_2 could be determined. The data at wavelengths longer than 920nm was omitted in order to reduce the effect of errors in the μ_a and μ_s' of Intralipid on the error in SO_2. The uncertainty in SO_2 was calculated from the standard deviation of the spectral measurements and the uncertainties of the fixed parameters used in the FEM.

3. RESULTS

A Bland-Altmann plot of the difference of the photoacoustically determined SO_2 in the three capillaries and CO-oximeter SO_2 measurements is shown in Figure 4.

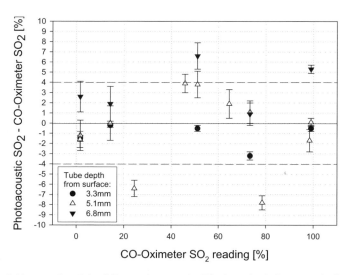

Figure 4 Bland-Altmann plot of the difference between the SO_2 determined photoacoustically and CO-oximeter SO_2 measurements. The graph shows the results obtained from capillaries at different depths and across the range of oxygenation levels that was studied.

The photoacoustically determined SO_2 was typically within ±4% of the CO-oximeter for capillary depths of less than 6mm across the entire range of blood oxygen saturations studied. At greater depths, the accuracy in SO_2 decreased to ±8%, while the uncertainty in the determined values increased due to a lower signal-to-noise ratio.

4. DISCUSSION

This study has demonstrated for the first time that photoacoustic spectroscopy is suitable for making quantitative, non-invasive, spatially resolved measurements of blood SO_2. It has been found that the accuracy in SO_2 decreases with increasing depth of the blood vessel and this is mainly due to lower signal-to-noise ratios. This was exacerbated by errors in the optical and geometrical parameters used in the FEM. Small inaccuracies in the absolute values and the wavelength-dependence of the optical coefficients of the medium surrounding the blood vessels cause significant deviations in the modelled photoacoustic spectrum, and hence the determined SO_2, especially at greater depths. There is the potential to overcome this limitation. Since the photoacoustic signal carries information on the optical properties of the tissue at any location of the illuminated volume, absorption and scattering coefficients can be determined from those parts of the signal that originate in the surrounding medium. The extracted optical coefficients could then be included in the FEM to achieve a higher accuracy in measuring SO_2 in deeper capillaries.

A very promising aspect of this work is that the principle of photoacoustic spectroscopy can be incorporated into photoacoustic imaging. The combination of the two techniques would enable to monitor, for example, absolute blood SO_2 along the length of an individual blood vessel at cm depths with 50-100µm resolution and to provide three-dimensional maps of SO_2, which may be of interest in the study of the development of the microvasculature in tumours.

5. REFERENCES

1. J. Laufer, C. Elwell, D. Delpy and P. Beard, Pulsed near-infrared photoacoustic spectroscopy of blood, *Proc. SPIE*, **5320**, 57-68 (2004).
2. X. Wang, Y. Pang, G. Ku, X. Xie, G. Stoica, and L. Wang, Noninvasive laser-induced photoacoustic tomography for structural and functional in vivo imaging of the brain, *Nature Biotechnology*, **21(7)**, 803-806 (2003).
3. P.C.Beard, Photoacoustic imaging of blood vessel equivalent phantoms, *Proc. SPIE*, **4618**, 54-62 (2002).
4. X. Wang, G. Ku, X. Xie, Y. Wang, G. Stoica, L. Wang, Non-invasive functional photoacoustic tomography of blood oxygen saturation in the brain, *Proc. SPIE*, **5320**, 69-76 (2004).
5. T.A. Troy, D.L. Page, E.M. Sevic-Mucraca, Optical properties of normal and diseased breast tissues: prognosis for optical mammography, *J Biomed Opt*, **1(3)**, 342-355 (1996).
6. C.R. Simpson, M. Kohl, M. Essenpreis, M. Cope, Near-infrared optical of ex-vivo human skin and subcutaneous tissues measured using the Monte Carlo inversion technique, *Phys Med Biol*, **43**, 2465-2478 (1998).
7. P.C. Beard, F. Pérennès, T.N. Mills, Transduction mechanisms of the Fabry-Perot polymer film sensing concept for wideband ultrasound detection, *IEEE Transactions on Ultrasonics, Ferroelectrics, and Frequency Control*, **46(6)**, 1575-1582 (1999).
8. S.R. Arridge, M. Schweiger, M. Hiraoka, D.T. Delpy, A finite element method for modelling photon transport in tissue, *Medical Physics*, **20(2)**, 299-309 (1993).
9. W. M. Star, in *Optical-thermal response of laser-irradiated tissue*, edited by A.J. Welch and M.J.C. van Gemert (Plenum Press, New York and London, 1995), pp.138-148.

26.

DIABETES-INDUCED DECREASE IN RENAL OXYGEN TENSION: EFFECTS OF AN ALTERED METABOLISM

Fredrik Palm, Per-Ola Carlsson, Angelica Fasching, Peter Hansell, and Per Liss[*]

1. INTRODUCTION

The metabolism within different parts of the kidney is highly heterogeneous and is likely to reflect the local energy demand and milieu in that specific region. The metabolism in the renal cortex has been found to be highly dependent on the availability of oxygen, i.e. aerobic metabolism. Glucose oxidation in the renal cortex is relatively low compared to that of the renal medulla. The metabolism within the renal medulla is also heterogeneous, with high glucose oxidation and high oxygen consumption in the outer part of the medulla, while the deeper situated inner medulla is highly dependent on anaerobic metabolism, i.e. low oxygen consumption and high glycolytic rate[1,2].

The blood flow to the renal medulla is derived through the *vasa recta*, acting as a counter current system in order to maintain the high osmolar gradient necessary for the formation of concentrated urine. While electrolytes are re-circulated from the ascending to the closely located descending *vasa recta*, oxygen is shunted in the opposite direction, resulting in a low delivery of oxygen to the medullary region[3,4]. Conditions which alter the shunting of oxygen, with concomitant alteration in medullary oxygen delivery, have the potential to influence the medullary oxygen tension (pO_2)[5].

Long-term hyperglycemia, e.g. diabetes mellitus, is known to significantly increase the risk of developing progressive renal dysfunction[6]. The exact mechanism accounting for the increased risk is not known, but it has been suggested that aggravated low pO_2 in the renal medulla may cause progression of nephropathy during certain pathological conditions[7,8].

[*] FP, POC, AF, PH, Department of Medical Cell Biology, Uppsala University, Biomedical Center, box 571, SE 751 23 Uppsala, Sweden. PL, Department of Oncology, Radiology and Clinical Immunology, Uppsala University, University Hospital, SE 751 85 Uppsala, Sweden.

2. DIABETES-INDUCED ALTERATION IN THE RENAL METABOLISM AND INTERSTITIAL pH

Diabetes-induced activation of the polyol pathway has been demonstrated in numerous tissues[9-12], including the renal medulla[13]. Increased activity through the polyol pathway alters the cellular redox-state, mainly due to increased $NADH/NAD^+$ ratio[14]. This will concomitantly shift the equilibrium between pyruvate and lactate, resulting in an increased lactate/pyruvate ratio, predominately as a result of increased lactate concentration[15]. Long-term diabetes has been shown to increase the lactate/pyruvate ratio in both the renal cortex and in the medulla, but inhibition of the polyol pathway only prevents the increase in the medullary region (Fig. 1)[13]. This finding is consistent with the almost exclusive presence of the enzyme aldose reductase in the medullary region[16].

It is important to note that the increased formation of lactate in our previous study[13] was not a result of hypoxia, since the levels of purine-base metabolites (adenosine, inosine and hypoxanthine) did not increase after the onset of hyperglycemia. Increased lactate/pyruvate ratio is known to occur during sustained hyperglycemia even though the oxygen supply is sufficient for full mitochondrial respiration. This state is commonly referred to as "pseudohypoxia" and is a result of the altered intracellular redox status[15].

The increased formation of lactate found in the renal medulla of diabetic animals resulted in a significantly lower interstitial pH in this region (Fig. 2)[13]. The increased lactate concentration will decrease the pH and the protons will re-circulate, in the same manner as electrolytes, in the medullary structures due to the counter-current mechanism in the *vasa recta*.

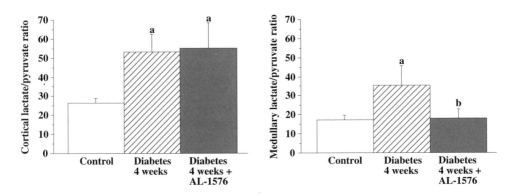

Figure 1. Lactate/pyruvate ratios in the renal cortex and the medulla in control and diabetic animals with and without treatment with the specific aldose reductase inhibitor AL-1576. a denotes p<0.05 versus non-diabetic control animals, whereas b denotes p<0.05 versus diabetic animals. Values are presented as mean ± SEM. Modified from data originally published by Palm *et al.*[13].

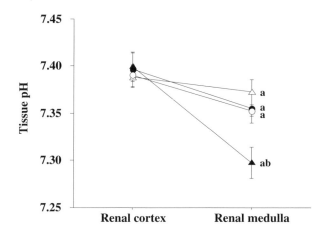

Figure 2. Renal interstitial pH in control (circles) and diabetic animals (triangles) with and without treatment of an aldose reductase inhibitor (empty and filled symbols, respectively). a denotes $p<0.05$ versus corresponding cortex value within the same group, whereas b denotes $p<0.05$ versus control animals within the same region. Values are presented as mean±SEM. Figure originally published in *Diabetologia, Polyol-pathway-dependent disturbances in renal medullary metabolism in experimental insulin-deficient diabetes mellitus in rats, Palm et al., 47:1223-1231, Copyright Springer Verlag, published with permission.*

3. INFLUENCE OF DIABETES ON RENAL OXYGEN TENSION

The *in vivo* pO_2 in any tissue of an animal is the result of the net delivery of oxygen and the oxygen consumption within that specific tissue. Altering any of these two parameters will undoubtedly affect the pO_2 in the tissue.

The diabetes-induced increase in renal medullary hydrogen ion concentration will increase the Bohr effect when acidic blood in the ascending *vasa recta* comes in the vicinity of the arterial blood in the descending *vasa recta*. The shunting of oxygen (from descending to ascending vessels) will increase and the net result will be an even further reduced oxygen delivery to the renal medulla. Decreased oxygen delivery to the medullary structures will occur despite the fact that the total blood perfusion might be unaffected, or even increased.

Brownlee and co-workers[17] have shown that hyperglycemia induces excessive formation of radical oxygen species (ROS) from the electron transport chain in the mitochondrial membrane. The hyperglycemia-induced increase in substrate available for the electron transport chain will increase the electrochemical potential gradient over the mitochondrial membrane. This will stabilize superoxide-generating intermediates in the electrode transport chain, resulting in increased formation of ROS[17]. A direct link between increased formation of ROS and decreased renal oxygen tension has been established[18]. In the same study, treatment of diabetic animals with the free radical scavenger α-tocopherol fully prevented the diabetes-induced decrease in renal pO_2. Furthermore, it was shown that increased levels of ROS were accompanied by increased mitochondrial respiration. Nitric oxide (NO) is known to inhibit mitochondrial respiration in a dose-dependent manner[19, 20]. The mechanism, by which ROS increases the cellular oxygen consumption, is probably by increasing the degradation of NO[21]. The inhibitory influence of NO on oxygen consumption will decrease and, thus, increase the mitochondrial respiration rate.

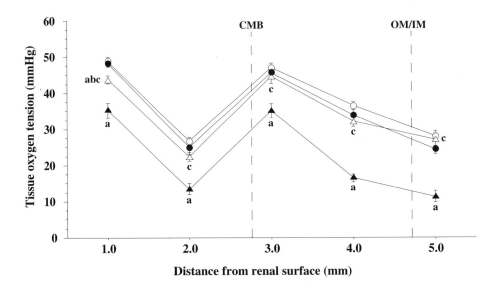

Figure 3. Tissue oxygen tension at different depths from the renal surface in non-diabetic control (filled circles), non-diabetic animals fed vitamin E (empty circles), diabetic animals (filled triangles) and diabetic animals fed vitamin E (empty triangles). a denotes $p<0.05$ versus non-diabetic control animals, b denotes $p<0.05$ non-diabetic fed vitamin E versus diabetic animals fed vitamin E and c denotes $p<0.05$ diabetic animals fed vitamin E versus non-diabetic animals. Values are presented as mean±SEM (CMB – anatomical border between cortex and medulla, OM/IM – anatomical border between outer and inner medulla). Figure originally published in *Diabetologia, Reactive oxygen species cause diabetes-induced decrease in renal oxygen tension, Palm et al., 46:1153-1160, Copyright Springer Verlag, published with permission.*

It is well known that glomerular hyperfiltration occurs during the early onset of hyperglycemia, both in diabetic patients[22] and in animal models of experimental diabetes[23, 24]. Increased glomerular filtration will increase the tubular load of electrolytes, resulting in increased tubular reabsorption and increased oxygen consumption. This does not appear to be a contributing mechanism to the diabetes-induced decrease in renal oxygen tension since alterations in the oxygen tension occur independently of the glomerular filtration rate[13, 18].

4. SUMMARY

During conditions with experimental diabetes mellitus, it is evident that several alterations in renal oxygen metabolism occur, including increased mitochondrial respiration and increased lactate accumulation in the renal tissue. Consequently, these alterations will contribute to decrease the interstitial pO_2, preferentially in the renal medulla of animals with sustained long-term hyperglycemia.

5. ACKNOWLEDGEMENTS

The skilled technical assistance of Astrid Nordin is gratefully acknowledged. This study was supported by grants from the Swedish Medical Research Council (project no. 10840, 15040 and 15043), the Swedish Diabetes Association, the Swedish Juvenile Diabetes Foundation, the Royal Swedish Research Society, the Selander Foundation, the Swedish Medical Society, and the Lars Hierta Memorial Foundation.

6. REFERENCES

1. J. J. Cohen, and D. E. Kamm, Renal metabolism: Relation to renal function., in: *The Kidney,* edited by M. B. Brenner, and F. C. Rector (W.B. Saunders, Philadelphia, 1981), pp. 144-248.
2. S. Klahr, and M. Hammarman, Renal metabolism, in: *The Kidney: Physiology and Pathophysiology,* edited by D. W. Seldin, and G. Giebisch (Raven Press, New York, 1985), pp. 699-718.
3. K. Aukland, and J. Krog, Renal oxygen tension, *Nature* **188**, 671 (1960).
4. M. N. Levy, and E. S. Imperial, Oxygen shunting in renal cortical and medullary capillaries, *Am J Physiol* **200**, 159-162 (1961).
5. K. Aukland, Studies on Intrarenal Circulation with Special Reference to Gas Exchange, *J Oslo City Hosp* **14**, 115-46 (1964).
6. The Diabetes Control and Complications Trail Research Group, The effect of intensive treatment of diabetes on the development and progression of long-term complications in insulin-dependent diabetes mellitus., *N Engl J Med* **329**, 977-86. (1993).
7. M. Brezis, S. Rosen, P. Silva, and F. H. Epstein, Renal ischemia: a new perspective, *Kidney Int* **26**, 375-383 (1984).
8. M. Brezis, and S. Rosen, Hypoxia of the renal medulla - its implication for disease, *New Engl J Med* **332**, 647-655 (1995).
9. G. Pugliese, R. G. Tilton, and J. R. Williamson, Glucose-induced metabolic imbalances in the pathogenesis of diabetic vascular disease, *Diabetes Metab Rev* **7**, 35-59. (1991).
10. P. J. Dyck, A. Windebank, H. Yasuda, F. J. Service, R. Rizza, and B. Zimmerman, Diabetic neuropathy, *Adv Exp Med Biol* **189**, 305-320 (1985).
11. M. J. Stevens, J. Dananberg, E. L. Feldman, S. A. Lattimer, M. Kamijo, T. P. Thomas, H. Shindo, A. A. Sima, and D. A. Greene, The linked roles of nitric oxide, aldose reductase and, (Na+,K+)-ATPase in the slowing of nerve conduction in the streptozotocin diabetic rat, *J Clin Invest* **94**, 853-859 (1994).
12. M. K. Van den Enden, J. R. Nyengaard, E. Ostrow, J. H. Burgan, and J. R. Williamson, Elevated glucose levels increase retinal glycolysis and sorbitol pathway metabolism. Implications for diabetic retinopathy, *Invest Ophthalmol Vis Sci* **36**, 1675-85 (1995).
13. F. Palm, P. Hansell, G. Ronquist, A. Waldenstrom, P. Liss, and P. O. Carlsson, Polyol-pathway-dependent disturbances in renal medullary metabolism in experimental insulin-deficient diabetes mellitus in rats, *Diabetologia* **47**, 1223-31 (2004).
14. R. G. Tilton, L. D. Baier, J. E. Harlow, S. R. Smith, E. Ostrow, and J. R. Williamson, Diabetes-induced glomerular dysfunction: links to a more reduced cytosolic ratio of NADH/NAD+, *Kidney Int* **41**, 778-788 (1992).
15. J. R. Williamson, K. Chang, M. Frangos, K. S. Hasan, Y. Ido, T. Kawamura, J. R. Nyengaard, M. van den Enden, C. Kilo, and R. G. Tilton, Hyperglycemic pseudohypoxia and diabetic complications, *Diabetes* **42**, 801-813 (1993).
16. R. I. Dorin, V. O. Shah, D. L. Kaplan, B. S. Vela, and P. G. Zager, Regulation of aldose reductase gene expression in renal cortex and medulla of rats, *Diabetologia* **38**, 46-54 (1995).
17. T. Nishikawa, D. Edelstein, X. L. Du, S. Yamagishi, T. Matsumura, Y. Kaneda, M. A. Yorek, D. Beebe, P. J. Oates, H. P. Hammes, I. Giardino, and M. Brownlee, Normalizing mitochondrial superoxide production blocks three pathways of hyperglycaemic damage, *Nature* **404**, 787-790 (2000).
18. F. Palm, J. Cederberg, P. Hansell, P. Liss, and P. O. Carlsson, Reactive oxygen species cause diabetes-induced decrease in renal oxygen tension, *Diabetologia* **46**, 1153-1160 (2003).
19. A. Koivisto, A. Matthias, G. Bronnikov, and J. Nedergaard, Kinetics of the inhibition of mitochondrial respiration by NO, *FEBS Letters* **417**, 75-80 (1997).
20. A. Koivisto, J. Pittner, M. Froelich, and A. E. Persson, Oxygen-dependent inhibition of respiration in isolated renal tubules by nitric oxide, *Kidney Int* **55**, 2368-2375 (1999).

21. C. G. Schnackenberg, Physiological and pathophysiological roles of oxygen radicals in the renal microvasculature, *Am J Physiol Regul Integr Comp Physiol* **282**, R335-R342 (2002).
22. C. E. Mogensen, Glomerular filtration rate and renal plasma flow in normal and diabetic man during elevation of blood sugar levels, *Scand J Clin Lab Invest* **28**, 177-82 (1971).
23. F. Palm, P. Liss, A. Fasching, P. Hansell, and P. O. Carlsson, Transient glomerular hyperfiltration in the streptozotocin-diabetic Wistar Furth rat, *Ups J Med Sci* **106**, 175-182 (2001).
24. M. P. O'Donnell, B. L. Kasiske, and W. F. Keane, Glomerular hemodynamic and structural alterations in experimental diabetes mellitus, *Faseb J* **2**, 2339-2347 (1988).

27

VASTUS LATERALIS METABOLIC RESPONSE TO EXPLOSIVE MAXIMAL ISOMETRIC LEG PRESS EXERCISE

Valentina Quaresima,[1,2] Stefano Crisostomi,[3] Leonardo Mottola,[1] Massimo Angelozzi,[2] Antonio Franco,[3] Cristiana Corsica,[3] Vittorio Calvisi,[2,4] and Marco Ferrari[1,2]*

1. INTRODUCTION

Explosive force production is considered as an additional class of strength tests.[1] An important rationale for testing explosive force production has been the short time available for production in various athletic and other activities. The most applied test is the rate of force development (RFD) during which the subject exerts his maximal force in an explosive way. RFD can be assessed as the maximal slope of the recorded force-time curve or as the slope after a fixed time following the initiation of the contraction.[2] Leg press is one of the common core exercises that are utilized by athletes to enhance performance in sport. In particular, leg press exercise develops the upper leg, because it works the quadriceps, the hamstrings and the gluteus maximum. This exercise is normally performed on a machine where the legs press against a weighted platform. Muscle bioenergetics during explosive muscle strength has not been clarified yet.[3]

Near infrared spectroscopy (NIRS) is becoming a widely used instrument for measuring tissue O_2 status.[4] In fact, NIRS is a non-invasive and relatively low cost optical technique that offers the advantage of being less restrictive (no limitations on the type of exercise to be performed), more comfortable and suitable than ^{31}P-MRS for monitoring (with high temporal resolution, up to 100 Hz) oxygenation (then indirectly oxidative metabolism) of multiple muscle groups.[5]

* [1]Dipartimento di Scienze e Tecnologie Biomediche, [2]Centro Interdipartimentale di Scienza dello Sport, [3]Scuola di Specializzazione in Medicina dello Sport, [4]Dipartimento di Scienze Chirurgiche, Università di L'Aquila, 67100 L'Aquila, Italy; e-mail: vale@univaq.it.

This study aimed at measuring *vastus lateralis* (VL) muscle O_2 saturation (SmO_2) response to a single very short-duration static (maximal voluntary force, MVF) leg press exercise.

2. METHODS

Seven male volunteers (23 ± 2 years; 79 ± 9 kg) were recruited for the experiments. The experimental procedures were explained, and all the subjects gave their informed consent. The subjects were physically active (although none were engaged in daily, intensive or specific training programs) with no history of serious lower extremity (specifically ankle or knee joint) injury. Leg press strength was measured on a commercial leg press equipped with strain gauge to convert analog force signal, sampled at 100 Hz, to digital signal and stored on a PC. The subjects were positioned on the sledge of the leg press with the knee ankle adjusted at 110° (Fig. 1). The waist was fixed and the subjects were allowed to stabilize their upper body by holding on to handles attached to the leg press. The position of each subject was documented so that it was identical for the duration of the protocol. Testing was only performed on the dominant leg.

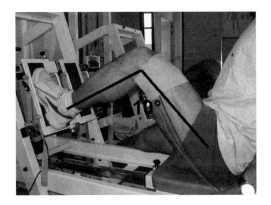

Figure 1. Experimental setup.

A warm-up period of 10 min on a treadmill preceded the testing session itself. Each subject (accustomed to the testing procedure) performed five static leg press exercises with his maximal voluntary effort. For each trial, subjects were thoroughly instructed to act "as forcefully and as fast as possible". Interval between the bouts was about 15 min. During later offline analysis the trial with the maximum static leg press strength was selected. The following force parameters were considered: maximum isometric force output, duration of leg press exercise, and RFD. RFD was calculated using the maximal slope of the force time curve ($\Delta force/\Delta time$). Normalized force output values were determined as force relative to maximum force (expressed as % of MVF). The reported data are referred to the bout associated with the best developed force.

NIRS measurements were performed (sampling rate: 6 Hz) with a NIRO-300 oximeter (Hamamatsu Photonics, Japan). The optical probe (consisting of one emitter and one detector 4.5 cm apart), supported by a rigid rubber shell, was firmly attached to the skin of the main body of VL by a double-sided adhesive sheet. The rigid rubber shell, in turn, was secured by a soft and elastic bandage. To identify the exact site for the positioning of the rubber shell, each subject performed preliminary leg contractions. Pen-marks were made on the skin to check for any sliding of the probe during the exercise. No sliding of the probe was observed at the end of the measurements in any subject. NIRS data were collected and transferred on-line to a computer for storage and subsequent analysis. SmO_2 was measured as tissue oxygenation index (TOI, %). TOI reflects the local balance between O_2 supply and O_2 consumption. Blood volume changes were measured as total hemoglobin changes (tHb, μM^*cm). Adipose tissue thickness (ATT) underlying the monitored VL area was measured with a skinfold caliper. Adipose tissue thickness was 4.2 ± 0.9 mm. Considering that ATT was less than 6 mm, it can be assumed that TOI variations reflect the metabolic changes occurring mainly at the muscle level.

3. RESULTS

Static leg press exercise duration and maximal force output (corrected for body mass) were 3.3 ± 0.5 s and 18 ± 2 N/kg, respectively (Table 1). The normalized profile of force output for all subjects is reported in Fig. 2.

Table 1. Force output and *vastus lateralis* TOI values measured over explosive static leg press exercise.

Subject	EXERCISE			TOI (%)			Time to min TOI (s)
	Duration (s)	Max Force (N)	RFD (N/s)	Baseline	End exercise	Minimum	
1	4	1050	2416	71	57	35	4
2	3	1195	4413	73	74	56	7
3	3	1264	5286	75	76	49	10
4	3	1328	6651	71	68	47	7
5	3	1224	2822	65	64	43	9
6	3	1190	8855	70	60	44	10
7	4	1445	3433	66	56	19	3
Mean±SD	3.3±0.5	1242±123	4839±2299	70±4	65±8	42±12*#	7±3

RFD: rate of force development. Time to min TOI: the time (after the end of exercise) at which TOI reached its minimum value. *: significantly different from baseline (P= 0.0003); #: significantly different from the end of exercise (P= 0.0001).

Figure 3 shows the VL oxygenation pattern observed in 2 out the 7 subjects (#2, 3). TOI was unchanged over the 3-s exercise and started to drop immediately after the exercise end. In the remaining 5 subjects, TOI was stable only over the first 1.5-2 s of the exercise; thereafter, TOI started to decline (5.3 ± 2.3 %/s) (Fig. 4). In all the subjects, TOI was reduced by 40 ± 16 % (in correspondence to TOI minimum value: 42 ± 12 %) in 7 ± 3 s after the end of the exercise (Table 1).

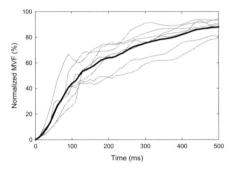

Figure 2. Time course of normalized maximal voluntary force (MVF, %) at the onset of the static leg press exercise. The solid line represents the average over the 7 subjects.

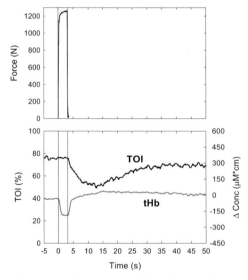

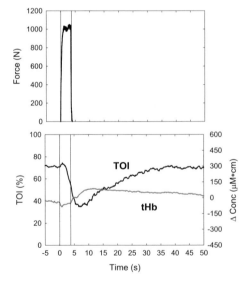

Figure 3. Time course of leg force output (upper panel), and *vastus lateralis* TOI and tHb (lower panel) before, during, and after static leg press exercise (subject #3).

Figure 4. Time course of leg force output (upper panel), and *vastus lateralis* TOI and tHb (lower panel) before, during, and after static leg press exercise (subject #1).

4. DISCUSSION

To the best of our knowledge, this is the first time that NIRS has been employed to assess SmO_2 in leg muscle during single very short trials of static leg press exercise. This non-invasive optical method has been found suitable for evaluating the metabolic response in exercising muscles.[5] It is well known that there are 3 distinct yet closely integrated processes that operate together to satisfy the energy requirements of skeletal muscle. The anaerobic energy system is divided into alactic and lactic components, referring to the processes involved in the splitting of the stored phosphagens, ATP and phosphocreatine (PCr), and the nonaerobic breakdown of carbohydrate to lactic acid through glycolysis. The aerobic energy system refers to the combustion of carbohydrates and fats in the presence of O_2. The interaction and relative contribution of the energy systems during single bouts of maximal exercise has not been clarified yet. Most recent research suggests that energy is derived from each of the energy-producing pathways during almost all exercise activities.[3] The duration of maximal exercise at which equal contributions are derived from the anaerobic and aerobic energy systems appears to occur between 1 to 2 min and most probably around 75 s, a time that is considerably earlier than has traditionally been suggested.[3] TOI data reported in Fig. 3 suggest that during the 3-s static leg press exercise, the main contributor of the energy system to the exercising VL muscle was represented by the anaerobic-alactic one. On the other hand, the initial decline of TOI within the first 3-s of exercise (Fig. 4) suggests that the lactic component of the anaerobic system was already involved, and/or O_2 was even utilized by VL. The aerobic pathway (oxidative metabolic response) starting after the end of the exercise period was evident in both the observed oxygenation patterns (Fig. 3 and 4). In this case TOI dropped to reach its minimum value and returned slowly to the pre-exercise value as a consequence of O_2 utilization for regenerating ATP, then for replenishing the PCr stores. The variability of the VL metabolic response, observed amongst the subjects, could be in part explained by the diverse performance of the subjects (Fig. 2).

In conclusion, this study suggests that NIRS could be used to: 1) profile in each muscle group the aerobic and anaerobic energy system contribution during a single bout of maximal exercise; 2) follow alteration of the profile as function of specific aerobic or anaerobic training or rehabilitation programs.

5. REFERENCES

1. G. J. Wilson and A. J. Murphy, The use of isometric tests of muscular function in athletic assessment, *Sports Med.* **22**(1), 19-37 (1996).
2. P. Aagaard, E. B. Simonsen, J. L. Andersen, P. Magnusson, and P. Dyhre-Poulsen, Increased rate of force development and neural drive of human skeletal muscle following resistance training, *J. Appl. Physiol.* **93**(4), 1318-1326 (2002).
3. P. B. Gastin, Energy system interaction and relative contribution during maximal exercise, *Sports Med.* **31**(10), 725-741 (2001).
4. M. Ferrari, L. Mottola, and V. Quaresima, Principles, techniques, and limitations of near infrared spectroscopy, *Can. J. Appl. Physiol.* **29**(4), 463-87 (2004).
5. V. Quaresima, R. Lepanto, and M. Ferrari, The use of near infrared spectroscopy in sports medicine, *J. Sports Med. Phys. Fitness* **43**(1), 1-13 (2003).

28.

THERMODYNAMIC APPROACH TO OPTIMIZE IMMOBILIZED METAL AFFINITY CHROMATOGRAPHY PURIFICATION OF PROTEIN C

James J. Lee[†] and Duane F. Bruley[‡]

1. INTRODUCTION

Deficiency of protein C (PC) reduces the body's natural ability to prevent coagulation and thrombosis and increase the possibility of blood clot formation, preventing the transport of oxygen, nutrient, and metabolic byproduct to tissue. If large quantities of PC can be produced at a low cost, it could be used as a safer alternative to coumadin and heparin for treatment, which can result in bleeding and skin necrosis from long-term use. Currently, PC is purified using immunoaffinity chromatography (IAC), which is a very expensive process, due to the difficulty of separating PC among other homologous vitamin K-dependent (VKD) proteins by other method. However, the relative high specificity of Immobilized Metal Affinity Chromatography (IMAC) has shown a high potential of separating PC from the PC homologues at a lower cost.[1, 2]

The aim of this study is to optimize IMAC/PC purification process for high yield, purity, and bioactivity by utilizing thermodynamic principles involved in the adsorption and elution processes of PC and other VKD proteins in IMAC. Thermodynamic parameters, such as, adsorption equilibrium constant (K_a) and the Gibb's free energy (ΔG_r°) of the adsorption/elution are being analyzed to study how various chromatographic operating conditions affect PC purification. This paper presents the scheme to obtain K_a and ΔG_r° of the IMAC process for purifying PC.

[†] James J. Lee, jjlee004@gwise.louisville.edu, University of Louisville, Louisville, KY
[‡] Duane F. Bruley, bruley@umbc.edu, University of Maryland Baltimore County, Baltimore, MD, Synthesizer Inc.

2. BACKGROUND

2.1. Protein C

Protein C has been chosen as the model molecule for this study for its importance as a therapeutic protein, our long-term interest, and availability of PC. PC is an anticoagulant, antithrombotic, and anti-inflammatory,[3, 4] and is produced in the liver.[5, 6] The current clinical method for treating PC deficiency is by coumadin and heparin. However, long-term applications of these drugs may cause complications, such as, bleeding or skin necrosis. PC has promising clinical potential for patients diagnosed with genetic deficiency of PC, septic shock, coumadin-induced necrosis, and heparin-induced thrombocytopenia, pregnant women who are PC deficient, patients undergoing hip and knee replacement, and patients undergoing angioplasty. PC has no known clinical side effects, and should be an ideal therapeutic for PC deficiency. PC is one of several VKD proteins that have a 60-71% homology in the amino acid sequence. These VKD proteins include coagulants, Factors II (prothrombin), VII, IX, and X.[5, 6] PC has fifteen surface histidine residues, which are the largest number, and therefore, possibly higher affinity in IMAC among VKD proteins.[7, 8]

2.2. IMAC for PC Purification

Purification of PC from human plasma is not possible by traditional ion exchange chromatography because of the many homologues in the plasma as previously stated. Immunoaffinity chromatography (IAC) yields a high affinity and specificity, but the high cost of monoclonal antibodies (MAbs) that are used as ligand for IAC is a significant disadvantage. The cost of MAb is approximately $1,200/mg, while the cost of Cu^{2+} used for IMAC is only $0.16/g. Therefore, it can be estimated that IMAC is roughly 100,000 times less expensive than IAC based on ligand cost. Results from our research group demonstrated the effectiveness of the IMAC with the combination of Cu^{2+} and iminodiacetic acid (IDA) for PC separation from prothrombin, a thrombotic VKD protein and known to be the most difficult to separate.[1] The PC purification by IMAC has also shown 83% PC recovery from inexpensive Cohn Fraction IV-1derived from blood plasma.[2]

In IMAC the adsorption affinity of PC and other homologous VKDs depends largely on the number of surface histidine residues.[7] It is also contributed by the surrounding hydrophobic residues which aid/hinder the protein histidine affinity to the metal ion.[8, 9] Therefore, the operational conditions of IMAC need to be designed with the consideration of these various binding forces for favoring PC adsorption.[8, 10] Some operation conditions influencing the binding forces include salt concentration, pH, buffer concentration, and temperature.[11, 12]

2.3. Thermodynamics of PC Adsorption

The energy required for protein adsorption are dependent on intra- and inter-molecular interactions between the protein molecules, the environment the proteins are in, and the site of protein adsorption. The equilibrium binding constant (K_a) and standard Gibb's free energy of a reaction (ΔG_r°) are two thermodynamic parameters that could describe protein adsorption in IMAC at equilibrium. ΔG_r^o is a measure of the difference in chemical potentials, or driving force to form substances on either side of the equilibrium. Relationship of these parameters is shown in Eq. (1), where R is the gas constant (8.3214 J/mol K) and T is the temperature (K).

$$\Delta G_r^o = -RT \ln K_a \qquad (1)$$

The equilibrium favoring PC adsorption increases as the value of ΔG_r^o decreases, and therefore, adsorption of PC to Cu^{2+} would be indicated by negative ΔG_r^o values under equilibrium.[13]

Figures 1a & 1b show the relationship between ΔG_r^o and K_a and the method to measure these values by a calorimetric or batch equilibrium experiments. Either a calorimetric method or a batch equilibrium experiment can be used to obtain K_a.[14] Once K_a is obtained, ΔG_r^o can be calculated from Eq. (1). The adsorption condition between PC and other homologous VKD proteins in IMAC can be examined by these thermodynamic parameters, as shown below for varying pH and salt concentrations.

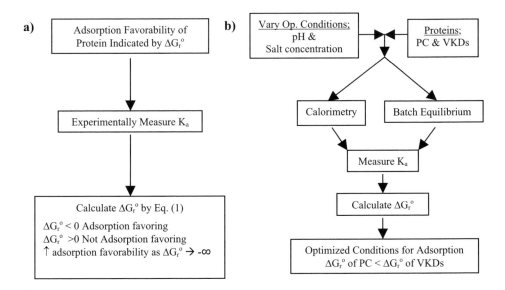

Figure 1. Schematic Diagram of the Proposed Thermodynamic Approach for Optimizing IMAC Process, a) general approach for obtaining ΔG_r^o, b) proposed method to obtain ΔG_r^o at various operating conditions and VDK proteins

3.   EXPERIMENTAL APPROACH FOR DETERMINING K_a AND ΔG_r^o

3.1. Isothermal Titration Calorimetry

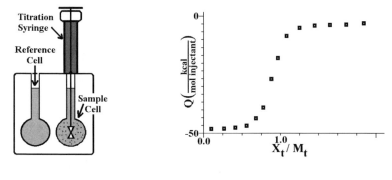

Figure 2. MicroCal	Figure 3. Schematic Diagram of an
VP-ITC Cell	Exothermic Adsorption

Isothermal titration calorimetry (ITC) measures reaction heat, and it has been used to study the effects of protein adsorption to immobilized metal ions in various conditions, such as, salt concentration, pH, temperature, and buffer composition.[11, 12] The VP-ITC instrument from MicroCal Inc. (Northampton, MA) has the sensitivity of 0.1 μcal. Adsorption heat from the sample cell (1.3 ml) is measured relative to a reference cell (Figure 2). In our ITC experiment, samples at a known PC concentration would be titrated into the cell at set containing a known amount of Cu^{2+} chelated matrix at a set time interval at a fixed volume, while the sample is homogeneously mixed in the sample cell. Experiments determine the adsorption energy of PC to Cu^{2+}. The control experiment measures heat of dilution and any non-specific binding of PC to the unchelated matrix.

Figure 3 shows an example of the output reaction heat. The reaction heat (Q; kcal/mol injectant) verses the ratio X_t/M_t is illustrated in Figure 3, where X_t (mM) and M_t (mM) respectively are the total amount of PC and the total amount of Cu^{2+} adsorption sites of the matrix in the cell. Equation (3) expresses the heat of adsorption measured in the sample cell at the end of the ith titration. K_a, molar heat of PC adsorption (ΔH^o, J/mol-PC), and stoichiometry of the adsorption reaction (n) can be determined by fitting a full thermogram to Eq. (2) via a Marquardt non-least squares analysis to get K_a and ΔG_r^o.

$$Q = \frac{nM_t \Delta H^o V_o}{2}\left[1 + \frac{X_t}{nM_t} + \frac{1}{nK_aM_t} - \sqrt{\left(1 + \frac{X_t}{nM_t} + \frac{1}{nK_aM_t}\right)^2 - \frac{4X_t}{nM_t}}\right] \tag{2}$$

3.2. Batch Equilibrium

Adsorption isotherms can also be used to obtain K_a and ΔG_r^o. The interaction and heterogeneous nature of the adsorption of a protein to metal ions is best described by the Langmuir-Freundlich equation (Eq. 3), where q* (μg/mg gel) is the PC concentration of the protein in the solid phase (matrix), $q_{m(LF)}$ ($\mu g/ml$) is the maximum PC concentration in the solid phase, C^* is the PC concentration in the liquid phase, and n_{LF} is the Langmuir-Freundlich number.[15]

$$q^* = \frac{q_{m(LF)}(C^*)^{n_{LF}}}{\frac{1}{K_a} + (C^*)^{n_{LF}}} \qquad (3)$$

This model has described the PC adsorption phenomena better than the other adsorption models, such as, the Langmuir, Temkin, or Freundlich model.[16]

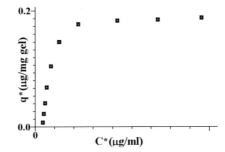

Figure 4. Illustration of Batch Equilibrium Data

Batch equilibrium experiments can be performed with samples containing an equal volume of the Cu^{2+} chelated IDA gel matrix mixed with various amounts of PC. Each batch is gently shaken for one hour to allow a full adsorption equilibrium state between PC and the metal ion. The suspensions are then centrifuged, and the amount of PC in the supernatant in each batch is quantified. This amount of bound PC/mg of the matrix (q^*) is determined by mass balance. Fitting the experimental data to the Langmuir-Freundlich equation (Figure 5) should provide K_a to calculate ΔG_r° [Eq. (1)].

4.   ON-GOING AND FUTURE STUDY

Currently ITC and batch equilibrium experiments are being performed for obtaining K_a and ΔG_r° for PC and prothrombin (one of the most problematic VKD proteins) adsorption in MAC at various NaCl and pH values. In the optimized adsorption process, ΔG_r° for PC should be less than that for prothrombin. K_a for PC should be greater than that for prothrombin. The other operating condition that needs to be studied is the elution of proteins using varying imidazole concentrations. The results of this approach for optimizing IMAC can be applied to other purification techniques and molecules.

5. AKNOWLEDGEMENTS

Research was made possible by funding through NSF-Grants CTS-0090749, and material support by the American Red Cross. The Whitaker Foundation Special Opportunity Award made the initial research possible. The Hematology Engineering Lab research group, Dr. Annamarie Ralston, and Dr. Fred Schwartz of the Centers for Advanced Research in Biology (CARB) in Rockville, MD are acknowledged for their assistance and support with this project.

6. REFERENCES

1. H. Wu and D. F. Bruley, Homologous human blood protein separation using immobilized metal affinity chromatography: protein C separation from prothrombin with application to the separation of factor IX and prothrombin, *Biotechnol. Progr.* **15**, 928 (1999).
2. H. Wu, D. F. Bruley, K. A. Kang, Protein C Separation form human plasma Cohn fraction IV-1 using immobilized metal affinity chromatography, *Adv. Exp. Med. & Bio.: Oxy. Trans. to Tis. XX,* **454**, 697-504 (Plenum. 1998).
3. C. T. Esmon, The Anticoagulant and Anti-Inflammatory Roles of the Protein C Anticoagulant Pathway, *J. Autoimmun.* **15**, 113-116 (2000).
4. C. T. Esmon, Protein C anticoagulant pathway and its role in controlling microvascular thrombosis and inflammation, *Crit. Care Med.* **29**(7), 48-51 (2001).
5. D. F. Bruley and W. N. Drohan, *Advances in Applied Biotechnology Series; Protein C and Related Anticoagulants*, **11** (Portfolio Pub. Co, Woodlands, TX 1990).
6. R. M. Bertina, *Protein C and Related Proteins; Biochemical and Clinical Aspects*, (Churchill Livingstone, Edinburgh London, 1998).
7. E. E. Thiessen, and D. F. Bruley, Theoretical Studies of IMAC Interfacial Phenomena for the Purification of Protein C, *Adv. Exp. Med. & Bio.: Oxy. Trans. to Tis. XXV*, **540**, 183-190 (Kluwer/Plenum, NY, 2003).
8. J. J. Lee, E. E. Thiessen and D. F. Bruley, Protein C Production; Metal Ion/Protein Interfacial Interaction in Immobilized Metal Affinity Chromatography, *Adv. Exp. Med. & Bio.: Oxy. Trans. to Tis. XXVI*, (Kluwer/Plenum, NY), In Press.
9. E. K .M. Ueda, P. W. Goutt and L. Moranti., Rev. of "Current and prospective application of metal ion-protein binding," *J. of Chromatography A* **988**, 1-23 (2003).
10. J. Porath, J. Carlsson, I. Olsson and G. Belfrage, Metal chelate affinity chromatography, a new approach to protein fractionation, *Nature* **258**, 598-599 (1975).
11. F. Y. Lin, W. Y. Chen and L. C. Sang, Microcalorimetric Studies of the Interactions of Lysozyme with Immobilized Metal Ions: Effects of Ion, pH Value, and Salt Concentration, *J. of Colloid and Interface Sci.* **214**, 373-379 (1999).
12. W. C. Chen, J. F. Lee, C. F. Wu and H. K. Tsao, Microcalorimetric Studies of the Interactions of Lysozyme with Immobilized Cu(II): Effects of pH Value and Salt Concentration, *J. of Colloid and Interface Sci.***190**, 49-54 (1997).
13. G. A. Holdgate, Making Cool Drugs Hot: The Use of Isothermal Titration Calorimetry as a Tool to Study Binding Energetics, *Biotechniques*, **31**, 164-184 (2001).
14. C. Jones, B. Mulloy and A.H. Thomas, *Methods in Molecular Biology, Volume 22; Microscopy, Optical Spectroscopy and Macroscopic Techniques*, Humana Press, NJ (1994).
15. S. Sharma and G. P. Agarwal, Interaction of proteins with immobilized metal ions: A comparative analysis using various isotherm models, *Anal. Biochem.* **288**, 126 (2001).
16. R. Nandakumar, H. Afshari and D. F. Bruley, Analysis of Equilibrium Adsorption Isotherms for Human Protein C Purification by Immobilized Metal Affinity Chromatography (IMAC), *Adv. Exp. Med. & Bio.: Oxy. Trans. to Tis. XXV*, **540**, 191-199 (Kluwer/Plenum, NY, 2002).

29

FLUORESCENCE ENHANCERS FOR FLUOROPHORE MEDIATED BIOSENSORS FOR CARDIOVASCULAR DISEASE DIAGNOSIS

Bin Hong and Kyung A. Kang*

1. INTRODUCTION

When heart muscles are damaged due to lack of oxygen, biomolecules, such as myoglobin (MG), C-reactive protein (CRP), cardiac troponin I (cTnI), and B-type natriuretic peptide (BNP), are released into the circulation system.[1-4] Clinically significant ranges of these biomarkers blood plasma are: 4.1 ~ 41 nM for MG[1]; 5.6 ~ 56 nM for CRP[2,3]; 31 ~ 310 pM for cTnI[3]; and 26 ~ 260 pM for BNP.[3,4] A fiber-optic, fluorophore-mediated, multi-analyte immuno-biosensing system has been under development for rapid, accurate, and simultaneous monitoring of these biomarkers in blood. The target sensing ranges of MG and CRP are sufficient for the fiber optic biosensor. The ranges for BNP and cTnI, however, are too low for regular sensing methods, requiring an additional mechanism of increasing sensitivity.

A plasmon rich *nanogold particle* (NGP), when located at an appropriate distance from a fluorophore, can reroute the electrons that contribute to self-quenching and, therefore, enhance the fluorescence (Fig. 1a).[5] This enhancement effect is highly dependent on the distance between a fluorophore and an NGP. If it is too close from a fluorophore, an NGP quenches the entire fluorescence (Fig. 1b), and if too far, an NGP will not affect the fluorescence (Fig. 1c). It was found that the size of an NGP also affects the fluorescence enhancement, according to our study results.

We have also found a few biocompatible solvents significantly enhancing fluorescence. This enhancement is possibly caused by the shift of excitation/emission spectrums for the fluorophore[6], and/or by isomerization of the fluorophore[7], or by more fluorescence retrieval when the fluorophore-tagged, sandwich protein complex shrinks in the solvent and gets closer to the sensor surface. This study investigates the effects of the NGP and the solvent on the performance improvement of the cTnI and BNP sensors.

---

* Bin Hong and Kyung A. Kang, Department of Chemical Engineering, University of Louisville, Louisville, KY 40292.

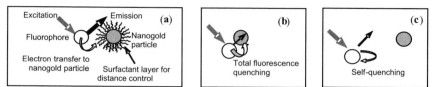

Figure 1. The effect of the distance between a fluorophore and a nanogold particle on the emission fluorescence. (**a**) At an appropriate distance: a fluorophore receives the excitation light and an NGP receives the electrons used for self-quenching, resulting in more emission light; (**b**) Too close: a fluorophore receives excitation light but the emission light is quenched, because the NGP attracts most of the electrons including the ones for fluorescence generation; (**c**) Too far: the plasmon field around the NGP does not reach the fluorophore.

2. MATERIALS, INSTRUMENTS AND METHODS

2.1 Materials and Instruments

The fluorometer (Analyte 2000) and the quartz fibers used for sensors were purchased from Research International (Monroe, WA). Protein C (PC) and PC specific monoclonal antibodies were donated by the American Red Cross (Rockville, MD). BNP (Bachem; Torrance, CA), anti-BNP monoclonal antibodies (Strategic Biosolutions; Newark, DE), cTnI, and anti-cTnI monoclonal antibodies (Fitzgerald; Concord, MA) were used to prepare BNP or cTnI sensors. Nanogold particles (2, 5 and 10 nm) linked with tannic acid (3 nm) were obtained from Ted Pella (Redding, CA). L-glutathion (1 nm) and 16-mercaptohexadecanoic acid (2 nm) used for *self-assembled monolayer* (SAM) formation on the NGP were from Sigma/Aldrich (St. Louis, MO). The fluorophores, Fluorolink™ Cyanine 5 (Cy5; 649/670 nm for excitation/emission) and Alexa Fluor® 647 (AF647; 649/666 nm for excitation/emission), were purchased from Amersham Pharmacia Biotech (Uppsala, Sweden) and Molecular Probes (Eugene, OR), respectively.

2.2 Methods

All sensors and sensing samples were prepared following the protocols established by Kang, *et al.*[8-10] As a means to adjust the distance between a fluorophore and an NGP, SAMs at particular thicknesses were immobilized onto the NGP surface. The resulting NGPs linked with SAMs (NGP-SAMs) were purified using a DispoDialyzer® dialysis tube (Spectrum Laboratories; Rancho Dominguez, CA) before use.

The protocols for BNP and cTnI biosensing were based on the PC sensing, described by Spiker and Kang.[8] When the enhancers (NGP-SAMs, solvent, or the mixture) were used, two more steps were added to the original protocol: 1) The new baseline with the enhancer, and 2) applying the enhancer after the washing step following the surface immuno-reactions between the analyte and the fluorophore-linked second antibody.

3. RESULTS AND DISCUSSION

3.1 Effect of NGP-SAM on Fluorescence Enhancement

When a fluorophore is excited by photons, the fluorescence is usually emitted with a quantum yield (QY) lower than one, mainly because of its self-quenching. An NGP at an appropriate distance from a fluorophore can reroute the electrons that are normally used for self-quenching, by absorbing these electrons in its strong plasmon field. Therefore, the QY of a fluorophore can be artificially increased by an NGP. The rerouting capacity of an NGP depends not only on the distance between an NGP and a fluorophore, but also on the NGP size because the plasmon field strength differs according to the size.

The effects of the distance between an NGP and a fluorophore and the NGP size on the fluorescence enhancement were investigated in free fluorophore solution (Fig. 2a) and also for the fiber optic immuno-biosensing (Fig. 2b). The free form measurement was performed using a quartz fiber tip polished at 45°, and the immuno-sensing was conducted in a model sensing system, protein C (PC; sensing range, 4 ~ 40 nM) biosensor, since the PC sensor has been studied extensively in our research group.[8-10] The concentration ranges of the NGP-SAMs and the fluorophore in the samples were 10^{-8}~10^{-7} and 10^{-10}~10^{-7} M, respectively. *The enhancement is defined as the increase in the fluorescence divided by the control (without the enhancer) fluorescence.*

Figure 3a shows the effect of the SAM thickness on the fluorescence enhancement for 5 nm NGPs linked with 1, 2 and 3 nm SAMs. All three SAM thicknesses show the enhancements for both free fluorophore and the PC sensing. Among the three SAM linked 5 nm NGPs, 2 nm SAM linked one provided the highest enhancement (44 and 215%, free form and PC sensing, respectively). 1 nm appeared to be too short to attract the electrons used only for self-quenching, while 3 nm, too long for effective electron rerouting. The effect of the NGP size (2, 5, 10 nm) was also investigated at a constant SAM thickness of (3 nm; Fig. 3b). The enhancements decrease as the NGP size increases, probably due to the different plasmon density for different sized NGPs. In both studies, i.e. the SAM thickness and the NGP size, the enhancement levels in PC sensing are much higher than that in free fluorophore.

Among the NGP sizes and the SAM thicknesses tested, 5 nm NGP linked with 2 nm SAMs (5nmNGP-SAM2nm) was found to be the best fluorescence enhancer with the enhancements of 44 and 215%, for the free fluorophore and the PC sensing, respectively.

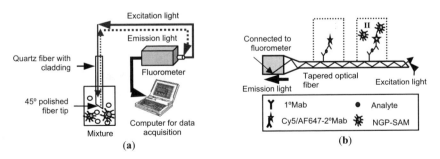

Figure 2 Schematic diagrams of fluorescence measurements: (**a**) Free fluorophores in solution using a tip of a quartz fiber, and (**b**) PC biosensing: **I**. During the immunosensing, the (1°Mab)-(PC)-(Cy5-2°Mab) complex is formed on the surface of a quartz fiber; **II**. NGP-SAMs are applied for fluorescence enhancement.

3.2 Effect of a Solvent on Fluorescence Enhancement

The quantum yield of a fluorophore is solvent sensitive since the excitation/emission spectrums may exhibit a red or blue shift in a different solvent.[6] The trans/cis isomerization of the fluorophore in the solvent may also cause the fluorescence enhancement.[7] Therefore, fluorescence enhancement may occur in a suitable solvent.

The fluorescence of free Cy5 was measured in the phosphate buffered saline (PBS) buffer and also a biocompatible solvent (Fig. 4). The fluorescence signals in the solvent were doubled (Fig. 4a) for the free Cy5, and in the PC sensing four times the enhancement was obtained (Fig. 4b). The difference in enhancement is probably due to the difference between the freely floating fluorophore and surface bound fluorophore. For the fluorophore mediated immuno-sensing, Cy5 was immobilized on the sensor surface by the sandwich protein complex. Therefore, the conformation change of the proteins in the solvent may also affect the light retrieval and fluorescence signal.

3.3 Biosensing of cTnI and BNP with Enhancers

cTnI and BNP were first measured in the fiber optic biosensors without enhancers. The initial biosensing conditions for cTnI and BNP sensors were: 12 cm in sensor size and 10/10 minutes for the sample and the AF647-2°Mab incubation. The sensors provided very low signals in the sensing range (< 50 pA; Fig. 5, △), probably not sufficient to accurately quantify these two markers in the plasma.

For testing the enhancer, the sensor size for BNP and cTnI sensors were reduced by half, estimating from our successful enhancement results in PC sensing with enhancers. Also, the sample and AF647-2°Mab incubation times were reduced to 5/5 minutes. First, 5nmNGP-SAM2nm enhancer was tested in the 6 cm BNP and cTnI biosensors. 7 ~ 56% and 33 ~ 110% signal enhancements were achieved for BNP and cTnI biosensing, respectively (Fig. 5, ▲), despite the reductions in the sensor size and the assay time.

The bio-compatible solvent was also tested in the 6 cm BNP and cTnI sensors for its fluorescence enhancement effects (Fig. 5, □) and 110 ~ 140% and 230 ~ 330% fluorescence enhancements were acquired, respectively. Also, the application of this solvent did not affect the sensor's reusability (7 times) probably because the conformation change of the protein complex was reversible in our experimental conditions.[11]

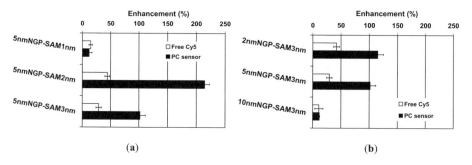

(a)   (b)

Figure 3 The effect of (**a**) the SAM thickness and (**b**) the NGP size on fluorescence enhancement for free Cy5 and in Cy5 mediated PC sensing. [Experimental conditions for PC sensing: sensor size, 6 cm; PC concentration, 1 μg/ml-plasma].

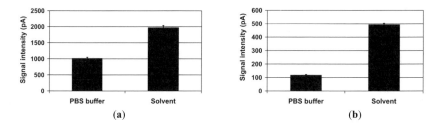

Figure 4. The solvent effect on fluorescence enhancement (**a**) for free Cy5 and (**b**) in Cy5 mediated PC

In order to maximize the fluorescence enhancement effect, 5nmNGP-SAM2nm was mixed with the solvent to produce the _nanogold particle reagent_ (NGPR), and tested for the 6 cm BNP and cTnI sensors. The fluorescence signal intensities were increased by 123 ~ 200% and 350 ~ 450% for BNP and cTnI biosensing, respectively (Fig. 5, ■). The enhancements by the NGPR appeared to be the cumulative effect of the NGP-SAMs and the solvent. Fig. 5 confirms that the sensors with even a half length and a half assay time can provide sufficient sensitivity with these enhancers.

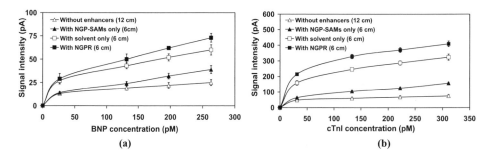

Figure 5. Biosensing (**a**) BNP and (**b**) cTnI without enhancers using 12 cm sensors with 10/10 min incubations and with various enhancers using 6 cm sensors with 5/5 min incubations.

4. CONCLUSIONS

The fluorophore mediated fiber optic immuno-biosensing of four cardiac markers, MG, CRP, BNP and cTnI, can be an important tool for rapid and accurate diagnosis of cardiovascular disease. The low sensing signals of BNP and cTnI for their target ranges need to be improved for the purpose of simultaneous quantification of four-markers. A nanogold particle fluorescence enhancer, NGP-SAMs, was able to improve the sensing performance by rerouting the electrons used for self-quenching in the fluorophore. The study results showed that this enhancement was dependent on the SAM thickness and the NGP size. Among the NGP-SAMs tested, 5nmNGP-SAM2nm was the best fluorescence enhancer with 215% enhancement in PC sensing. Another fluorescence enhancer, a biosensor compatible solvent, was also able to enhance the fluorescence significantly (318%). 5nmNGP-SAM2nm and solvent were tested for the BNP and the cTnI biosensors. Both enhancers were able to enhance the fluorescence and improve the sensitivity of the sensors significantly. The mixture of 5nmNGP-SAM2nm and the solvent (NGPR) improved BNP and cTnI sensing performances even more effectively and the enhancement level by NGPR was found to be the cumulative results of the NGP-SAM and of the solvent. An accurate and sensitive quantification of BNP and cTnI can now be achieved using the NGPR by 6 cm sensor and 10 minutes of assay time with an average signal-to-noise ratio of 18.

Future studies will focus on the enhancement mechanisms by the NGP-SAMs and the solvent by testing various combinations of the NGP size, the SAM thickness, the fluorophore, and the solvent.

5. ACKNOWLEDGEMENT

Authors would like to thank Kentucky Science and Engineering Foundation for the financial support (KSEF-148-502-03-55), National Science Foundation for the partial financial support (BES-0330075) of cardiac marker biosensing, and American Red Cross (Rockville, MD) for the donation of PC and two types of antibodies against PC.

6. REFERENCES

1. J. P. Chapelle, A. Albert, J. P. Smeets, J. Boland, C. Heusghem, and H. E. Kulbertus, Serum myoglobin determinations in the assessment of acute myocardial infarction, *Eur. Heart J.* **3** (2), 122-129 (1982).
2. H. Riese, T. G. M. Vrijkotte, P. Meijer, C. Kluft, and E. J. C. de Geus, Diagnostic strategies for C-reactive protein, *BMC Cardiovasc. Disord.* **2** (1), 9 (2002).
3. M. S. Sabatine , D. A. Morrow, J. A. de Lemos, C. M. Gibson, S. A. Murphy, N. Rifai, C. McCabe, E. M. Antman, C. P. Cannon, and E. Braunwald, Multimarker approach to risk stratification in non-ST elevation acute coronary syndromes: Simultaneous assessment of troponin I, c-reactive protein, and b-type natriuretic peptide, *Circ.* **105**, 1760-1763 (2002).
4. A. S. Maisel, P. Krishnaswamy, R. M. Nowak, J. McCord, J. E. Hollander, P. Duc, T. Omland, A. B. Storrow, W. T. Abraham, A. H. B. Wu, P. Clopton, P. G. Steg, A. Westheim, C. W. Knudsen, A. Perez, R. Kazanegra, H. C. Herrmann, and P. A. McCullough, Rapid measurement of B-type natriuretic peptide in the emergency diagnosis of heart failure, *New Engl. J. Med.* **347**, 161-167 (2002).
5. K. G. Thomas, and P. V. Kamat, Making gold nanoparticles glow: enhanced emission from a surface-bound fluoroprobe, *J. Am. Chem. Soc.* **122**, 2655 -2656 (2000).
6. V. Buschmann, K. D. Weston, and M. Sauer, Spectroscopic study and evaluation of red-absorbing fluorescent dyes, *Bioconjug. Chem.* **14**, 195 - 204 (2003).
7. J. Rodríguez, D. Scherlis, D. Estrin, P. F. Aramendía, and R. M. Negri, AM1 study of the ground and excited state potential energy surfaces of symmetric carbocyanines, *J. Phys. Chem. A.* **101**, 6998 -7006 (1997).
8. J. O. Spiker, K. A. Kang, W. N. Drohan, and D. F. Bruley, Preliminary Study of Biosensor Optimization for the Detection of Protein C, *Advances in Experimental Medicine and Biology* **454**, 681-688, Plenum, New York, (1998).
9. H. J. Kwon, H. I. Balcer, and K. A. Kang, Protein C biosensor sensitivity for biological samples and sensor reusability, *Comp. Biochem. Physiol. Part A* **132**, 231-238 (2002).
10. L. Tang, and K. A. Kang, Preliminary study of simultaneous multi-anticoagulant deficiency diagnosis by a fiber optic multi-analyte biosensor, *Proceedings of the 31st ISOTT Annual Meeting*, Aug. 16-20, Rochester, NY (2003).
11. A. L. Jacobson, and P. J. Krueger, Infrared spectroscopic studies of solvent-induced conformational changes in globular proteins, *Biochim. Biophys. Acta* **393**, 271-283 (1975).

30.

NIRS-DETECTED CHANGES IN THE MOTOR CORTEX DURING MENTAL REHEARSAL OF PHYSICAL ACTIVITY (IMAGINARY EXERCISE)

Chris E. Cooper, Duncan Pryor, Caroline Hall, & Murray Griffin [*]

1. INTRODUCTION

Mental rehearsal of a physical task (or motor ideation) has been studied since the 1930's [1]. Initial studies looking at the electrical activity in the muscle (EMG) date back to Jacobson [2] who demonstrated that the muscles used in mentally rehearsing a task are the same as those used to perform the task. Since this initial report a range of studies has provided evidence for [3-5] or against [6-8] muscle activation during mental imagery. The nature of the mental rehearsal EMG patterns do not always mirror those produced during physical activity [9] and it has recently been suggested [10] that EMG changes are due to a practice effect and not mental imagery.

The advent of modern imagery techniques have clearly demonstrated that mental imagery of physical activity does activate the brain, however. SPECT [11], PET [12], EEG [13], MEG [14] and fMRI [15] have shown that imagery activates a large fraction of the neural networks that are involved in motor performance. However, although most studies report changes in the motor cortex during motor ideation, as with EMG the detailed nature of the changes can differ between real and imaginary exercise [16, 17]. The physical confines of an MRI or PET study put severe limitations on the kind of psychological experiments that can be performed. In the case of athletes, for example, the imagery process is most successful when the nature and surroundings are closely linked to the physical act of exercise being imagined. In this study we tested whether NIRS could be used to measure motor ideation in the motor cortex, with the ultimate aim of developing a more realistic and more readily reproducible assay for the kind of motor ideation that is used in practice by professional sportspeople.

[*] Department of Biological Sciences, University of Essex, Wivenhoe Park, Colchester CO4 3SQ, England.

2. METHODS

Six subjects (5 male and 1 female, age 40 ± 15 years, mean $\pm$ SD) participated in the study. Subjects consented to the study, which complied with University of Essex ethics committee regulations. Subjects were seated with their dominant arm resting on a desk top throughout the study. A three wavelength spectrometer (NIRO200 Hamamatsu, Japan) was used with a 4 cm source:detector separation to measure changes in the oxygenation of hemoglobin in the motor cortex. An optical pathlength of 25 cm was used to convert optical density changes on μM oxyhaemoglobin concentration changes. A 1 Hz acquisition rate was used. Only one of the two NIRS probes in the NIRO200 was used in this study. The probe was positioned over the motor cortex (20% of the distance from the centre of the head to the ear on the opposite side to the dominant hand being exercised in the study). The hair was parted to allow optimum probe contact (as judged by a maximum signal:noise ratio). The subject then performed a preliminary standard "finger tapping" protocol to confirm that changes in the motor cortex could be detected.

The experimental protocol lasted 29 minutes. Two minutes of baseline readings were followed by four minutes of finger tapping (4 x 30s on, 30s off). There was then a five minute baseline prior to four minutes of exercise. During the exercise subjects squeezed and released the bulb of a sphygmomanometer continuously over a repeated five second bout. This bout consisted of a one second isometric contraction (squeeze) followed by release (one second) followed by a three second break. A further five minutes rest was followed by four minutes of "imaginary exercise", where the subjects attempted to visualise the exercise they had just performed. The protocol concluded with five minutes of rest.

3. RESULTS

Figure 1 shows the average oxygenation change for the six subjects during the finger tapping protocol (two minutes to four minutes). The increase in oxygenation during the motor task clearly demonstrates that the spectrometer is measuring oxygenation changes over the motor cortex.

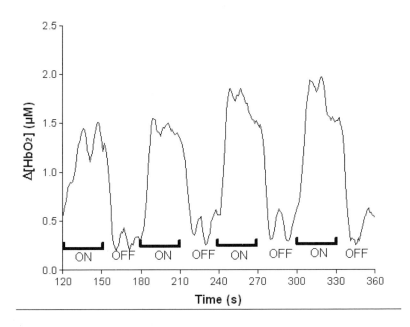

Figure 1. Changes in brain hemoglobin oxygenation during finger tapping protocol. Data averaged for all six subjects. Changes in oxyhemoglobin concentration are relative to that at the beginning of the study (t = 0s).

Figure 2 illustrates the separate pattern of oxygenation changes for each subject over the whole 29 minute protocol. In all cases a clear increase in oxygenation/deoxygenation can be seen coinciding with the on/off cycle of finger tapping. During the ball squeezing a similar magnitude of increase in oxygenation is observed, and is generally maintained over the whole four minute exercise period, decreasing during the subsequent rest period. All six subjects also show a detectable increase in oxygenation during mental rehearsal of the exercise, although in some cases the decrease to baseline in the subsequent rest period is not obvious.

The data for all the subjects are averaged in Figure 3. The magnitude of work in the muscle is of the order ball squeezing > finger tapping > mental imagery. Interestingly, though the maximum oxygenation change is identical for all three protocols. The oxygenation change observed will be a sum of the amount of the brain activated in the NIRS field of view and the amplitude of that activation. One, very simplistic, explanation of the data is therefore that equal parts of the brain are activated for the three protocols and that these parts are all maximally activated (or at least have identical hemodynamic responses to that activation). More complex interpretations are, of course, possible.

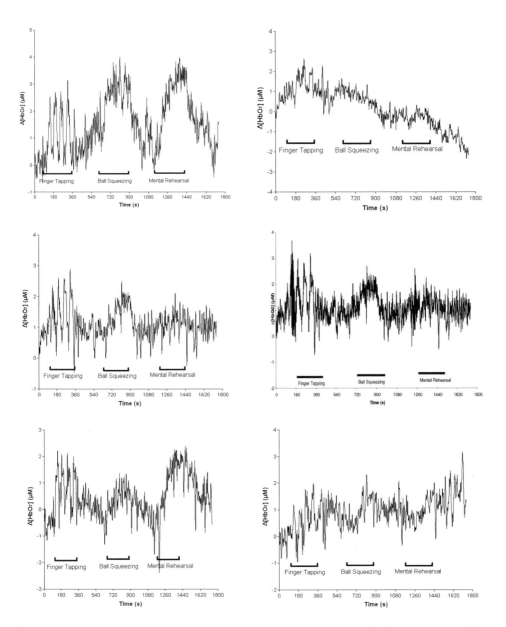

Figure 2. Changes in muscle hemoglobin oxygenation for the six subjects during the 29 minute protocol. Bars indicate the time periods for the finger tapping (four cycles), ball squeezing or mental rehearsal of ball squeezing.

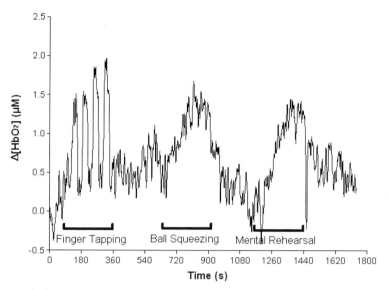

Figure 3. Magnitude of muscle hemoglobin oxygenation changes during the protocol. Data from Figure 2 averaged for all six subjects.

4. DISCUSSION

In this study we have demonstrated that it is possible to use NIRS to detect changes in the motor cortex following exercise *and* mental rehearsal (imagery) of that exercise. In principle this opens up the possibility of a range of studies of motor ideation, free from the confines of the confined and artificial environments required for MRI and PET. In practice there are a number of studies that need to be done before NIRS can be generally applied to exercise effects on the brain. First there are likely to be changes in heart rate and cardiac output in exercise that are generally more severe than that used for traditional motor cortex activation e.g. finger tapping. Smaller changes could, in principle, also occur during imagery. In the absence of topographic and tomographic data the possibility exists that systemic factors could be the cause of the increase in blood flow (and hence oxygenation) during exercise and/or imagery. Simultaneous measures of blood pressure and/or measurements over "control" regions of the brain would alleviate these concerns, either making use of dual probe or topographic NIRS systems.

We are also interested in simultaneous measurements in the muscle and brain [18]. It has been suggested that improvements in muscular strength can be driven neurally (via practice effects on central motor programming) without a requirement for repeated muscle activation. Experiments suggesting this include those from Yue and co-workers who demonstrated increases in voluntary abduction force of the human finger by mental imagery alone [19], even if the finger was in a plaster cast [20]. It is unclear, however, to what blood flow changes in the finger could be occurring despite the presence of the plaster cast. NIRS measurements in the muscle and brain could help determine the relative extent of central and local training effects.

5. ACKNOWLEDGEMENTS

We thank the Wellcome Trust (Showcase Award), the MRC (Discipline Hopping Award) and the University of Essex (Research Promotion Fund) for financial assistance. We are grateful to Clare Elwell and Julian Henty (Medical Physics and Bioengineering, University College London) for technical assistance with setting up the measurements of brain functional activation.

6. REFERENCES

1. M. Jeannerod and V. Frak, Mental imaging of motor activity in humans, *Curr. Opin. Neurobiol.* **9**, 735-739. (1999).
2. E. Jacobson, Electrical measurement of neuromuscular states during mental activity, *Amer. J. Physiol.* **94**, 22-34 (1930).
3. R. M. Suinn, in: *Psychology in sports: Methods and applications*, edited by R. Suinn (Burgess International, Minneapolis, 1980), pp. 26-36.
4. E. I. Bird, EMG quantification of mental practice, *Percept. Mot. Skills* **59**, 899-906 (1984).
5. M. Bonnet, J. Decety, M. Jeannerod, and J. Requin, Mental simulation of an action modulates the excitability of spinal reflex pathways in man, *Brain Res. Cogn. Brain Res.* **5**, 221-228. (1997).
6. B. D. Hale, The effects of internal and external imagery on muscular and ocular concomitants, *J. Sport Psychol.* **4**, 379-387 (1982).
7. T. Wehner, S. Vogt, and M. Stadler, Task-specific EMG-characteristics during mental training, *Psych. Res.* **46**, 389-401 (1984).
8. S. Yahagi, K. Shimura, and T. Kasai, An increase in cortical excitability with no change in spinal excitability during motor imagery, *Percept. Mot. Skills* **83**, 288-290. (1996).
9. J. M. Slade, D. M. Landers, and P. E. Martin, Muscular activity during real and imagined movements: A test of inflow explanations, *J. Sport Exerc. Psychol.* **24**, 151-167 (2002).
10. B. S. Hale, J. S. Raglin, and D. M. Koceja, Effect of mental imagery of a motor task on the Hoffmann reflex, *Behav. Brain Res.* **142**, 81-87 (2003).
11. J. Decety, H. Sjoholm, E. Ryding, G. Stenberg, and D. H. Ingvar, The cerebellum participates in mental activity: tomographic measurements of regional cerebral blood flow, *Brain Res.* **535**, 313-317 (1990).
12. J. Decety, D. Perani, M. Jeannerod, V. Bettinardi, B. Tadary, R. Woods, J. C. Mazziotta, and F. Fazio, Mapping motor representations with positron emission tomography, *Nature* **371**, 600-602 (1994).
13. D. F. Marks and A. R. Isaac, Topographical Distribution of Eeg Activity Accompanying Visual and Motor Imagery in Vivid and Non-Vivid Imagers, *Br. J. Psychol.* **86**, 271-282 (1995).
14. A. Schnitzler, S. Salenius, R. Salmelin, V. Jousmaki, and R. Hari, Involvement of primary motor cortex in motor imagery: a neuromagnetic study, *Neuroimage* **6**, 201-208 (1997).
15. C. A. Porro, M. P. Francescato, V. Cettolo, M. E. Diamond, C. Baraldi, C. Zuiani, M. Bazzocchi, and P. E. di Prampero, Primary motor and sensory cortex activation during motor performance and motor imagery: a functional magnetic resonance imaging study, *J. Neorosci.* **16**, 7688-7698. (1996).
16. J. Decety, Do imagined and executed actions share the same neural substrate?, *Brain Res. Cogn. Brain Res.* **3**, 87-93. (1996).
17. M. P. Deiber, V. Ibanez, M. Honda, N. Sadato, R. Raman, and M. Hallett, Cerebral processes related to visuomotor imagery and generation of simple finger movements studied with positron emission tomography, *Neuroimage* **7**, 73-85 (1998).
18. C. E. Cooper, M. Blannin, C. Hall, and M. Griffin, NIRS-detected changes in the arm during mental rehearsal of physical activity (imaginary exercise), *Adv. Exp. Biol. Med.* **This issue**, (2005).
19. G. Yue and K. J. Cole, Strength increases from the motor program: comparison of training with maximal voluntary and imagined muscle contractions, *J. Neurophysiol.* **67**, 1114-1123. (1992).
20. G. H. Yue, S. L. Wilson, K. J. Cole, W. G. Darling, and W. T. C. Yuh, Imagined muscle contraction training increases voluntary neural drive to muscle, *J. Psychophysiol.* **10**, 198-208 (1996).

31.

NIRS-DETECTED CHANGES IN THE ARM DURING MENTAL REHEARSAL OF PHYSICAL ACTIVITY (IMAGINARY EXERCISE)

Chris E. Cooper, Mark Blannin, Caroline Hall and Murray Griffin[*]

1. INTRODUCTION

Using imagery to perform a conscious mental rehearsal of a physical task, designed to aid performance [1], has proved beneficial in a variety of tasks, ranging from physical therapy to flight training to sporting performance [2]. Yet the physical/psychological basis underpinning the success of mental rehearsal is still unclear [3]. A number of theories have been proposed. One, psychoneuromuscular theory, suggests that imagery results in a neuromuscular pattern that is identical to the patterns used in actual movements [4]. Mental rehearsal is therefore likened to exercise with the gain turned down. Consistent with this theory functional magnetic resonance studies have shown that imagery activates a large fraction of the neural networks that are involved in motor performance [5-7]. However, there is conflicting evidence as to whether this effect extends to the activation of the site of physical activity itself [8, 9].

Previous studies testing the psychoneuromuscular theory of mental rehearsal have looked at changes in muscle EMG with conflicting findings supporting [8] or disputing [9] effects at the muscle. Blood volume and blood flow increase in response to exercise, to increase the flow of oxygen to the working muscles [10]. There have been no previous studies comparing the effect of these hemodynamic parameters on mental imagery of a physical activity. Near infrared spectroscopy (NIRS) is an easy and non-invasive way of measuring continuous blood volume changes during exercise. It has the advantage for studies of imagery that, once placed on the muscle, the probes and measuring device are light and non-intrusive. There is therefore minimal psychological distraction to the subject in performing the imagery test as opposed to more invasive or "bulky" measuring devices.

---

[*] Department of Biological Sciences, University of Essex, Wivenhoe Park, Colchester CO4 3SQ, England.

2. METHODS

Fifteen subjects (13 male and 2 females, age 25.1 ± 6.0, mean $\pm$ SD) participated in the study. Subjects consented to the study, which complied with University of Essex ethics committee regulations. Subjects were seated with their dominant arm resting on a desk top throughout the study. A dual wavelength spectrometer (Micro RunMan, NirSales Inc. Philadelphia, PA, USA) was used with a 3 cm source:detector separation to measure changes in the concentration of hemoglobin in the muscles above the dominant forearm [11]. As originally designed the algorithm used by the RunMan measured arbitrary changes in hemoglobin concentration and (hemoglobin + myoglobin) oxygenation calculated from changes in light attenuation at 780 and 850 nm (using 20 nm bandwidth filters). We have adapted this algorithm to measure quantified relative changes in these parameters (in µM chromophore using *in vitro* optical extinction coefficients [12] and a modified version of the Beer-Lambert Law, allowing for the spectral distortions [13] and additional pathlength [14] due to the multiple scattering of light by tissue). This is essentially the same algorithm as used in the Hamamatsu NIRO range of NIR spectrometers (although the reduced number of wavelengths and the much broader spectral bandwidth could lead to some differences in the absolute values). The probe was securely strapped to the arm to maintain its stability and position throughout the test (note that the exercise protocol was such that there was minimal movement of the arm during the protocol). A 1 Hz acquisition rate was used. The light source was positioned along the longitudinal axis of the anterior forearm, approximately $2 - 4$ cm from the antecubital fossa. Subjects performed two protocols and they were always in the same order. Immediately following probe placement and calibration there was a two minute baseline period, followed by a four minute exercise period and then a four minute recovery period. During the four minutes of exercise subjects squeezed and released the bulb of a sphygmomanometer continuously over a repeated five second bout. This bout consisted of a 1 second isometric contraction (squeeze) followed by release (one second) followed by a 3 second break. This rhythm was repeated for the 4 minute exercise. During the baseline and recovery period the subject was asked to sit in a quiescent state.

The probe was then removed and the subject left to their own devices as long as they did not engage in exercise (other than walking). Following a minimum of one hour's rest, the probe was replaced in the same position as before. In the second protocol the same procedure was followed (calibration, two minute baseline, four minute activity, four minute recovery) only this time the subjects were asked to mentally rehearse the exercise following the same rhythm as before.

All subjects completed a standard Vividness of Movement Imagery Questionnaire VMIQ to assess the impact of their ability to image the physiological response [15]. The VMIQ is a 48 item questionnaire which measures imagery on five-point Likert scales in both internal ("feeling" oneself perform) and external ("seeing" oneself perform) modalities (24 questions for each). It is designed to measure the vividness of a subjects' movement imagery and has a test-retest reliability coefficient of .76 [15]. The lowest and therefore "best" possible score is 48 (24 in each modality) whereas the highest and "worse" score is 240 (120 in each modality). The VMIQ has been used in various studies assessing the impact of vividness of imagery on sporting performance [16, 17].

3. RESULTS

The fifteen subjects showed on average a significant increase in blood volume during the exercise (Figure 1a). A large initial rise was followed by a slower increase till the end of the test. After the end of the test the blood volume decreased slowly. During the imagery protocol there was a slow continuous rise to a somewhat lower level than that seen for the exercise activity itself (Figure 1b). Subtracting the exercise-induced change from that induced by imagery (Figure 1c) reveals that the initial rapid blood volume increase is unique to exercise, but the slower increase is identical in both procedures (from half way into the exercise test till the end there is no change in blood volume between exercise and imagery). The initial exercise-induced rise is readily reversed at the end of the protocol. The primary event in the muscle seems to be an increase in blood volume as the hand is moved. However, secondary blood volume changes can be readily mimicked by the conscious mental rehearsal of the task.

Note that on average there is an increase in blood volume seen in the two minutes prior to both exercise and imagery. Separate studies (results not shown) on an optical phantom revealed that there is no change in the lamp power and/or detector sensitivity during this period (completely flat baselines were observed). This effect is likely to be due to the well-characterised anticipatory effect when a subject knows they are about to be required to "perform" a protocol [18].

The overall blood volume change induced by imagery is substantial, amounting to 65% of that induced by the exercise being imaged. Superficially this agrees with the idea that imagery can be likened to exercise with the gain turned down. However, this averaged data masks a wide range of biological variation. When the data from all the individuals are compared it is clear that, whilst all subjects show an exercise-induced blood volume increase, there is a larger variation amongst the imagery protocols, with some subjects even showing decreases in blood volume This biological variation is not random – it relates to the subject's ability to image motor tasks. The value of the VMIQ test (lower = better at imagery) significantly positively correlates with how much larger the exercise-induced blood volume change is compared to that induced by imagery (Figure 2).

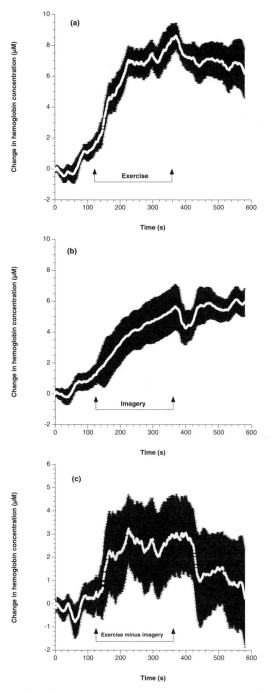

Figure 1. Changes in muscle hemoglobin concentration during isometric contractions of the forearm (a), imagery of that exercise (b) and the difference between exercise and imagery (c). The period during which the subjects performed exercise or imagery is indicated. Data are reported as mean ± SEM.

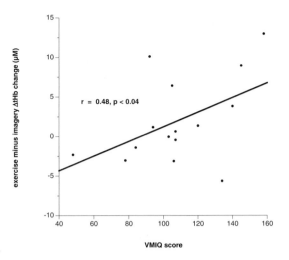

Figure 2. Magnitude of muscle hemoglobin concentration changes correlate with ability to image. Data from Vividness of Movement Imagery Questionnaire (VMIQ) plotted against how much larger the hemoglobin concentration change is for each individual during exercise compared to imagery (exercise minus imagery, taken from the data shown in Figure 1). A one-tailed t-test. (chosen on the null hypothesis that those with a low VMIQ should be better at inducing a blood volume change during imagery) showed a significant correlation.

4. DISCUSSION

Mental rehearsal is known to improve performance in physical skills. Most noticeably elite athletes use extensive imagery in their training. It has been shown that specific physical training combined with mental rehearsal is more effective than physical training alone [19]. A range of psychological theories exist to explain the advantages of imagery [20] e.g. symbolic learning [21], dual coding [22] and bio informational theory [23]. These theories tend to focus on effects of imagery on psychological models of the motor skills learning process. Yet the performance-enhancing effects of imagery are greatest the more the subject is familiar with the task, when cognitive activity should be less important [24]. Psychoneuromuscular theory claims that when mentally rehearsing a task, the same neurological pathways as when actually performing the task are used but with – as it were - the gain turned down. Functional magnetic resonance studies have clearly demonstrated that the same regional hemodynamic changes in the brain are activated during exercise and motor imagery performance [5-7]. In this study we have used the sensitivity of Near Infrared Spectroscopy to small changes in blood volume to extend this finding to the hemodynamics of the muscle itself. This study has tested the psychoneuromuscular theory of mental imagery and has found significant evidence to support it. Furthermore these finding may have implications for exercise rehabilitation. As increasing blood flow to injured muscles promotes healing this study would suggest that imagery of a motor task in an injured limb may be a way of promoting flow and healing in a limb that would otherwise be dormant [25].

5. ACKNOWLEDGEMENTS

We thank the Wellcome Trust (Showcase Award) and the University of Essex (Research Promotion Fund) for financial assistance.

6. REFERENCES

1. M. Jeannerod and V. Frak, Mental imaging of motor activity in humans, *Curr. Opin. Neurobiol.* **9**, 735-739. (1999).
2. M. Denis, Visual imagery and the use of mental practice in the development of motor skills, *Can. J. Appl. Sport Sci.* **10**, 4S-16S. (1985).
3. S. M. Murphy, Imagery interventions in sport, *Med. Sci. Sports Exerc.* **26**, 486-494. (1994).
4. E. Jacobson, Electrical measurement of neuromuscular states during mental activity, *Amer. J. Physiol.* **94**, 22-34 (1930).
5. J. Decety, Do imagined and executed actions share the same neural substrate?, *Brain Res. Cogn. Brain Res.* **3**, 87-93. (1996).
6. C. A. Porro, M. P. Francescato, V. Cettolo, M. E. Diamond, P. Baraldi, C. Zuiani, M. Bazzocchi, and P. E. di Prampero, Primary motor and sensory cortex activation during motor performance and motor imagery: a functional magnetic resonance imaging study, *J Neurosci* **16**, 7688-7698. (1996).
7. M. Lotze, P. Montoya, M. Erb, E. Hulsmann, H. Flor, U. Klose, N. Birbaumer, and W. Grodd, Activation of cortical and cerebellar motor areas during executed and imagined hand movements: an fMRI study, *J. Cogn. Neurosci.* **11**, 491-501. (1999).
8. M. Bonnet, J. Decety, M. Jeannerod, and J. Requin, Mental simulation of an action modulates the excitability of spinal reflex pathways in man, *Brain Res. Cogn. Brain Res.* **5**, 221-228. (1997).
9. S. Yahagi, K. Shimura, and T. Kasai, An increase in cortical excitability with no change in spinal excitability during motor imagery, *Percept. Mot. Skills* **83**, 288-290. (1996).
10. R. A. De Blasi, M. Ferrari, A. Natali, G. Conti, A. Mega, and A. Gasparetto, Noninvasive measurement of forearm blood flow and oxygen consumption by near-infrared spectroscopy, *J Appl. Physiol.* **76**, 1388-1393 (1994).
11. D. M. Mancini, L. Bolinger, H. Li, K. Kendrick, B. Chance, and J. R. Wilson, Validation Of Near-Infrared Spectroscopy In Humans, *J. Appl. Physiol.* **77**, 2740-2747 (1994).
12. S. J. Matcher, C. E. Elwell, C. E. Cooper, M. Cope, and D. T. Delpy, Performance Comparison of Several Published Tissue Near-Infrared Spectroscopy Algorithms, *Anal. Biochem.* **227**, 54-68 (1995).
13. M. Essenpreis, C. E. Elwell, P. van der Zee, S. R. Arridge, and D. T. Delpy, Spectral dependance of temporal point spread functions in human tissues, *Applied Optics* **32**, 418-425 (1993).
14. A. Duncan, J. H. Meek, M. Clemence, C. E. Elwell, L. Tyszczuk, M. Cope, and D. T. Delpy, Optical pathlength measurements on adult head, calf and forearm and the head of the newborn infant using phase resolved spectroscopy, *Phys. Med. Biol.* **40**, 295-304 (1995).
15. A. Isaac, D. Marks, and D. Russell, An instrument for assessing imagery of movement: the Vividness of Movement Imagery Questionnaire (VMIQ), *J. Ment. Imag.* **10**, 23-30 (1986).
16. A. R. Isaac, Mental practice: Does it work in the field?, *The Sports Psychologist* **6**, 192-198 (1992).
17. S. M. Murphy and K. A. Martin, in: *Advances in Sport Psychology*, edited by T. S. Horn (Human Kinetics., Champaign, IL, 2002), pp. 405-439.
18. R. J. Roberts and T. C. Weerts, Forearm Blood-Flow Responding Prior to Voluntary Isometric Contraction, *Psychophysiol.* **21**, 363-370 (1984).
19. J. E. Driskell, C. Copper, and A. Moran, Does Mental Practice Enhance Performance, *J. Appl. Psychol.* **79**, 481-492 (1994).
20. S. M. Murphy and D. P. Jowdy, in: *Advances in Sport Psychology*, edited by T. S. Horn (Human Kinetics, Champaign, Il., 1992), pp. 222-250.
21. R. S. Sackett, The influences of symbolic rehearsal upon the retention of a maze habit, *J. Gen. Psychol.* **10**, 376-395 (1930).
22. J. Annett, Motor Imagery - Perception or Action, *Neuropsychologia* **33**, 1395-1417 (1995).
23. J. Hecker and L. Kazcor, Application of imagery theory to sport psychology: some preliminary findings, *J. Sport Exerc. Psychol.* **10**, 363-373 (1986).
24. A. Richardson, *Mental Imagery* (Springer, New York, 1969).
25. P. L. Jackson, M. F. Lafleur, F. Malouin, C. Richards, and J. Doyon, Potential role of mental practice using motor imagery in neurologic rehabilitation, *Arch. Phys. Med. Rehabi.l* **82**, 1133-1141. (2001).

32

APPLICATION OF NIRS IN MICE: A STUDY COMPARING THE OXYGENATION OF CEREBRAL BLOOD AND MAIN TISSUE OXYGENATION OF MICE AND RAT

De Visscher Geofrey[*][♥], Verreth Wim[*][♥], Blockx Helga[*], van Rossem Koen[♦], Holvoet Paul[*], Flameng Willem[*]

1. INTRODUCTION

In modern neuroscience, insight is gained in important pathophysiological processes underlying Alzheimer and Parkinson disease by the development of transgenic mice models[1-2]. Obtaining on-line information on the cerebral oxygenation, hemodynamics, cerebrovascular reactivity and mitochondrial function is crucial for the validation of these models, the understanding of the pathology itself, and the search for potential therapies. In addition, it would be very interesting to study cerebral blood flow and mitochondrial function in transgenic mice models of atherosclerosis[3] or cerebrovascular deformations leading to neurodegeneration[4].

A suitable technique for such measurement does not exist at the moment. Therefore, we tested if multi-wavelength near infrared spectroscopy (NIRS) technology, a minimally invasive technique, can be used as a tool to examine cerebral oxygenation and hemodynamics in the anaesthetised mouse.

In the first study we compare basic NIRS variables measured in mice with those obtained in rats and establish the possibility of NIRS measurement of cerebral blood and cellular oxygenation in mice. In a second study we determine the optimal dose of indocyanine green (ICG) providing a sufficiently large signal in the brain of mice and estimate the effect of the tracer injection upon the other NIRS variables

---

[♥] Authors contributed equally to this article
[♦] Barrier Therapeutics NV, Geel, Belgium
[*] Laboratory of Cardiovascular Research, CEHA, Catholic University of Leuven, Leuven, Belgium

Address for correspondence:
   Dr. De Visscher G., CEHA, KULeuven, Minderbroedersstraat 17, B3000 Leuven, Belgium,
   tel: +32 16 337852 or +32 16 337298, fax: +32 16 337855

2. MATERIALS AND METHODS

2.1. Animal Preparation and Experimental Protocol

Animal housing and treatment conditions complied with the European Union directive # 86/609 on animal welfare. The study was performed on 10 male Sprague-Dawley rats (Charles River, Sulzfeld, Germany) weighing 380-430 g, 10 C57BL6 mice (CEHA, KULeuven, Leuven, Belgium) weighing 24-26 g and 12 NMRI mice (ANIMALIUM, St. Rafael Proefdierencentrum, Leuven, Belgium) weighing 24-26 g. They were allowed free access to food and water (12/12-hour day-night cycle). The rat preparation for NIRS has previously been described in detail[5]. In short, anaesthetic induction was achieved by 4% isoflurane in 30/70% O_2/N_2O. Instead of endotracheal intubation, standard in rats, the C57BL6 mice were kept on 2% isoflurane by means of an anaesthesia mask and the NMRI mice underwent tracheotomy and were mechanically ventilated (7µl/g body weight at 150 strokes/min). In the NMRI mice a PE50 cannula was inserted into the right external jugular vein towards the vena cava for ICG injection and the aorta descendens was cannulated to monitor mean arterial blood pressure (MABP) and heart rate (HR). All animals were positioned on a rectal temperature controlled heating pad and fixed into a stereotaxic apparatus (David Kopf, Tujunga, CA, USA). The skin was removed and combined with either temporal muscle resection or retraction for rats or mice, respectively. Possible sites of bleeding were cauterised and the skull was painted black. The receiving and emitting optode, identical in both studies, were placed onto the temporal bones by means of micromanipulators. The space between the skull and optodes was filled with optical coupling gel (R.P. Cargille Laboratories Inc., Cedar Grove, New Jersey, USA) and the skull was covered with black modelling clay (Eberhard Faber GmbH., Neumarkt, Germany). Anaesthesia was then lowered to 1.5% isoflurane.

After finalisation of the animal preparation transillumination of the brain was started. The light entrance slit of the spectrograph was set at 350µm and 160µm for rat and mice, respectively. The measured variables were: concentration of deoxyhemoglobin ([Hb]), oxyhemoglobin ([HbO_2]), total hemoglobin ([HbT]) and oxidised cytochrome oxidase (Cu_A) as well as the optical path lengths at 740, 840 and 960nm. For the dose-finding study the same variables were measured along with MABP and HR, and in each animal, boli of 0.5, 1, 2.5, 5 and 10µl ICG were given at 5, 10, 15, 20 and 25 min after the onset of the measurements, respectively. All animals were killed by terminal anoxia.

2.2. NIRS Equipment

The NIRS system and algorithms have been developed at University College London and described in detail before[6-8]. The system allows the on-line assessment of absolute values of [Hb] and absolute changes in [HbO_2] and Cu_A. Terminal anoxia at the end of an in vivo experiment allows back-calculation of the absolute [HbO_2] and Cu_A. The [HbT] at each time point (= [Hb]+[HbO_2]) and the oxygen saturation of the hemoglobin in the cerebrum ($SmcO_2$ = ([HbO_2]/([Hb]+[HbO_2]))x100) can be calculated.

A 1mg/ml ICG (IR-125, laser grade; Acros, Geel, Belgium) solution was used containing 5% bovine serum albumin (BSA fraction V; Sigma, Bornem, Belgium) to bind the ICG. The solution was sterilized by filtration with a 0.22 µm filter unit.

NIR spectra were collected contiguously with a period of 100ms and 10 spectra were averaged to obtain a time resolution of 1 second (1 Hz) throughout the entire experiment.

2.3. Data and Statistical Analysis

For each animal the data of the NIRS measured variables ($[HbO_2]$, $[Hb]$, $[HbT]$, Cu_A and $SmcO_2$) were averaged over a 5-min period starting 2 min after the onset of the measurement for the first study and over a 3-min period starting 1min after the onset of the measurement for the dose-finding study. For the first study we calculated the differential path length factor (DPF) by dividing path length by the interoptode distance[9].

ΔICG was calculated from the ICG bolus transit curve by subtraction of the baseline value before injection from the peak value after injection. The rise time (Δt) was measured from the first value where the ICG signal deviates from the baseline towards the peak value. A blood flow index (BFI) was then calculated by dividing ΔICG by Δt[10,5].

In order to evaluate cross-talk between the ICG signal and other NIRS signals during and after the ICG transit, we examined the duration of the disturbances of the different NIRS variables. We thus measured the time interval between the peak of the disturbance and the point of return to 10% of its amplitude (t_{90})[5] for $[Hb]$, $[HbO_2]$, Cu_A and ICG.

Results are expressed as median [95% confidence interval (CI)]. A two-sided Wilcoxon-Mann-Whitney rank-sum test and a two-sided Wilcoxon signed-ranks test were used for comparison of the mice and rat data and the dose-finding data, respectively. Two-sided probability values of less than 0.05 were regarded as statistically significant.

3. RESULTS

3.1. Comparison Study

Table 1 shows the results of the optical path lengths and DPF at 740, 840 and 960nm. Considering the absolute path lengths, large differences were observed at every wavelength, clearly indicating a larger tissue volume measured in the rats. To normalize for these differences we calculated the DPF, thereby correcting for the interoptode distance, corresponding to the width of the skull. The DPF at 740nm is lower in mouse brain whereas the DPF at 840nm and 960nm are not different from rat.

Table 1. Near infrared spectroscopy assessed optical path length and DPF data

	Mouse (n = 10)	Rat (n = 10)	p-value
Path length 740nm (cm)	2.25 [1.74, 2.62]	6.31 [6.07, 6.69]	< 0.001
Path length 840nm (cm)	2.62 [2.38, 2.73]	5.41 [4.98, 5.78]	< 0.001
Path length 960nm (cm)	2.15 [2.01, 2.26]	4.21 [4.06, 4.38]	< 0.001
DPF 740nm	2.51 [1.93, 2.91]	3.50 [3.37, 3.71]	< 0.001
DPF 840nm	2.91 [2.64, 3.04]	3.01 [2.77, 3.21]	0.393
DPF 960nm	2.39 [2.23, 2.51]	2.34 [2.26, 2.43]	0.280

Values are expressed as median [95% CI].

3.2. Dose-finding Study

The physiological data derived from the NMRI mice on the onset of the dose-finding study were stable (MABP: 71.9 [67.1, 89.5] mmHg; HR: 592 [543, 618] beats per min; T_{rectal}: 36.8 [36.4, 37.0] °C; T_{ear}: 33.4 [33.0, 33.9] °C).

Figure 1 shows the path lengths at 740, 840 and 960nm in the mice brain. These data indicate that after larger ICG injection volumes the 960nm path length significantly increases and both 740 and 840nm significantly decrease. This is in accordance with the increased absorption of light in the lower end of the spectra where ICG is a major contributor to the general absorption spectrum.

Table 2 presents a summary of the t_{90} data. After 0.5μl ICG bolus injection, the change in the signal of [HbO$_2$] was too small to have a sufficient signal to noise ratio. T_{90} values of all variables rose with larger ICG bolus, except for t_{90} of Hb after 1μl ICG bolus injection. T_{90} of [Hb] values of 2.5, 5 and 10μl are significantly different from 0.5μl value, t_{90} of [HbO$_2$] of the 10μl bolus is significantly different from 1μl and t_{90} values of [Cu$_A$] and ICG all significantly increased dose dependently.

To compare variability we divided BFI by the injected volume (figure 2). Significant differences were found between 0.5 and 1μl, 2.5 and 5μl, 2.5 and 10μl and 5 and 10μl.

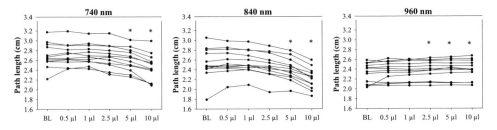

Figure 1. Path length at 740, 840 and 960nm after the different ICG boli. Statistical significance was assessed by Wilcoxon signed-ranks test. *: significantly different from baseline values ($\alpha = 0.05$). BL = baseline

Table 2. t_{90} (s) of NIRS signals after i.v. boli of different ICG dosages

	ICG 0.5μl	ICG 1μl	ICG 2.5μl	ICG 5μl	ICG 10μl
[Hb] (μM)	3.58 [1.10, 4.66]	3.22 [1.89, 9.22]	16.9[10.4, 55.2]*	68.6 [24.0, 155.4]*	91.8 [65.9, 115.0]*
[HbO$_2$] (μM)	ND	1.84 [0.90, 2.88]	3.17 [1.98, 9.32]	8.82 [1.89, 70.3]	10.1 [3.66, 32.7]°
Cu$_A$ (μM)	1.59 [0.79, 3.08]	2.18 [1.18, 3.57] *	39.8 [9.71, 93.1] *	113.5 [90.5, 246.0] *	235.0 [173.6, 308.1] *
ICG (μM)	11.8 [2.18, 60.0]	68.4 [52.1, 96.1] *	109.9 [82.1, 134.2] *	119.6 [96.4, 131.0] *	141.5 [112.4, 179.1] *

Values are expressed as median [95% CI]. ND: not determinable. *: significantly different from 0.5μl values ($\alpha = 0.05$). °: significantly different from 1μl values ($\alpha = 0.05$). t_{90} of Cu$_A$ and ICG were all significantly different.

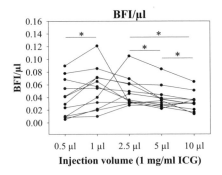

Figure 2. BFI/μl. Blood flow index divided by the amount of ICG injected. Statistical significance was assessed by Wilcoxon Signed Ranks test. *: significantly different from each other ($\alpha = 0.05$)

4. DISCUSSION

At present, NIRS has been shown an accurate research tool for the study of cerebral hemodynamics and oxygenation in several species such as pig[10], piglet[8] and rat[11]. However, further miniaturisation of the system in order to obtain measurements in mice, is not evident. To our knowledge the present study is a first attempt to evaluate brain oxygenation in mice with multi-wavelength NIR spectroscopy.

A first prerequisite to allow absolute quantification of [Hb], [HbO$_2$] and Cu$_A$ is the determination of the optical path length[6]. Path lengths at 740, 840 and 960nm are significantly smaller in mice compared to rat. The path lengths are the actual average distance the light travels between the emitting and receiving optodes and are not very useful to compare between species since they are inter-optode distance dependent[9], (9mm for mice and 18mm for rat). Calculation of the DPF enables comparison of both species. We showed a decreased DPF at 740nm in mice compared to rat but equal DPF at 840 and 960nm. Although shortening of the interoptode space theoretically shortens DPF and increases DPF in vivo[12], we find contradictory results. At this time we do not have a proper explanation for these data. The strong absorption peak of water at 960nm[6] can be used to calculate and compare the actual DPF under the assumption that the water content of the brain is 80% in both mice[13] and rat[14]. Since the DPF at 960nm are comparable between both species, we can conclude that they have similar scattering properties and that the algorithms allow to obtain absolute values for [Hb], [HbO$_2$] and Cu$_A$ in mice.

Mice showed a significantly lower [HbT] than rats (data not shown). Only 72.3% of rat [HbT] was measured in mice, this compares to the value of 78.1% calculated from the average hemoglobin concentration published for rat and mice (14.2 and 11.1g/100ml respectively[15]). The lower hemoglobin concentration was consistently found for oxy- and deoxyhaemoglobin (71.2% and 79.1% respectively), resulting in comparable SmcO$_2$ between both species. For Cu$_A$ in mice brain equivalent concentrations were obtained as in references 16 and 8, but validation still needs to be performed.

In the second study we investigated the optimal ICG bolus to qualitatively measure cerebral blood flow. The aim was a sufficiently large signal in the brain of mice with only minor cross-talk effects. We evaluated the effect of ICG on the 3 path lengths (figure 1). Due to significant changes in path length after large ICG boli, back-calculation of NIRS variables was incorrect. It is therefore not possible to simultaneously measure BFI and [Hb], [HbO$_2$] and Cu$_A$. Most likely ICG accumulates in the blood and dominates the NIR spectrum. In rat this problem does not occur due to the small bolus to blood volume ratio (5µl ICG/±20ml blood) whereas the most appropriate volume for BFI calculation in mice, 2.5µl, is only diluted in 2ml of blood. To estimate the impact of ICG boli on the NIRS variables without taking the change in path length into account, we calculated their t$_{90}$. T$_{90}$ of all variables rose with larger ICG bolus (table 2), demonstrating the need to minimise bolus volume. On the contrary, while investigating variability by dividing BFI by the injection volume (figure 2), we found that small injection volumes results in a large variability, indicating the need to increase the injected bolus.

To best fit all criteria, minimise bolus for minor disturbances in path length and t$_{90}$ and maximise bolus to minimise variability, we find at present the optimal bolus to measure BFI in mice to be 2.5µl ICG. Further attempts to increase reproducibility of the 1µl bolus injection are a technical challenge but will be attempted.

We conclude that multi-wavelength NIRS can be used in mice to assess oxygenation of cerebral blood and the concentration of oxidised cytochrome oxidase in brain tissue. To our knowledge no previous report showing this possibility has been published. Monitoring of the transit of an indocyanine green bolus with NIRS can be used to qualitatively estimate cerebral blood flow in mice. A bolus with a volume of 2.5µl of a 1mg/ml ICG solution gives a good signal-to-noise ratio with acceptable disturbances of the other NIRS measurements and minor variability of the blood flow index. At present combination of the assessment oxygenation of cerebral blood and the concentration of oxidised cytochrome oxidase with cerebral blood flow estimation is not yet possible due to technical difficulties. We consider the application of this technique in the study of transgenic mice models as a major step forward.

5. ACKNOWLEDGEMENTS

The authors would like to thank Prof. Dr. David T. Delpy from the Dept. of Medical Physics & Bioengineering of the University College London for his useful comment when preparing the manuscript.

6. REFERENCES

1. Lo Bianco C, Ridet JL, Schneider BL, Deglon N, Aebischer P. alpha -Synucleinopathy and selective dopaminergic neuron loss in a rat lentiviral-based model of Parkinson's disease. Proc.Natl.Acad.Sci.U.S.A, 2002; 99: 10813-10818.
2. Aliev G, Seyidova D, Lamb BT, Obrenovich ME, Siedlak SL, Vinters HV, Friedland RP, LaManna JC, Smith MA, Perry G. Mitochondria and vascular lesions as a central target for the development of Alzheimer's disease and Alzheimer disease-like pathology in transgenic mice. Neurol.Res., 2003; 25: 665-674.
3. Quarck R, De Geest B, Stengel D, Mertens A, Lox M, Theilmeier G, Michiels C, Raes M, Bult H, Collen D, Van Veldhoven P, Ninio E, Holvoet P. Adenovirus-mediated gene transfer of human platelet-activating factor-acetylhydrolase prevents injury-induced neointima formation and reduces spontaneous atherosclerosis in apolipoprotein E-deficient mice. Circulation, 2001; 103: 2495-2500.
4. Storkebaum E, Carmeliet P. VEGF: a critical player in neurodegeneration. J.Clin.Invest, 2004; 113: 14-18.
5. De Visscher G, van Rossem K, Van Reempts J, Borgers M, Flameng W, Reneman RS. Cerebral blood flow assessment with indocyanine green bolus transit detection by near-infrared spectroscopy in the rat. Comp Biochem.Physiol, 2002; 132: 87-95.
6. Cope M, Delpy DT, Wray S, Wyatt JS, Reynolds EO. A CCD spectrophotometer to quantitate the concentration of chromophores in living tissue utilising the absorption peak of water at 975nm. Adv.Exp.Med.Biol., 1989; 248: 33-40.
7. Matcher SJ, Cooper CE. Absolute quantification of deoxyhaemoglobin concentration in tissue near infrared spectroscopy. Phys.Med.Biol., 1994; 39: 1295-1312.
8. Springett R, Wylezinska M, Cady EB, Cope M, Delpy DT. Oxygen dependency of cerebral oxidative phosphorylation in newborn piglets. J.Cereb.Blood Flow Metab, 2000a; 20: 280-289.
9. Delpy DT, Cope M, van der Zee P, Arridge S, Wray S, Wyatt J. Estimation of optical pathlength through tissue from direct time of flight measurement. Phys.Med.Biol., 1988; 33: 1433-1442.
10. Kuebler WM, Sckell A, Habler O, Kleen M, Kuhnle GE, Welte M, Messmer K, Goetz AE. Noninvasive measurement of regional cerebral blood flow by near-infrared spectroscopy and indocyanine green. J.Cereb.Blood Flow Metab., 1998; 18: 445-456.
11. De Visscher G, Springett R, Delpy DT, Van Reempts J, Borgers M, van Rossem K. Nitric oxide does not inhibit cerebral cytochrome oxidase in vivo or in the reactive hyperemic phase after brief anoxia in the adult rat. J.Cereb.Blood Flow Metab, 2002; 22: 515-519.
12. Arridge SR, Cope M, Delpy DT. The theoretical basis for the determination of optical pathlengths in tissue: temporal and frequency analysis. Phys.Med.Biol., 1992; 37: 1531-1560.
13. Murakami K, Kondo T, Yang G, Chen SF, Morita-Fujimura Y, Chan PH. Cold injury in mice: a model to study mechanisms of brain edema and neuronal apoptosis. Prog.Neurobiol., 1999; 57: 289-299.
14. Dobbing J, Sands J. Growth and development of the brain and spinal cord of the guinea pig. Brain Res., 1970; 17: 115-123.
15. Gross DR. Animal models in cardiovascular research, 2 ed. Kluwer Academic Publishers: 1994: 1-494.
16. Cooper CE, Cope M, Springett R, Amess PN, Penrice J, Tyszczuk L, Punwani S, Ordidge R, Wyatt J, Delpy DT. Use of mitochondrial inhibitors to demonstrate that cytochrome oxidase near-infrared spectroscopy can measure mitochondrial dysfunction noninvasively in the brain. J.Cereb.Blood Flow Metab, 1999; 19: 27-38.

33.

NEAR INFRARED SPECTROSCOPY AS A NON-INVASIVE ASSESSMENT OF CORTICAL ABNORMALITY IN MIGRAINE?

Murray Griffin,* Duncan Prior,* Chris E. Cooper,* Arnold J. Wilkins,[†] and Clare E. Elwell[‡]

1. INTRODUCTION

Migraine is a brain disorder associated with debilitating episodic head pain that afflicts approximately 7% of men and 20% women. It has been estimated to cost US employers about $13 billion per annum.[1, 2] There is an abnormal haemodynamic response to strong visual stimuli in migraine that has been measured using functional magnetic resonance imaging (fMRI).[3] An alternative optical technique, near infrared spectroscopy (NIRS), is more suitable than fMRI for mass studies because of its low cost. Its simplicity and portability offer the potential for widespread use in outpatient clinics. We report preliminary work designed, ultimately, to evaluate the use of NIRS as a screening and therapy evaluation tool in migraine.

2. BACKGROUND: MIGRAINE

2.1. Strong Patterns

Patterns of stripes (gratings) vary predictably in only one dimension and have been widely used in vision research. The gratings that have parameters optimal for visibility at low contrast[4] are also those that at high contrast interfere maximally with the visibility of other stimuli with which they are combined (in studies of visual masking[5]). Patterns with the same characteristics induce the highest amplitude evoked potentials[6] and are

* Department of Biological Sciences, University of Essex, Wivenhoe Park, Colchester CO4 3SQ, England.
† Corresponding author: Department of Psychology, University of Essex, Wivenhoe Park, Colchester CO4 3SQ, England. arnold@essex.ac.uk
‡ Department of Medical Physics and Bioengineering, University College London. 1st Floor Shropshire House, 11-20 Capper Street, London WC1E 6JA.

associated with the greatest fMRI blood oxygenation level dependent (BOLD) signal[7]. In brief, they are stimuli to which the visual system is in a general sense maximally sensitive, and will be referred to here as *strong* patterns.

2.2. Aversion, Perceptual Distortions and Seizures in Response to Strong Patterns

Strong patterns can induce a variety of perceptual distortions – illusions of colour, shape and motion. The origin of the illusions is unknown, despite a century of study, although various mechanisms have been proposed.[8] Some individuals are far more susceptible than others to these distortions, usually individuals with frequent headaches.[8] The illusion susceptibility increases in the 24 hours prior to a headache.[9] If the headaches are unilateral, the distortions predominate in one lateral visual field.[8] Individuals with migraine find the patterns aversive[10] and these individuals are particularly susceptible to the perceptual distortions that the patterns evoke.[8] In migraine with aura, the distortions tend to occur interictally in the visual field affected by the aura.[8]

Strong patterns can induce seizures in patients with photosensitive epilepsy.[11, 12] The patterns that induce paroxysmal epileptiform EEG activity in patients with photosensitive epilepsy are very similar indeed to the characteristics of patterns that provoke perceptual distortions in normal observers and, to a greater extent, those with migraine.[8]

2.3. Cortical Abnormality in Migraine

The cerebral cortex responds abnormally in migraine. The abnormality is poorly understood, although there are several disparate but convergent lines of evidence, recently reviewed,[13] consistent with the hypothesis that the cortex is hyperexcitable: migraine and epilepsy are co-morbid conditions[14] and several anticonvulsant drugs have been shown to prevent migraine in randomised controlled trials[15]. In migraineurs (1) magnetic stimulation of the visual cortex stimulates phosphenes more readily,[16] (2) the evoked potential fails to show the usual habituation with repeated stimulation,[17, 18] (3) *strong* (epileptogenic) patterns are aversive,[10] (4) the fMRI BOLD response to *strong* patterns is selectively elevated.[3]

We hypothesise that the perceptual distortions occur because a spread of excitation causes visual neurons to fire inappropriately. According to this hypothesis, the degree of susceptibility to distortions increases with, and reflects, the degree of cortical hyperexcitability.

2.4. Triggering of Headaches

When they are asked, about 40% of patients with migraine will report visually provoked attacks.[19] A substantial proportion report that flicker induces attacks, and a smaller proportion are aware that patterns of stripes can also be a problem. The role of vision in the induction of migraine attacks has received only one investigation in a dozen epidemiologic studies of migraine with a sample size of 1,000-20,000 patients.[19]

The possibility that many headaches are visually provoked has been suggested by (1) double-masked studies that have shown the imperceptible high frequency flicker from fluorescent lighting to be responsible for many headaches experienced by office workers,[20] and (2) the success of recent ophthalmic techniques for headache prevention.[21]

3. BACKGROUND: NEAR INFRARED SPECTROSCOPY

Near infrared spectroscopy exploits the relative transparency of biological tissue to light between 700-1000nm. The most important tissue chromophores within this region are oxyhaemoglobin (HbO2), deoxyhaemoglobin (HHb) and oxidised cytochrome oxidase (CytOx), which display significantly different absorption characteristics, enabling spectroscopic measurements of their relative concentrations in the adult brain.[22] A major advantage of the technique is its non invasive and continuous nature - the light intensity levels used being well below those associated with damage. Systems are portable and measurements can be made easily and repeatedly, without the need for a dedicated laboratory. This makes the technique of potential use in outpatient clinics. Recent technical advances have led to the development of NIRS systems that can measure a variety of oxygenation and haemodynamic parameters in quantitative absolute units, with high temporal resolution and a degree of localisation. Many previous studies have demonstrated the suitability NIRS for measurements of functional activation in adults, particularly over the visual cortex.[23, 24]

4. PURPOSE OF THE STUDY

The purpose of the present study was to determine whether we would be able in normal volunteers to measure the haemodynamic response of the visual cortex to patterns of stripes, both those associated with illusions (*strong* patterns) and those of higher spatial frequency. According to the data obtained by fMRI BOLD,[3] any difference between the response to strong patterns and those of higher spatial frequency should be marginal in normal volunteers but greater in migraineurs.

5. METHODS

5.1. Participants

Six male and two female students and staff of the University of Essex aged 20-57 served as unpaid volunteers. All had normal or corrected-to-normal visual acuity. The volunteers gave written, informed consent, and the study was approved by the University of Essex Ethics committee.

5.2. Procedure

A commercially available NIRO 200 dual-channel spectrometer (Hamamatsu Photonics K.K) was used. Probes with interoptode spacing of 40mm incorporating the emitting and detecting fibres were fixed to the scalp. The optodes were placed with horizontal symmetry either side of two locations: O1 and 10mm above FP4 (measured according to the International 10-20 system of electrode placement.[25]). Optical attenuation data were acquired simultaneously from the frontal and occipital regions at 6Hz.

Horizontal square wave gratings, circular in outline, subtending 18 degrees, with spatial frequencies of 2.5 and 8.5 cpd (Michelson contrast >90%) were presented on the LCD TFT screen of a personal computer. The patterns were each presented for 27s with a 37s interval between presentations, during which the screen was uniformly illuminated with the same space-averaged luminance (48 cd.m^{-2}). Two test sessions were given with 5-20min rest between each. The patterns were presented in random order, 5 times each within a session, with the constraint that no pattern appeared more than twice in succession. The order of presentation in the second session was the reverse of that in the first. Each pattern was therefore presented a total of 10 times.

The recorded attenuation data were converted into changes in concentration of the chromophores HbO$_2$ and HHb.

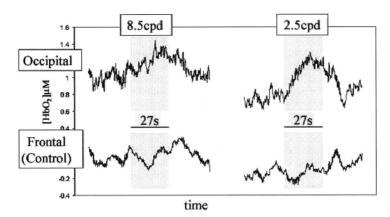

Figure 1. Micromolar oxyhaemglobin concentration changes (from arbitrary baseline) in response to a 8.5 cpd grating (left) and 2.5 cpd grating (right). Upper traces - probes over occipital region; lower traces - probes over frontal region (control). Average of ten sweeps in one representative participant. Periods of 27s stimulation (shown by horizontal bar) were preceded and followed by periods of 37s uniform screen (27s shown) with the same space-averaged reflectance as the stimulus.

6. RESULTS

Figure 1 shows, for a representative participant, the micromolar oxyhaemoglobin concentration changes over time (about an arbitrary baseline) before, during and after the presentation of a grating. The curves show the average of 10 sweeps and have been filtered using a running average of 1s. The curves on the left of the figure show the data for the grating with spatial frequency of 8.5 cycles per degree and the curves on the right the data for the *strong* grating with spatial frequency of 2.5 cycles per degree. The concentration changes are greater over the occipital region (upper curves) than over the frontal region (lower curves). Note the slow decrease in signal following the pattern offset.

Table 1 shows the mean micromolar oxyhaemglobin concentration changes during the 27s presentation of the patterns. The data in Table 1 are differences in the signal obtained during the stimulus and during the latter 27s of each of the 20 periods lasting 37s that immediately preceded the stimuli (during which the screen was a uniform grey). (The 10s immediately following a stimulus was not included to avoid the slow offset of the haemodynamic response.) Over the occipital region both grating stimuli gave a significant increase in signal compared with the grey screen (t(7)=1.88, p=0.05, one-tail, for the 8.5cpd grating; t(7)=2.44, p=0.02, one-tail, for the 2.5cpd grating). The differences over the frontal region, which served as a control for systemic effects, did not approach significance. The occipital response to the 2.5cpd grating was marginally larger than that to the 8.5cpd grating (t(7)=1.16, p=0.14, one-tail).

Note that the data analysed here are raw and unfiltered. The relatively high standard deviations reflect the large differences between individuals. Some of the differences may be due to variation in optode placement relative to cortical structures, but according to the fMRI BOLD findings, the variation may also reflect stable differences in the cortical haemodynamic response from one individual to another.

Table 1. Micromolar oxyhaemoglobin concentration changes.

Participant	Occipital region		Frontal region	
	8.5cpd	2.5cpd	8.5cpd	2.5cpd
1	0.122	0.090	-0.014	-0.020
2	0.102	0.993	-0.039	0.363
3	-0.042	0.007	-0.134	-0.336
4	0.079	-0.026	0.034	-0.309
5	0.303	0.173	0.105	-0.094
6	-0.014	0.237	-0.217	-0.081
7	0.125	0.329	0.040	-0.007
8	0.855	0.828	-0.114	-0.026
Average	0.191	0.329	-0.042	-0.064
SD	(0.288)	(0.380)	(0.107)	(0.215)

7. DISCUSSION

The change in micromolar oxyhaemglobin concentration as a result of a grating stimulus was clearly measured using NIRS. As anticipated on the basis of fMRI BOLD findings,[3] the response to the two gratings was similar, though slightly larger for the *strong* stimulus. It now remains to be determined, in a much larger cohort of subjects, whether the individual differences can be explained at least in part by diagnostic category, and whether the differences between the pattern stimuli will be greater in participants with migraine, as the fMRI BOLD findings would predict. The present findings indicate that NIRS has the potential to provide a non-invasive, portable and inexpensive monitor of visual disturbance in migraine.

8. REFERENCES

1. Hu, XH, *et al.*, Burden of migraine in the United States: disability and economic costs. *Archives of Internal Medicine.*, 1999. **159**(8): p. 813-8.
2. Gerth, WC, *et al.*, The multinational impact of migraine symptoms on healthcare utilisation and work loss. *Pharmacoeconomics*, 2001. **19**(2): p. 197-206.
3. Huang, J, *et al.*, Visual distortion provoked by a stimulus in migraine associated with hyperneuronal activity. *Headache*, 2003. **43**(6): p. 664-71.
4. De Valois, RL, *et al.*, *Spatial Vision.*(Oxford University Press: Oxford. 1990.
5. Chronicle, EP, *et al.*, Gratings that induce distortions mask superimposed targets. *Perception*, 1996. **25**: p. 661-8.
6. Plant, GT, *et al.*, Transient visually evoked potentials to the pattern reversal and onset of sinusoidal gratings. *Electroencephalography and Clinical Neurophysiology*, 1983. **52**(2): p. 147-58.
7. Singh, KD, *et al.*, Spatiotemporal frequency and direction sensitivities of human visual areas measured using fMRI. *Neuroimage*, 2000. **12**(5): p. 550-64.
8. Wilkins, AJ, *et al.*, A neurological basis for visual discomfort. *Brain*, 1984. **107**: p. 989-1017.
9. Nulty, D, *et al.*, Mood, pattern sensitivity and headache: a longitudinal study. *Psychol. Med*, 1987. **17**: p. 705-13.
10. Marcus, DA, *et al.*, Migraine and stripe-induced visual discomfort. *Achives of Neurology*, 1989. **46**(10): p. 1129-32.
11. Wilkins, AJ, *et al.*, Neurophysiological aspects of pattern-sensitive epilepsy. *Brain*, 1979. **102**: p. 1-25.
12. Wilkins, AJ, *et al.*, Visually-induced seizures. *Progress in Neurobiology*, 1980. **15**: p. 86-117.
13. Welch, KM, Contemporary concepts of migraine pathogenesis. *Neurology*, 2003. **61**((Suppl 4)): p. S2-S8.
14. Lipton, RB, *et al.*, Comorbidity of migraine: the connection between migraine and epilepsy. *Neurology*, 1994. **44**(10 Suppl 7): p. S28-S32.
15. Diener, HC, *et al.*, Advances in pharmacological treatment of migraine. *Expert Opin. Investig Drugs*, 2001. **10**(10): p. 1831-45.
16. Aurora, SK, *et al.*, Brain excitability in migraine: evidence from transcranial magnetic stimulation studies. *Curr. Opin. Neurol.*, 1998. **113**(Jun): p. 205-9.
17. Schoenen, J, Deficient habituation of evoked cortical potentials in migraine: a link betgween brain biology, behavior and trigeminovascular activations? *Biomed. Pharmocother.*, 1996. **50**: p. 71-8.
18. Kropp, P, *et al.*, Contingent negative variation during migraine attack and interval: evidence for nomalization of slow cortical potentials during the attack. *Cephalalgia*, 1995. **15**: p. 123-8.
19. Hay, KM, *et al.*, 1044 women with migraine: the effect of environmental stimuli. *Headache*, 1994. **34**(3): p. 166-8.
20. Wilkins, AJ, *et al.*, Fluorescent lighting, headaches and eye-strain. *Lighting Research and Technology*, 1989. **21**(1): p. 11-8.
21. Wilkins, AJ, *et al.*, Tinted spectacles and visually sensitive migraine. *Cephalalgia*, 2002. **22**: p. 711-9.
22. Elwell, CE,Measurement and data analysis techniques for the investigation of adult cerebral haemodynamics using near infrared spectroscopy.,PhD thesis,University of London,1995
23. Meek, JH, *et al.*, Regional changes in cerebral haemodynamics due to visual stimulus measured by near infrared spectroscopy. *Proc. R. Soc. Lond. B. Biol. Sci.*, 1995. **261**: p. 351-6.
24. Villringer, A, *et al.*, Near infrared spectroscopy (NIRS) - a new tool to study hemodynamic changes during activation of brain function in human adults. *Neuroscience Letters*, 1993. **154**: p. 101-4.
25. Klem, GH, *et al.*, The ten-twenty electrode system of the International Federation. *Electroencephalogr Clin Neurophysiol, Supplement*, 1999. **52**: p. 3-6.

34.

INVESTIGATION OF OXYGEN SATURATION DERIVED FROM CARDIAC PULSATIONS MEASURED ON THE ADULT HEAD USING NIR SPECTROSCOPY

Terence S. Leung[*], Ilias Tachtsidis[*], Praideepan Velayuthan*, Caroline Oliver[#], Julian R. Henty[*], Holly Jones[#], Martin Smith[#], Clare E. Elwell[*], and David T. Delpy[*]

1. INTRODUCTION

Cardiac related pulsatile signals can be detected in different parts of the human body, including the finger, ear lobe and forehead [1] by using near infrared (NIR) monitoring. These pulsatile signals are due to attenuation of light by the increase of arterial blood volume during systole in the cardiac cycle. Pulse oximetry exploits these pulsatile signals to calculate oxygen saturation (S_pO_2) [2]. There are two types of pulse oximetry: (1) the transmission type where the light source and detector are facing each other across the measurement site (e.g. ear lobe or finger), and (2) the reflectance type where both the light source and detector are in the same plane (e.g. forehead or forearm) [1] with the source detector spacing typically less than 1 cm. With more sensitive optical instruments, it has been shown that these pulsatile signals can be measured on the forehead at a greater spacing using either a CCD spectrometer [3] or a phase resolved system [4]. Both of these two methods can be considered as operating in reflectance mode with large source detector spacings of 3 or 3.5 cm. These pulsatile signals were also thought to be mainly caused by the change in arterial blood volume. However, at source detector spacing larger than ~2 cm, NIR light can penetrate through the skull into the brain and the measured pulsatile signals are likely partly to include components caused by brain movement [5] as well as arterial blood pulsations in the scalp and the brain. The objective of this paper is to compare three algorithms used to calculate oxygen saturation from the head pulsatile signals, (signified by SpO_2^h to distinguish it from SpO_2 measured at other sites) with large source detector spacing. Two of the algorithms implicitly allow the possibility of venous blood contributing to SpO_2^h. We will show the SpO_2^h calculated by three algorithms in 8 adult subjects during normoxia and hypoxia. Examples of phase differences between the oxy and deoxy-haemoglobin (ΔHbO_2 and ΔHHb) signals, which could imply a venous contribution, will also be presented.

[*]Department of Medical Physics & Bioengineering, University College London, London, WC1E 6JA, U.K.
[#]Dept. of Neuroanaesthesia, The Nat. Hosp. for Neurology & Neurosurgery

2. METHODS

2.1 Experimental Study

Eight adult subjects (mean age 31 ± 3 years) participated in this study which was approved by the UCL Hospital Ethics Committee. An optical probe with a source detector spacing of 3.5 cm was placed on the left side of the foreheads of subjects. The light source was provided by a tungsten halogen lamp (Model 77501, Oriel Instruments) via an optical fibre bundle. The transmitted light was collected by another fibre bundle linked to an imaging spectrograph (SPEX 270M, JY Optical Systems Instruments SA, Inc.) which dispersed the light on to a cooled CCD detector (Wright Instruments). Intensity spectra were collected between 670 and 990 nm with a spectral resolution of 5 nm and exposure time of 50 ms. A pulse oximeter probe (Novametrix 500) operating in beat-to-beat mode was attached to subjects' ear lobes to monitor SpO_2. In the first part of the experiment when the subjects were resting, they breathed room air through a face mask. In the second part of the experiment, the fraction of inspired oxygen (FiO_2) was reduced from 21% to 10-15% such that SpO_2 fell from 98% to 90%. At this point the subjects were given 100% oxygen for 5 breaths which caused SpO_2 to rapidly return to 98-100%, then they returned to air breathing. The same manoeuvre was repeated three times.

2.2 The Three Algorithms to Calculate $SpO_2{}^h$

The intensity spectra were converted to attenuation spectra with respect to a reference spectrum which was the average of 200 spectra (10 seconds) and further smoothed by a 3rd order Savitsky-Golay filter. The attenuation spectra calculated in this way can be interpreted as changes in attenuation $\Delta A(\lambda)$ from a nominal baseline and were used in the following analyses. For the conversion from $\Delta A(\lambda)$ to ΔHbO_2 and ΔHHb, the wavelength range $746 - 906$ nm was used. The wavelength dependence of the differential pathlength was taken into account. Of the three methods described below, method B and C were developed by the authors.

Method A   A flow diagram for this method is shown in Fig.1(a). This method is that used by [3,4] who calculated $SpO_2{}^h$ using data from a CCD multi-wavelength spectrometer and a phase resolved system, respectively. At each wavelength λ_i, $\Delta A(\lambda_i, t_j)$ a block of 256 samples (12.8 s) was Fourier transformed (FT) and the subsequent FTs were carried out at 1 s time intervals. For each FT spectrum, the spectral peak around the heart rate frequency (~ 1 Hz) and its 1st harmonic were identified and the sum of their magnitudes, which was taken as the energy of the cardiac pulsations, over a bandwidth of 0.8 Hz was calculated. With the sum of magnitudes calculated at all wavelengths, an attenuation spectrum for the cardiac pulsations was thus obtained, i.e. $\Delta A_p(\lambda)$ which was then converted to ΔHbO_2 and ΔHHb by least square fitting to the specific extinction coefficient spectra. The $SpO_2{}^h$ was approximated by:

$$SpO_2{}^h = \frac{\Delta HbO_2}{\Delta HbO_2 + \Delta HHb} \times 100\% \tag{1}$$

The $SpO_2{}^h$ calculated is therefore a running average over 12.8 s (256 samples).

(a) Method A

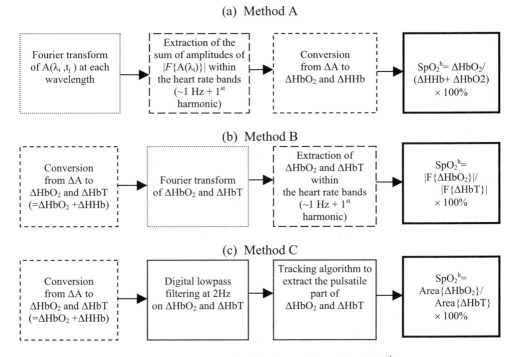

Figure 1. The three methods implemented to calculate SpO$_2$^{h}

Method B   As shown in Fig.1(b), the basic elements are the same as method A but in a different sequence. For each $\Delta A(\lambda,t_j)$ spectrum at t_j, one value of ΔHbO_2 and ΔHHb were obtained respectively by least square fitting. A FT was then performed on 256 samples of ΔHbO_2 and ΔHbT (= ΔHbO_2 + ΔHHb) separately with the time interval between subsequent FTs of 1 s. The corresponding spectral peaks around the heart rate frequency were identified as well as its first harmonic. The sums of the magnitudes around the spectral peaks were then calculated which were regarded as the energy of the cardiac pulsations in ΔHbO_2 and ΔHbT. The SpO$_2$^{h} was calculated as

$$SpO_2^{\ h} = \frac{\sum_i \left| F\{\Delta HbO_2\} \right|_i}{\sum_i \left| F\{\Delta HbT\} \right|_i} \times 100\% \tag{2}$$

where $F\{\ \}$ signifies the FT and i includes the indexes of the frequency bins within the heart rate bandwidth and that of its 1st harmonic. As in Method A, the calculated SpO$_2$^{h} is a running average over 12.8 s.

Method C   As depicted in Fig.1(c), each $\Delta A(\lambda,t_j)$ spectrum at t_j was first converted to one value of ΔHbO_2 and ΔHHb, respectively. Subsequently, a digital lowpass filter (5th order Butterworth with a cut-off frequency of 2Hz) was used to remove high frequency noise from the ΔHbO_2 and ΔHbT signals. A tracking algorithm based on trough detection and cubic spline fitting was implemented to track the baseline movement and to isolate the pulsatile component of the ΔHbO_2 and ΔHbT signals. The SpO$_2$^{h} was then calculated on a beat-to-beat basis from the areas under each cardiac pulsation in ΔHbO_2 and ΔHbT:

$$SpO_2{}^h = \frac{Area\{\Delta HbO_2\}}{Area\{\Delta HbT\}} \times 100\% \qquad (3)$$

4. RESULTS

4.1 Normoxia

During normoxia, 60 seconds worth of data were analysed for each subject. The individual means ± standard deviations of $SpO_2{}^h$ for all 8 subjects calculated by methods A, B and C are summarised in Fig.2 alongside the group means ± standard deviations among the mean $SpO_2{}^h$ in 8 subjects, i.e. 90.0 ± 5.8 % (method A), 100.0 ± 9.1 % (method B) and 94.0 ± 9.5 % (method C). While the means of method A and B, and those of method B and C are statistically different (p<0.05), those of method A and C are not. The SpO_2 measured by the pulse oximeter was 98% for all subjects.

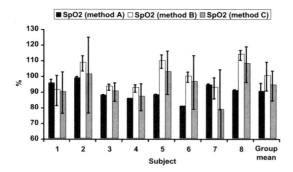

Figure 2. Individual and group means ± standard deviations of $SpO_2{}^h$ calculated by methods A, B and C

4.2 Hypoxia

An example of $SpO_2{}^h$ calculated by methods A, B and C, together with SpO_2 measured by the pulse oximeter as FiO_2 was changed is shown in Fig.3. The correlations between SpO_2 measured by the pulse oximeter and $SpO_2{}^h$ calculated by method A, B and C during approximately 700 seconds of FiO_2 variations are summarised in Table 1. For comparison purpose, $SpO_2{}^h$ calculated by method C and SpO_2 measured by the pulse oximeter have been averaged for 12.8s (same as methods A and B).

Table 1. Correlation coefficient (r) between SpO_2 measured by the pulse oximeter (P.O.) and $SpO_2{}^h$ calculated by methods A, B and C ($\sqrt{}$ for p<0.005)

Subject	r for Method A (p < 0.005)	r for Method B (p < 0.005)	r for Method C (p < 0.005)
1	0.42 ($\sqrt{}$)	0.31 ($\sqrt{}$)	0.66 ($\sqrt{}$)
2	0.67 ($\sqrt{}$)	0.58 ($\sqrt{}$)	0.73 ($\sqrt{}$)
3	0.49 ($\sqrt{}$)	0.37 ($\sqrt{}$)	0.66 ($\sqrt{}$)
4	0.18 ($\sqrt{}$)	0.45 ($\sqrt{}$)	0.68 ($\sqrt{}$)
5	0.35 ($\sqrt{}$)	0.45 ($\sqrt{}$)	0.50 ($\sqrt{}$)
6	0.56 ($\sqrt{}$)	0.31 ($\sqrt{}$)	0.48 ($\sqrt{}$)
7	-0.08 ($\times$)	0.05 ($\times$)	0.20 ($\times$)
8	0.29 ($\sqrt{}$)	0.64 ($\sqrt{}$)	0.76 ($\sqrt{}$)

5. DISCUSSION AND CONCLUSIONS

Method A involves performing FTs and taking the magnitudes of $\Delta A(\lambda)$ at each wavelength λ as the first procedure. Since magnitudes contain no information about the phase at each signal, method A intrinsically ignores any phase difference between ΔHbO_2 and ΔHHb, i.e. it assumes that both ΔHbO_2 and ΔHHb are in phase. This assumption is reasonable when the pulsatile signals are solely due to a change in arterial blood volume (a central assumption of the work carried out in [3,4]). The final values of ΔHbO_2 and ΔHHb are both mainly positive leading in most cases to $SpO_2^h < 100\%$. Each $\Delta A(\lambda,t_j)$ spectrum is subject to certain instrumental noise and using 256 $\Delta A(\lambda,t_j)$ spectra to calculate one value of ΔHbO_2 and ΔHHb improves the signal-to-noise ratio (SNR). This is shown in Fig.2 where SpO_2^h calculated by method A has the smallest inter and intra subject standard deviation of the three methods. Method B converts $\Delta A(\lambda)$ to ΔHbO_2 and ΔHHb as a first step and hence preserves the phase difference between ΔHbO_2 and ΔHHb. Since each value of ΔHbO_2 and ΔHHb is calculated from only one $\Delta A(\lambda)$ spectrum, the results are more noisy than those of method A. Each SpO_2^h is calculated from the FT of 256 samples of ΔHbO_2 and ΔHHb and thus can be considered as rolling average of 12.8 s. Method C also preserves the phase difference between the ΔHbO_2 and ΔHHb signals but suffers the same SNR problem as method B due to the use of only one $\Delta A(\lambda)$ spectrum for the conversion. Method C calculates SpO_2^h on a beat-to-beat basis but the algorithm may introduce certain errors while isolating the pulsatile part of ΔHbO_2 and ΔHHb. These errors contribute to the high inter and intra subject variation in the SpO_2^h values calculated by this method (Fig 2). With a rolling average of 12.8s, however, the resulting SpO_2^h have, in most cases, the highest correlation with SpO_2 from the pulse oximeter. From the results of the normoxia and hypoxia studies, method A seems to offer the most reasonable results; SpO_2^h has the lowest standard deviation and is generally less than 100% which fits with the idea of SpO_2^h being a ratio defined as $\Delta HbO_2 / \Delta HbT$ and $\Delta HbT > \Delta HbO_2$. Using method C for beat-to-beat analysis however, we found several instances where ΔHbO_2 and ΔHHb were out of phase and occasionally going in opposite directions (180° out of phase) during a cardiac cycle as shown in Fig.4(b). As a result, the ΔHbT is smaller than ΔHbO_2 and hence $SpO_2^h > 100\%$. This also happens in Method B which preserves phase change. This kind of phase change is unlikely to be caused by a change in arterial blood volume. Arterial HHb cannot decrease by more than 2% in a cardiac cycle (normal SpO_2 varies between 98 and 100%) and yet we see a larger drop (14% drop in ΔHHb with respect to ΔHbO_2) from Fig.4(b). Venous ΔHHb, on the other hand, can decrease noticeably during a cardiac cycle. Two possible mechanisms for the observed changes are: (1) the arterial to venous volume ratio increases during a cardiac cycle because the arterial compartment expands and presses against the venous compartment during systole, (2) the small rise in cerebral blood flow during systole increases the proportion of ΔHbO_2 (and hence reduces the proportion of ΔHHb) in the venous compartment. During normoxia the mean SpO_2^h calculated by methods A, B and C were different from the arterial SpO_2 measured by the pulse oximeter. Also, during hypoxia, SpO_2^h calculated by methods A and B were only partially correlated to the arterial SpO_2 as shown in Table 1 suggesting that the arterial contribution to SpO_2^h is only partial. To conclude, method A is only suitable if one is prepared to assume that the pulsatile signals from the head are totally arterial. Otherwise, other algorithms which preserve phase changes between ΔHbO_2 and ΔHHb, such as methods B and C should be used.

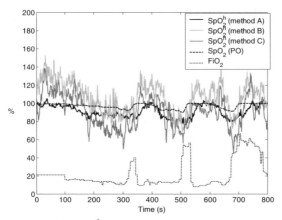

Figure 3. Changing FiO_2: examples of SpO_2^h calculated by methods A, B & C, and SpO_2 measured by the pulse oximeter

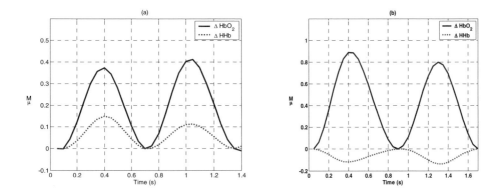

Figure 4. Examples of ΔHbO_2 and ΔHHb during a cardiac cycle: (a) both ΔHbO_2 and ΔHHb are positive, (b) ΔHbO_2 is positive but ΔHHb is negative

6. ACKNOWLEDGEMENTS

The authors would like to thank all the volunteers who participated in this study, and the Wellcome Trust and EPSRC who funded this work.

7. REFERENCES

1. D.G.Clayton, R.K.Webb, A.C.Ralston, D.Duthie, W.B.Runciman, Pulse oximeter probes. A comparison between finger, nose, ear and forehead probes under conditions of poor perfusion, *Anaesthesia*, **46**(4), 260-5 (1991)
2. I.Yoshiya, Y.Shimada, K.Tanaka, Spectrophotometric monitoring of arterial oxygen saturation in the fingertip, *Med Biol Eng Comput,* **18**, 27-32 (1980)

3. M.Kohl, C.Nolte, H.R.Heekeren, S.Horst, U.Scholz, H.Obrig, and A.Villringer, Determination of the wavelength dependence of the differential pathlength factor from near infrared pulse signals, *Physics in Medicine and Biology,* **43**, 1771-1782 (1998)

4. M.A.Franceschini, E.Gratton, S.Fantini, Noninvasive optical method of measuring tissue and arterial saturation: an application to absoulte pulse oximetry of the brain, *Optics Letter*, **24**(12), 829-831 (1999)

5. M.Firbank, E.Okada, D.T.Delpy, A theoretical study of the signal contribution of regions of the adult head to near infrared spectroscopy studies of visual evoked responses, *Neuroimaging* **8**, 69-78 (1998)

35.

A WIDE GAP SECOND DERIVATIVE NIR SPECTROSCOPIC METHOD FOR MEASURING TISSUE HEMOGLOBIN OXYGEN SATURATION

Dean E. Myers, Chris E. Cooper, Greg J. Beilman, John D. Mowlem, LeAnn D. Anderson, Roxanne P. Seifert, and Joseph P. Ortner *

1. INTRODUCTION

Absolute quantification of tissue chromophores can, in principle, be obtained by applying second derivative methods[1-4] to near-infrared spectroscopy. These methods transform optical attenuation measurements into second derivative units in order to provide robustness to wavelength dependent scattering, chromophores with constant absorption, baseline/instrumentation drift, and movement artifact.

Previous methods are more sensitive to deoxyhemoglobin than oxyhemoglobin because oxyhemoglobin lacks significant absorbance curvature within the wavelength region used to calculate a second derivative attenuation value. These methods have relied on broad spectrum wavelength measurements, spectral smoothing and component fitting. Excessive noise at high hemoglobin oxygen saturation is a problem.

A simple four wavelength method for quantifying tissue hemoglobin oxygen saturation (StO_2) having greatest spectral sensitivity occurring at high hemoglobin oxygen saturation (>50%) is described. Measurement performance was tested in isolated blood, isolated blood perfused animal organs, and healthy human volunteers with induced limb ischemia.

* Dean E. Myers, LeAnn D. Anderson, Roxanne P. Seifert, and Joseph P. Ortner all of Hutchinson Technology Inc., Hutchinson, MN 55350, USA. Chris E. Cooper, Department of Biological Sciences, Central Campus, University of Essex, Wivenhoe Park, Colchester CO4 3SQ, Essek, UK. Greg J. Beilman, Department of Surgery, University of Minnesota, Minneapolis, MN 55455, USA. John D. Mowlem, Emergency Services, Hutchinson Community Hospital, Hutchinson, MN 55350, USA.

2. METHODS

2.1 Equipment

Two spectrometer designs were used. A first full spectrum spectrometer, Biospectrometer-NB (Hutchinson Technology Inc, Hutchinson, MN), consisted of a tungsten halogen light source and a reflective grating model SD2000 CCD array photodetector (Ocean Optics, Tampa Bay, FL). Six 400 micron glass receive fibers were coupled to the grating in a slit pattern and resulted in a bandwidth resolution of 15 nm full width half maximum (FWHM). Fiber optic reflectance probes had numerous 200 micron optical send fibers to provide sufficient illumination of the tissue for probe spacings of 8 mm and 15 mm. Known transmission peaks of didymium glass were used to calibrate the CCD array pixels to a corresponding wavelength.

A second commercially available spectrometer, InSpectra^{TM} Tissue Spectrometer Model 325 (Hutchinson Technology Inc, Hutchinson, MN), consisted of four photomultiplier tube detectors coupled to interference filters having center wavelengths of 680, 720, 760, and 800 nm. All filters had a bandwidth of 10 nm FWHM. Reflectance probes consisted of a single 400 micron glass optical receive fiber and four similar fibers each coupled to a light emitting diode. A 12 inch length of 1.5 mm diameter plastic optical fiber coupled and mixed each fiber light source to the measurement sample. Probe spacings of 12 mm, 15 mm, 20 mm, and 25 mm were used.

A closed cell polyethylene foam light scattering calibrator, Plastazote® LD45 (Zotefoams Inc, Walton, KY), was used to capture reference light intensity at each wavelength prior to placing a reflectance probe on the tissue measurement sites.

An isolated blood perfusion apparatus (Figure 1) consisted of a Minimax^{TM} 3381 fiber membrane oxygenator (Medtronic, Minneapolis, MN) connected to a peristaltic pump and controlled temperature water bath. A custom flow cell, to mimic blood-perfused tissue, distributed the blood within the optical path of the measurement probes and was constructed to provide a variable thickness of blood flowing over a block of LD45 Plastazote foam of sufficient volume to contain all emitted light paths. Two 0.05 mm thick Mylar D plastic film layers, (DuPont Teijin Films, Hopewell, VA) separated flowing blood from the top mounted reflectance probes and the bottom mounted foam scattering layer. A co-oximeter was used to measure %SO$_2$ for sampled blood.

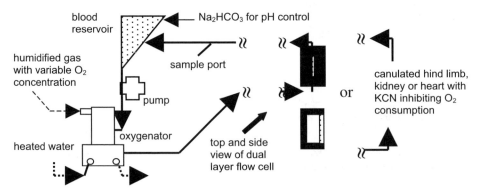

Figure 1. Schematic of isolated blood perfusion circuit for testing StO$_2$ performance.

2.2 StO$_2$ Algorithm

Tissue attenuation (A) measurements were calculated as log (reference intensity / sample intensity) at 680, 720, 760, and 800 nm. A wide 40 nm gap second derivative transformation (Figure 2) was used to obtain 2nd derivative attenuation measurements at 720 and 760 nm. These two 2nd derivative attenuation signals are related to the four measured attenuation wavelengths as follows:

$$2D_{720} = A_{760} - 2A_{720} + A_{680} \tag{1}$$

$$2D_{760} = A_{800} - 2A_{760} + A_{720} \tag{2}$$

For each tissue spectrum measurement a scaled $2D_{720}$ value was used to predict tissue %StO$_2$ from a predetermined empirical calibration relationship:

$$\text{scaled } 2D_{720} = 2D_{720} \ / \ 2D_{760} \tag{3}$$

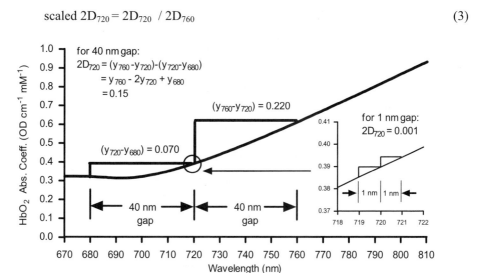

Figure 2. An example of how Eq. (1) is obtained. A 40 nm wavelength gap specifically amplifies the contribution of HbO$_2$ concentration to the 720 nm second derivative attenuation ($2D_{720}$). The wider wavelength gap spans the non-linear region of published HbO$_2$ absorbance. [5]

2.3 Performance Evaluation of StO$_2$

Scaled $2D_{720}$ (Eq. 3) was empirically correlated to blood hemoglobin oxygen saturation with the dual layer flowcell apparatus. The same apparatus was then used to test multiple devices with different probe spacings at different ranges of hemoglobin concentration and blood thickness. A diffusion theory light transport model for a single layer infinite slab[6] was used to verify the in vitro calibration curve. Isolated blood perfused animal organs were titrated with potassium cyanide to maintain the arterial/venous %SO$_2$ gradient to a difference of 5 or less %SO$_2$ units. Human limb ischemia was induced via inflation of a pneumatic tourniquet 50 mmHg above systolic blood pressure.

3. RESULTS AND DISCUSSION

Figure 3 shows that a wide 40 nm gap provides a scaled 2nd derivative spectrum with improved sensitivity to hemoglobin oxygen saturation. Figure 4 illustrates that the empirically derived calibration curve was dependent upon spectrometer optical resolution and agreed with modeled results from diffusion theory. The calibration curve (common to probe spacing) was robust to changes in blood layer thickness and total hemoglobin concentration. Figure 5 shows the correlation results comparing $\%StO_2$ to weighted arterial and venous $\%SO_2$ for the animal models tested. Mean error relative to the co-oximeter was +3.3, +10.1, and +6.7 $\%SO_2$ units respectively for hind limb, heart, and kidney. The described StO_2 method readily detected limb ischemia in 13 male and 13 female human volunteers for thenar hand and dorsal forearm muscle sites (Figure 6).

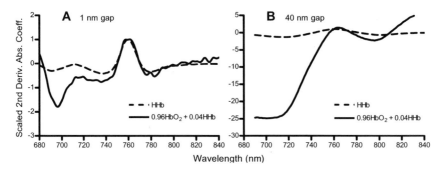

Figure 3. Transmitted pure HbO_2 and HHb spectra from published source[5]: **A)** 1 nm gap and corresponding **B)** 40 nm gap 2nd derivative absorption coefficient scaled to 760 nm. The 1 nm gap spectra required best fit smoothing to provide visually presentable 2nd derivative spectra. The 40 nm gap derivative spectrum does not require smoothing and is significantly more sensitive to oxyhemoglobin concentration.

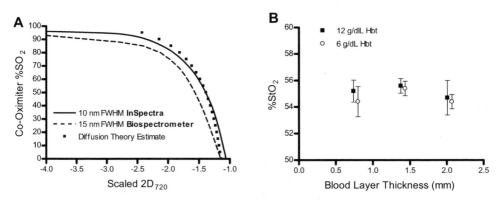

Figure 4. Isolated blood circuit experiments with the dual layer flowcell. **A)** Calibrated relationship between co-oximeter $\%SO_2$ and scaled $2D_{720}$. Average spectral relationship obtained at 8.5 g/dL Hbt with four InSpectra spectrometers using 12, 15, 20, and 25 mm probe spacings. Input parameters for diffusion theory estimate of the calibration curve were: 20 mm spacing, leg scattering profile[7], 0.15 mM Hbt, and 70% water. **B)** Effect of variable blood layer thickness and hemoglobin concentration on InSpectra $\%StO_2$ with blood controlled to 55 $\%SO_2$. The data represents mean +/- 1 std for five measurements at each condition measured with a 25 mm probe. To clarify results, the 12 g/dL and 6 g/dL Hbt summary data are slightly shifted along the x-axis.

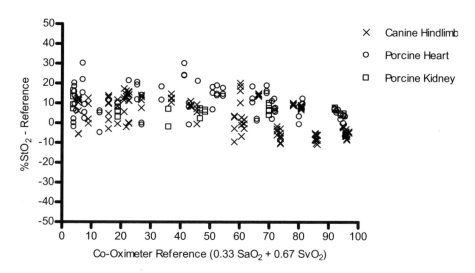

Figure 5. Correlation results for isolated blood perfused organs titrated with cyanide to maintain $\%SvO_2$ to within 5 units of the manipulated $\%SaO_2$. Biospectrometer-NB StO_2 results are from two isolated canine limb experiments using four spectrometers with 15 mm probes to obtain 34 paired measurements. Six isolated porcine hearts were measured with an 8 mm probe connected to a Biospectrometer-NB spectrometer to obtain 113 paired measurements. For two isolated kidneys, 25 paired readings were obtained with the same spectrometer system used for the isolated hearts.

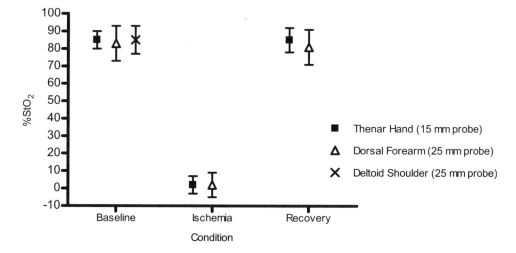

Figure 6. StO_2 measurements for 13 male and 13 female normotensive adult human volunteers with induced limb ischemia (only baseline data collected at deltoid site). Mean with ± 1 standard deviation limits for 26 individuals. Baseline and recovery were statistically significant from ischemia ($p < 0.01$) for thenar and dorsal sites. Three discrete wavelength spectrometers (InSpectra) were used for all probe spacings and tissue sites. Arterial cuff ischemia was less than 15 minutes to obtain approximate nadir StO_2 condition.

4. CONCLUSIONS

We describe a four wavelength continuous wave near-infrared algorithm for estimating hemoglobin oxygen saturation in tissue using single depth attenuation measurements. The 40 nm gap second derivative algorithm and in vitro calibration method provided StO_2 measurements of reasonable accuracy that were applied across a variety of tissues and probe spacings with no measured or assumed values for optical pathlength or optical scattering.

5. FUTURE WORK

Additional studies are planned to determine how robust the measurement method is to scattering changes and variable concentration of additional tissue absorbers including carboxyhemoglobin, methemoglobin, and water in relation to the total amount of hemoglobin present in the measurement sample.

6. FURTHER DETAILS

Address all correspondence to: Dean E. Myers, Principal Development Engineer, Hutchinson Technology Inc., 40 West Highland Park Drive NE, Hutchinson, MN 55350, USA. Tel: 320-587-1732; Fax: 320-587-1555; E-mail: dean.myers@hti.htch.com

7. ACKNOWLEDGEMENTS

Chris Cooper is grateful for a MRC Discipline Hopping Award. Hutchinson Technology Inc. funded all research activities including cooperative consultancies between the various author affiliations.

8. REFERENCES

1. S. J. Matcher and C. E. Cooper, Absolute quantification of deoxyhemoglobin concentration in tissue near infrared spectroscopy, *Phys Med Biol* **39**, 1295-1312 (1994).
2. C. E. Cooper, C. E. Elwell, J. H. Meek, S. J. Matcher, J. S. Wyatt, M. Cope, and D. T. Delpy, The noninvasive measurement of absolute cerebral deoxyhemoglobin concentration and mean optical path length in the neonatal brain by second derivative near infrared spectroscopy, *Pediatr Res* **39**, 32-8 (1996).
3. K. A. Schenkman, D. R. Marble, D. H. Burns, and E. O Feigl, Optical spectroscopic method for in vivo measurement of cardiac myoglobin oxygen saturation, *Applied Spectroscopy* **53**(3)**,** 332-338 (1999).
4. T. Binzoni, T. Leung, V. Hollis, S. Bianchi, J. H. Fasel, H. Bounameaux, E. Hiltbrand, and D. Delpy, Human tibia bone marrow: defining a model for the study of haemodynamics as a function of age by near infrared spectroscopy, *J Physiol Anthropol Appl Human Sci* **22**, 211-218 (2003).
5. S. J. Matcher, C. E. Elwell, C. E. Cooper, M. Cope, and D. T. Delpy, Performance comparison of several published tissue near-infrared spectroscopy algorithms, *Anal Biochem* **227**, 54-68 (1995).
6. S. R. Arridge, M. Cope, and D. T. Delpy, The theoretical basis for the determination of optical pathlengths in tissue: temporal and frequency analysis, Phys *Med Biol* **37**(7)**,** 1531-1560 (1992).
7. S. J. Matcher, P. Kirkpatrick, K. Nahid, M. Cope, and D. T. Delpy, Absolute quantification methods in tissue near infrared spectroscopy, *Proc SPIE* **2389**, 486-495 (1995).

36.

SIMULATION OF Mb/Hb IN NIRS AND OXYGEN GRADIENT IN THE HUMAN AND CANINE SKELETAL MUSCLES USING H-NMR AND NIRS

Shoko Nioka, Dah-Jyuu Wang, Joohee Im, Takahumi Hamaoka, Zhiyue J. Wang, John S. Leigh and Britton Chance*

1. INTRODUCTION

There has been great interest in the role of the oxygen carrying functions of myoglobin (Mb) and hemoglobin (Hb) in the skeletal muscles[1] which are thought to be associated with muscle performance[2], fiber types[3] and capillary density[4], indicating proportional use of oxygen carrying functions with oxygen demands. Mb and Hb have been shown to directly influence capillary, extra cellular and intracellular PO_2[5-6] to various degrees in different animals and organs[7-8]. Canine Gastrocnemius muscles consist of mostly fatigue resistant fibers and have a very high capillary density and high myoglobin[9], compared to human skeletal muscles[10]. They had a greater endurance to Mb desaturation in our pilot study, while human muscle exhibited about 50% of desaturation in the light exercise[11-12], and ischemia[13] using ^{1}H-NMR. In the ^{1}H-NMR, the deoxy Hb signal is shifted about 3 ppm from the Mb peak; however because of low visibility of the Hb signal, the contamination of Hb is negligible[14]. On the other hand, Near Infrared Spectroscopy (NIRS) can not distinguish between Hb and Mb because of their overlapped absorbance characteristics[15]. Thus the ^{1}H-NMR Mb determination is ideal to help distinguish Hb and Mb in the NIRS signal by simulating Hb/Mb contribution to study oxygen gradients between capillary and myocytes. We have demonstrated possible Hb/Mb ratio as well as PO_2 in the capillary and myocytes in dogs and humans.

* Dept. of Biochemistry and Radiology, University of Pennsylvania, DJW is presently in the Department of Radiology, The Children's Hospital of Philadelphia; Dr. Hamaoka is in the Dept. of Preventive Med and Public Health, Tokyo Med College; ZJW is presently in the Department of Diagnostic Imaging, Texas Children's Hospital, Baylor College of Medicine.

2. MATERIALS AND METHODS

The canine and the human protocols were approved by the University of Pennsylvania laboratory animal research committee. Four mongrel dogs (ca. 10kg) were anesthetized with pentobarbital, and the Achilles tendon was tied to a string gauge, the force measurement. The calf muscle was stimulated at a sciatic nerve to yield a submaximal force development. After obtaining a resting period baseline, animals received submaximal stimulation with 10% hypoxia for 3-6 minutes. Thereafter, the FiO2 was gradually reduced to 0% anoxia while stimulation was maintained. In the human study, the arm was cuffed about 10 minutes either in the magnet for the NMR study, or outside the magnet for the NIRS measurement. The forearm was positioned at the same level as the heart to avoid blood volume changes due to a static pressure effect.

The NMR and NIRS data acquisitions were simultaneously carried out in a 40 cm bore 2.8 Tesla magnet in the dog study. The pulse sequence used for ^{1}HNMR was similar to that reported previously[15]. A frequency domain NIRS (PM200, NIM Inc., Philadelphia) was used to measure HB + Mb signals. In brief, absorption coefficients of two wavelengths were obtained and used to calculate tissue oxygen saturation in the dog study. In the human study, CW NIRS imager using LED was used (NIM Inc.), and absolute saturation was calibrated based on the resting value of 50%[15] and 0% at the end of cuff ischemia.

Simulation of the Hb/Mb ratio in the NIRS was carried out to predict apparent Hb saturation using Hb and Mb P_{50} of 26.6, and 3.2 torr[16], and Hills coefficients of 2.7 and 1 respectively. Molar equivalent of Hb, Mb for optical absorption coefficients in our wavelengths is 4:1. NIRS measures changes in oxygen saturation in the oxygen transferable blood volume during the experiments. This is the compartment of our interest, and we assumed that saturation ranged from 50% at rest[15] to 0% at the end of each experiment, when there is no longer oxygen available (anoxia). The resting myocyte PO2 ($P_{myo}O_2$) is predicted as 18 torr[17]. The apparent Hb saturation, $S_{Hb}O_2$ is described as;

$$S_{Hb}O_2 \;=\; (1/(4\,f_{Hb}) + 3/4)\cdot S_{NIR}O_2 - (1/(4\,f_{Hb}) - 1/4)\cdot S_{Mb}O_2 \tag{1}$$

Where $S_{NIR}O_2$, $S_{Mb}O_2$ are saturations obtained from NIRS and NMR experimentally. f_{Hb}, f_{Mb}, are fractions of Hb and Mb ($f_{Hb} + f_{Mb} = 1$), assumed in the NIRS saturation signals respectively. The factors on the parameters in the Eq 1. came from translation of oxygen capacity from molar equivalent.

3. RESULTS

3.1. Human Cuff Ischemia Study with Simulation

From the measurements and ^{1}H-NMR and NIRS, we plotted the Mb and Mb+Hb saturation curve over time (Figure 1A). There was an earlier and greater deoxygenation seen in the NIRS in the first 3 minutes to 70%, while the ^{1}H-NMR signal had only 25% desaturation. After 3 to 6 minutes, there was a delayed rapid phase of desaturation in the ^{1}H-NMR signal to 85%, while NIRS desaturated slightly by 25% at that time. From the timing differences of fast phases in NIRS and NMR, we can imply roughly that NIRS is more sensitive to the Hb, which desaturated faster, while the Mb desaturation comes after 3 minutes.

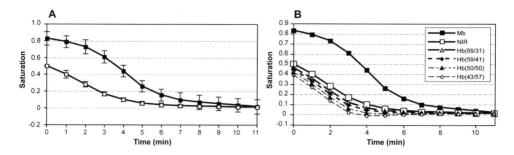

Figure 1. A; the time course of ¹H-NMR (closed rectangle) and NIRS (open rectangle) saturations during the arm cuff ischemia. Error bars indicate standard error. B; Saturations of Hb obtained from simulating $_{Hb}/f_{Mb}$ ratio from 100/0 (raw NIRS) to 47/57 (shown by the dotted lines), together with Mb saturation from NMR.

We then simulated the contribution of the Hb/Mb in the NIRS to be from 100%/0% (equivalent to raw data), to below 50/50 (Figure 1B). We can see as Hb/Mb is decreasing, (or the Mb contribution is increasing), apparent Hb saturation was lowered. Simulation suggests that below 50/50 of Hb/Mb estimation is not possible since the Hb at 4 minutes hit the lowest values below zero. On the other hand, Mb saturation was shifting down crossing the P_{50} values of 3.2 at about 4 minutes.

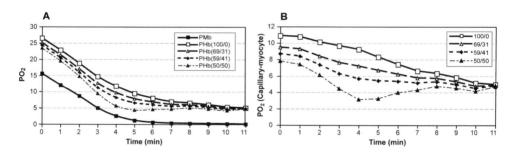

Figure 2. A; time course of PO_2 in myocyte (PMb, closed rectangle), and in the capillary (PHb) from the simulating various Hb/Mb (dotted lines) in human cuff ischemia. B; time course of PO_2 gradients between capillaries and myocytes.

We plotted the same data against PO_2 as Figure 1 (Figure 2A). The actual PO_2 depends on the Hb/Mb contribution on the NIRS. The possible actual capillary PO_2 (shown as PHb) was lowered as the contribution of Mb in the NIRS increased. The PO_2 gradients between capillary and myocytes is plotted in Figure 2B, which shows smaller PO_2 gradients between the capillaries and myocytes, if more Mb contributes in the NIRS during the time course, with possible f_{Mb} between 0% and 50%. In case of Hb/Mb 100/0, O_2 gradient starts from more than 10 torr at normal level, and ends about 5 torr at anoxia level. In the case of Hb/Mb contribution 50/50 in NIRS, the O_2 gradient was smaller (8 torr) at normoxia and decreased to 5 torr at anoxia, less different between the normoxia and hypoxia when the largest estimated Mb case was considered.

3.2. Canine Hypoxia Study with Sub-maximum Workload and Simulation

Preliminary dog studies showed that in order to obtain a full course of myoglobin desaturation in a relatively short time (within minutes), dog Gastrocnemius muscles required high submaximal workloads as well as limited supply of oxygen. In either one of these conditions, dog Gastrocnemius muscle did not show more than 30% desaturation. The FiO_2 at which half of the NIR and NMR signal becomes deoxygenated were 10 and 2.4%, respectively. These data suggested a much greater oxygen capacity in the dog than in the human skeletal muscles.

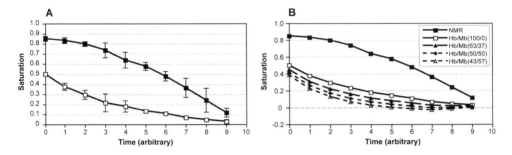

Figure 3. Saturation of Mb (NMR) and Mb+Hb (NIRS) were monitored during hypoxia to anoxia at submaximal workload (A). Simulated possible Hb/Mb and subsequently obtained Hb saturations were plotted as dotted lines (B).

In this study, in order to obtain whole spectral of desaturation in a relatively short time course, after a normal resting at 0 time, dogs were given low FiO_2 followed by anoxia while submaximal workloads were continued. Substantial Mb desaturation was seen when hypoxia was given at the submaximal stimulation (Figure 3A). In this figure, a relatively rapid phase of NIRS desaturation, at the beginning of the first 3 min was followed by later desaturation of NMR from 3 min to the end, similar to the human cuff study. However, NIRS desaturation was rather continuous in the dog study. We simulated possible Hb/Mb contribution in the NIR signals from Hb/Mb 100% (NIR signal is all due to Hb), to 43/57 (Mb is greater than Hb), and shown in Figure 3B. Note that in the case of 43/57 Hb/Mb, Hb saturation at a later time fell down below zero, which suggested the values were not adequately estimated and that the possible Hb/Mb contribution in the NIRS should be higher than 43/57. With increasing Mb contribution in the NIRS, the Hb desaturation curve was lower and steeper in the first 3 minutes. The simulation concluded that the maximum Mb contribution is near 50/% in the dog Gastrocnemius muscle, similar to the human muscle.

The PO_2 profiles from Mb and Hb saturation were plotted as myocyte and capillary PO_2, against time in the stimulated dog muscles (Figure 4A). The simulated Hb/Mb ranged from 100% to 43/57 are indicated in the dotted lines. We plotted PO_2 gradients between the capillary (PHb) and the myocytes (Pmyo) which were calculated from Hb and Mb saturation respectively, against time (Figure 4B). The possible oxygen gradients between capillary and muscle cells were larger with higher Hb contribution (less Mb contribution), ranging from 5 to 8.5 at rest and 4 and 9.5 torr at the highest (4-6 min) respectively in the dog muscles with hypoxia during submaximal workload.

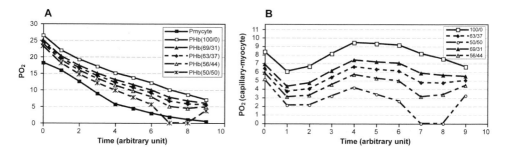

Figure 4. The PO_2 from Mb (Pmyo) and Hb saturation (PHb) were plotted against time in the stimulated dog muscles (A). The PO_2 values obtained from the simulated Hb/Mb ranged from 100% to 50/50 were shown as dotted lines. The same data are plotted as PO_2 gradients between capillaries and myocytes (B).

There appeared to be rather complicated trends of oxygen gradient with time, but the same trends were present in all cases of Hb/Mb. The trends consist of three different profiles of the oxygen gradient, even though both Mb and Hb curves against time were smoothly displayed in Figure 3. Here, the first profile, at time 0 and 1, shows that there is a reduction of oxygen gradient when dog muscle was under high workloads with mild hypoxia from the resting muscles. The second profile of oxygen gradients is that of increasing under an increasing amount of hypoxia during submaximal workload (from time 1 to time 4). And thirdly, when the oxygen gradient was already high at a high workload, it remained constant until oxygen was depleted and work no longer was achieved (after time 4 to end).

4. DISCUSSION

This study has demonstrated for the first time the possible oxygen gradients between capillaries and myocytes occurring in the human and dog muscles quantitatively, *in vivo*. The ^{1}H-NMR is used as a gold standard in providing an estimation of myocyte PO_2 and possible Hb/Mb contribution. Along with this knowledge, we can also estimate possible capillary PO_2 simultaneously using NIRS measured in the magnet.

The data showed that in the resting human arm muscle, limiting oxygen supply caused a linear reduction of O_2 gradient between capillary and myocyte. In addition, the greater the Mb contribution, the lesser the oxygen gradient. On the other hand, submaximally stimulated dog muscles with limited O_2 supply showed three O_2 muscle gradient profiles. Between resting to the submaximal working muscle, the O_2 gradient actually decreased. This was against our prediction, perhaps it occurs due to an excellent ability of increasing oxygen supply. The second profile showed an increased O_2 gradient during the submaximal workload, with more severe hypoxia at 3-6 arbitrary times, and it can be explained by limiting further oxygen supply. The last profile of the O_2 gradient was persistently constant in spite of more fetal hypoxia at the arbitrary time of 6 to 9 minutes. While oxygen supply was at risk, the work output was declining at the end of the time course.

With regard to the dog and human muscle responses to Mb and Hb desaturation, we have found that first, in order to desaturate 50% of the Mb, dog muscle needed to be in a more extreme condition (i.e. sub-maximal workload with hypoxia) while human muscle did not require as much (i.e. light exercise or cuff ischemia). This finding suggests a greater O_2 storing capacity in the dog muscle, which is also in agreement with the results from Mb/Hb quantification[9]. Secondly, we have found that both dog and human muscles have equal Hb/Mb contribution in the NIR suggesting that Hb and Mb work hand in hand in delivering oxygen from the capillary to the mitochondria. In summary, simulation can help identify the possible Hb saturation in the capillary and with in vivo ^{1}H-NMR Mb measurement combined, we can better understand the dynamics of oxygen gradients between capillary and myocyte.

5. ACKNOWLEDGEMENT

This study was supported by NIH HL 44125, and P41-RR02305.

6. REFERENCES

1. Conley KE, Ordway GA, and Richardson RS. Deciphering the mysteries of myoglobin n striated muscles. Acta. Physiol. Scand. 168:623-634. (2000).
2. Wagner PD. Algebraic analysis of the determination of VO2max. Respir. Physiol.1992. 93:221-237.
3. Nemeth PM, Lowry OH Myoglobin levels in individual human skeletal muscle fibers of different types. J. Histochem Cytochem. 32:1211-6. (1984).
4. Armstrong. RB. Essen-Gustavsson B, Hoppeler H, Jones JH, Kayar SR, Laughlin MH, Lindholm A. Tayler CR, Weibel ER. O2 delivery at VO$_2$max and oxidative capacity in muscles of standard bred horses. J. Appl. Physiol. 73:2274-2282. (1992).
5. Groebe K, Thews G. Calculated intra- and extra cellular PO$_2$ gradients in heavily working red muscle. Am J. Physiol. 259:H84-92. (1990).
6. Kindig CA, Howlett RA, Hogan MC. Effect of extra cellular PO$_2$ on the fall in intracellular PO$_2$ in contracting single myocytes. J. Appl Physiol. 94:1964-70. (2003).
7. O'brian. PJ, Shen H, McCutcheon LJ, O'Grady M, Byrne PJ, Ferguson HW, et al. Rapid, simple and sensitive micro assay for skeletal and cardiac muscle myoglobin and hemoglobin: use in various animals indicates functional role of myohemoproteins. Mol Cell Biochem. 112:45-52. (1992).
8. Conley KE, Jones C. Myoglobin content and oxygen diffusion: model analysis of horse and steer muscle. Am J. Physiol. Dec; 271(6 Pt 1):C2027-36. (1996).
9. Maxwell LC, Barclay JK, Mohrman DE, Faulkner JA. Physiological characteristics of skeletal muscles of dogs and cats. Am J. Physiol. Jul; 233(1):C14-8. (1977).
10. Hermansen L, Wachtlova M. Capillary density of skeletal muscles in well-trained and untrained men. J. Appl. Physiol. 30:860-863. (1971).
11. Richardson RS, Noyszewski EA, Kendrick KF, Leigh JS, Wagner PD. Myoglobin O$_2$ desaturation during exercise. Evidence of limited O$_2$ transport. J Clin Invest. 96:1916-26. (1995).
12. Mole PA, Chung Y, Tran TK, Sailasuta N, Hurd R, Jue T. Myoglobin desaturation with exercise intensity in human gastrocnemius muscle. Am J Physiol. 277:R173-80.9. (1999).
13. Wang ZY, Noyszewski EA, Leigh JS Jr. In vivo MRS measurement of deoxymyoglobin in human forearms. Magn Reson Med. Jun;14(3):562-7. (1990).
14. Wang DJ, Nioka S, Wang Z, Leigh JS, Chance B. NMR visibility studies of N-delta proton of proximal histidine in deoxyhemoglobin in lysed and intact red cells. Magn Reson Med. 1993. 30:759-63.
15. Hamaoka T, et al. Quantification of ischemic muscle deoxygenation by near infrared time-resolved spectroscopy. J. Biomed. Optics. 5:102-105. (2000).
16. Rossi-Fanelli A, Antonini E. Studies of the oxygen and carbon monooxide equilibria of human myoglobin. Arch. Biochem. Biophysics. 77:478-492. (1958).
17. Connett RJ, Gayeski TEJ, Honig CR. Lactate efflux is unrelated to intracellular PO$_2$ in a working red muscle in situ. J. Appl. Physiol. 61:402-408. (1986).

37.

EXPLORING PREFRONTAL CORTEX OXYGENATION IN SCHIZOPHRENIA BY FUNCTIONAL NEAR-INFRARED SPECTROSCOPY

Valentina Quaresima[1], Patricia Giosuè[2], Rita Roncone[2], Massimo Casacchia[2], and Marco Ferrari[1] *

1. INTRODUCTION

The specific nature of frontal lobe dysfunction in schizophrenia remains unclear. However, impairments in working memory have been proposed to underlie various cognitive and functional deficits in schizophrenia. So far, one of the most well-studied cognitive deficits in schizophrenia is that of impaired verbal fluency which some have argued is primarily due to dysfunction in access and/or retrieval of lexical information, while others have maintained that the primary defect is in the organization of the semantic memory. The verbal fluency task (VFT) is a neuropsychological task which permits assessment of the subject's ability to retrieve nouns based on a common criterion. The letter-fluency version is based on a phonological criterion, requiring the subject to pronounce as many words as possible beginning with a certain letter. Performances have been regarded to be mainly associated with frontal lobe function, particularly the left hemisphere. An attenuated frontal activation during a VFT has been demonstrated in patients with schizophrenia by functional magnetic resonance imaging (fMRI).[1]

Considering the several advantages of using functional near-infrared spectroscopy (fNIRS) over fMRI for performing cognitive studies, we aimed at assessing bilateral prefrontal cortex (PFC) oxygenation changes in response to VFT in schizophrenia patients and controls. We hypothesized that patients would have a PFC hypoactivation with respect to controls.

* [1]Dept. of Biomedical Sciences and Technologies, [2]Dept. of Experimental Medicine, University of L'Aquila, 67100 L'Aquila, Italy; e-mail: vale@univaq.it

2. METHODS

All the subjects participating in this study (approved by the Ethical Committee of the University of L'Aquila) were right-handed. The school education was 12.3±1.1 and 13.5±2.7 years for patients (n=9) and controls (n=9), respectively. The subjects with schizophrenia were diagnosed according to Diagnostic and Statistical Manual of Mental Disorders (fourth edition) established by a psychiatric evaluation. All patients were receiving medication; four (# 1, 5, 6 and 9) were receiving typical neuroleptic; four (# 2, 4, 7 and 8) were receiving atypical neuroleptic, and one (# 3) both typical and atypical neuroleptic. The dosage of antipsychotic drugs, calculated as equivalent to chlorpromazine, was 709±551 meq/die. The control subjects had no history of substance abuse or other medical, psychiatric or neurological disorder that might affect brain function.

A 2-min letter version of the VFT was adopted in this study.[2] Following a 2-min baseline, subjects were asked to produce aloud (overt speech) as many nouns as possible (within the allocated 120 s) beginning with the letters "S", "P", "F", "C" (acoustically presented at time 0, 30, 60, 90 s, respectively). No repetitions or proper nouns were permitted. Corrected responses were recorded for each subject. As control task, subjects were requested to listen to a 2-min story.

For fNIRS measurements the four-wavelength NIRO-300 oximeter (Hamamatsu Photonics, Japan) equipped with 2 channels was employed. Each optical probe (consisting of a source and a detector 5 cm apart) was placed and fixed (by double-sided adhesive sheet) over Fp1 (left) and Fp2 (right) of the 10-20 system for the EEG electrode placement. All measurements were performed on control subjects and patients seated in a comfortable chair in a quiet room. Concentration changes (expressed in $\Delta\mu M*cm$) of oxyhemoglobin (O_2Hb) and deoxyhemoglobin (HHb) were collected (sampling frequency: 6 Hz) before, during and after VFT execution, and transferred on-line to a computer for storage and subsequent analysis. ΔO_2Hb and ΔHHb values, expressed in μM, were obtained dividing ΔO_2Hb and ΔHHb ($\Delta\mu M*cm$) by the optical pathlength corrected for age. Simultaneous measurement of heart rate was obtained by pulse oximetry (N-200; Nellcor, Puritan Bennett, St. Louis, MO, USA) with the sensor attached to the index finger of the subject's right hand.

The amplitude of O_2Hb and HHb task-related changes in PFC was calculated by the difference between the rest condition (mean value over the 1.5-2 min baseline) and the end of each letter interval of VFT (mean value over the last 5 s). The mean and standard deviation of O_2Hb and HHb values within left and right PFC were determined separately and compared. The paired *t* test was used to compare the O_2Hb or HHb values in both hemispheres within each group and between groups. Comparisons between the performance of controls and patients in the VFT were performed with the two-tailed *t* test. Statistics was performed by a SPSS software version 10.0 (SPSS, Chicago, Il, USA). The criterion for significance was $P<0.05$.

3. RESULTS

No difference in education, age at the onset or duration of illness was found among medicated patients. During VFT, the controls achieved an average of 35.1±0.7 correct responses for the 4 letters, and the performance was almost constant over the 4-letter condition (S=8.3±1.7, P=8.1±2.1, F=8.9±2.8, C=9.8±2.1). The subjects with schizophrenia achieved an average of 20.2±0.4 correct responses for the 4 letters, with a similar performance over the 4-letter condition (S=5.4±2.1, P=5.1±1.7, F=4.6±2.7, C=5.1±2.1). Comparing the performance (mean value of correct responses) between the 2 groups, a lower performance (P=0.0014) was found in patients (5.1±1.9) than in controls (8.8±1.8).

The typical fNIRS activation response (characterized by O$_2$Hb increase and HHb smaller decrease) was observed in both PFCs of all control subjects during VFT execution. On the other hand, the pattern of PFC activation in response to VFT was found to differ between controls and patients.

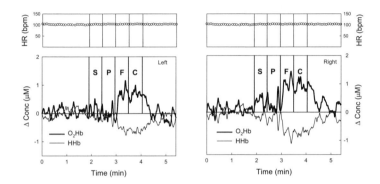

Figure 1. Typical time course of the right and the left prefrontal cortex ΔO$_2$Hb and ΔHHb during 2-min verbal fluency task. The vertical lines indicate the duration of words production beginning with each acoustically presented letter. The patient generated 7, 5, 5, 7 words for the letters S,P,F,C, respectively. The optical probes were centered at the Fp1 and Fp2 according to the international 10-20 system for the EEG electrode placement. Patient #6. HR: heart rate; bpm: beats per min

Figure 1 shows a typical response of ΔO$_2$Hb and ΔHHb, over the PFC areas, in patients. Unlike the controls, most of the patients showed a consistent O$_2$Hb increase after the first 30-60 s of the task execution. Thereafter O$_2$Hb continued to rise, reaching its maximum change within the end of VFT, whereas it decreased progressively after the end of the task. Concomitantly, HHb decreased (with a delayed starting point as O$_2$Hb) during VFT, and it returned to the pre-task value within the observation time. The heart rate was unchanged throughout the execution of VFT (Fig. 1, upper panels). The amplitude of O$_2$Hb and HHb changes for each subject with schizophrenia and controls are reported in Table 1 and 2, respectively. The consistent inter-patient variability was mainly due to the fact that 2 out of 9 patients had no oxygenation changes during VFT. Comparing the oxygenation event related response in both groups, a significant attenuated response to VFT was found in patients (Fig. 2). In all controls and patients O$_2$Hb and HHb tracings over the 2-min control task were not different from the corresponding baseline tracings.

Table 1.   Subjects with schizophrenia. Left (L) and right (R) prefrontal cortex concentration changes in O_2Hb and HHb induced by each letter over the 2-min VFT

Subj	Hem	"S" $\Delta[O_2Hb]$	"S" $\Delta[HHb]$	"P" $\Delta[O_2Hb]$	"P" $\Delta[HHb]$	"F" $\Delta[O_2Hb]$	"F" $\Delta[HHb]$	"C" $\Delta[O_2Hb]$	"C" $\Delta[HHb]$
1	L	0.9	-0.5	1.1	-0.3	1.7	-0.7	2.0	-1.2
	R	0.5	-0.2	1.0	-0.2	1.5	-0.7	1.4	-0.8
2	L	0.4	0.0	0.9	-0.2	1.3	-0.6	0.5	-0.1
	R	0.1	0.1	0.7	-0.4	0.6	-0.5	-0.1	-0.2
3	L	-0.5	-0.1	-1.0	-0.2	-1.3	-0.1	-0.9	-0.1
	R	1.2	1.0	-0.5	0.3	-0.5	0.1	-0.4	0
4	L	1.5	0	1.8	0.2	1.3	0.1	1.0	0
	R	2.0	0.3	2.2	0.4	2.3	0	1.9	0.3
5	L	0	-0.2	0.5	-0.3	0	-0.1	0.6	-0.4
	R	0.5	-0.3	1.0	-0.3	0.4	0	1.1	-0.5
6	L	-0.1	0	0.0	-0.2	0.8	-0.7	0.6	-0.6
	R	0.2	-0.1	0.5	-0.5	1.1	-1.0	0.7	-0.7
7	L	-0.2	-0.2	0.2	-0.1	0.3	-0.2	0.6	-0.4
	R	0.3	-0.4	0.5	-0.2	0.2	-0.4	0.8	-0.4
8	L	0	0	0	0	0	0	0	0
	R	0	0	0	0	0	0	0	0
9	L	0	0	0	0	0	0	0	0
	R	0	0	0	0	0	0	0	0
Mean ± SD	L	0.2±0.6	-0.1±0.2	0.4±0.8	-0.1±0.2	0.5±0.9	0.3±0.3	0.5±0.8	-0.3±0.4
	R	0.5±0.7	0.0±0.4	0.6±0.8	-0.1±0.3	0.6±0.9	-0.3±0.4	0.6±0.8	-0.3±0.4

Subj: subjects; Hem: hemisphere.

4. DISCUSSION

The most significant studies performed on subjects with different mental diseases by using fNIRS are summarized in Table 3. The attenuated PFC activation found in schizophrenia patients during VFT confirm the previous fMRI results,[1] and those obtained in a recent fNIRS study performed on subjects with schizophrenia.[18]

However, a large variability was found amongst patients. In particular, bilateral PFC oxygenation was unaffected by VFT in two patients (# 8 and 9) (Table 1). It has been recently hypothesized that functional hypofrontality in bipolar disorder may be at least partly caused by the dysfunction of vascular automatic regulation due to vascular factors such as altered signal transduction pathways.[17] This is not the case of patients #8 and 9, because in a separate study their bilateral PFC was activated during a visual spatial working memory test capable of specifically determining whether working memory was compromised.

In conclusion, fNIRS might be extensively employed to evaluate non-invasively "hypofrontality" in psychiatric patients, and to explain neural circuitry abnormalities that may change over the course of the illness by pharmacological treatment and/or cognitive rehabilitation.

Table 2. Control subjects. Left (L) and right (R) prefrontal cortex concentration changes in O_2Hb and HHb induced by each letter over the 2-min VFT

Subj	Hem	"S" $\Delta[O_2Hb]$	"S" $\Delta[HHb]$	"P" $\Delta[O_2Hb]$	"P" $\Delta[HHb]$	"F" $\Delta[O_2Hb]$	"F" $\Delta[HHb]$	"C" $\Delta[O_2Hb]$	"C" $\Delta[HHb]$
1	L	1.8	-0.6	1.7	-0.7	1.3	-0.8	1.3	-0.9
	R	2.2	-0.5	2.0	-0.6	1.8	-0.7	1.4	-0.8
2	L	2.4	-0.5	2.6	-0.4	3.9	-1.2	3.6	-1.4
	R	0.9	-1.1	0.5	-1.0	1.6	-1.9	1.5	-2.0
3	L	2.5	-1.0	2.2	-0.9	2.5	-1.3	1.7	-1.2
	R	2.4	-0.5	2.6	-0.5	2.7	-1.0	2.0	-0.9
4	L	0.6	-0.4	1.1	-0.3	1.8	-0.7	1.3	-0.5
	R	1.4	-0.7	2.2	-0.6	2.2	-1.1	1.8	-0.9
5	L	2.3	-0.2	3.4	-0.6	3.5	-0.6	3.4	-0.7
	R	2.1	-0.4	3.1	-0.5	2.5	-0.5	2.5	-0.6
6	L	1.7	-0.7	2.7	-0.9	2.5	-1.0	0.9	-0.4
	R	3.2	0.3	3.9	-0.1	4.1	-0.3	2.3	-0.3
7	L	1.1	-0.4	1.0	-0.3	1.3	-0.4	1.6	-0.7
	R	1.4	-0.4	1.2	-0.4	1.5	-0.7	1.8	-0.8
8	L	1.3	-0.8	1.5	-0.6	2.4	-0.8	1.2	-0.7
	R	0.5	-0.1	0.7	-0.1	1.5	-0.4	0.6	-0.1
9	L	2.0	-1.1	0.2	-0.5	0.8	-1.0	1.1	-0.5
	R	1.6	-0.6	0.2	-0.2	0.2	-0.3	1.5	-0.5
Mean ± SD	L	1.7±0.6	-0.6±0.3	1.8±1.0	-0.6±0.2	2.2±1.0	-0.9±0.3	1.8±1.0	-0.8±0.3
	R	1.7±0.8	-0.4±0.4	1.8±1.3	-0.4±0.3	2.0±1.1	-0.8±0.5	1.7±0.6	-0.8±0.5

Subj: subjects; Hem: hemisphere.

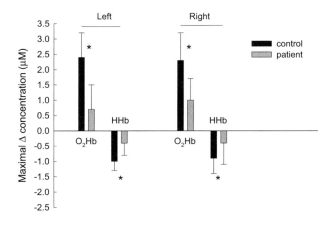

Figure 2. Comparison of the maximal concentration changes of O_2Hb and HHb between control and patient groups. *: $P<0.05$.

Table 3. Functional near infrared spectroscopy (fNIRS) studies in mental diseases.

#	Subj.	Dis.	Stimulus	Instrument	O$_2$Hb	HHb	Remarks
3	P: 38 CS: 38	S	MDT	Shimadzu prototype (Ch:2)	↑	↓	P: 17 atypical responses
4	P: 10 CS: 10	S	CPT	Critikon (Ch: 2)	↓	↑	Lack of lateralized activation
5	P: 13 CS: 10	S	RNG RC	NIRO-300, Hamamatsu (Ch:2)	↑ RNG ↓ RC ↑ RC$_{CS}$	↓ RNG = RC	RNG$_P$: smaller responses
6	P: 10+13 CS: 16	D S	VFT	ETG-100 Hitachi (Ch: 24)	↑	↓	P: smaller response
7	P: 36 CS: 36	D	MDT	Shimadzu (Ch: 2)	↑	↓	P: altered hemispheric differences
8	P: 9 CS: 10	D	VFT	HEO-200 Omron (Ch: 1)	CS: ↑ P: =	CS: ↓ P: =	
9	P: 12 CS: 12	D	MDT AC	Scanning prototype (Ch: 4)	↑	=	
10	P: 25 CS: 21	D BD	VFT	HEO-200 Omron (Ch: 1)	↑	↓	P: smaller response
11	P: 9 CS: 9	D	VFT	NIRO-300, Hamamatsu (Ch: 2)	↑	↓	P: smaller response
12	P: 8 CS: 26	PTSD	VFT	ETG-100 Hitachi (Ch: 24)	↑	↓	P: O$_2$Hb smaller response
13	P: 8 CS: 26	PTSD	traumatic images	ETG-100 Hitachi (Ch: 24)	CS: = P: ↑	CS: = P: ↓	
14	P: 23 CS: 31	PD	erotic, spider, images	NIRO-300, Hamamatsu (Ch: 2)	↑	↓	P: smaller response
15	P: 10 CS: 10	AD	VFT	Critikon 2020 (Ch: 2)	↑	↓	P: altered hemispheric differences
16	P: 19 CS: 19	AD	VFT	NIRO-500, Hamamatsu (Ch: 2)	↑	↓	P: O$_2$Hb decrease in parietal cortex
17	P: 9 CS: 9	BD	VFT	ETG-100 Hitachi (Ch: 24)	↑	↓	P: smaller response
18	P: 62 CS: 31	S	VFT LN	HEO-200 Omron (Ch: 1)	↑	↓	P: smaller response

#: reference number; AC: arithmetic calculation; AD: Alzheimer's dementia; BD: bipolar disorders; CS: control subjects; Ch: channel; CPT: continuous performance test; D: depression; Dis: disease; LN: letter number span test; MDT: mirror drawing task; P: patients; PD: panic disorder; PTSD: post traumatic stress disorder; Subj: subjects; RC: ruler-catching; RNG: random number generation; S: schizophrenia; VFT: verbal fluency task.

5. REFERENCES

1. V. A. Curtis, E. T. Bullmore, M. J. Brammer, I. C. Wright, S. C. Williams, R. G. Morris, T. S. Sharma, R. M. Murray, and P.K. McGuire, Attenuated frontal activation during a verbal fluency task in patients with schizophrenia, *Am. J. Psychiatry* **155**(8), 1056-1063 (1998).
2. M. J. Herrmann, A. C. Ehlis, and A. J. Fallgatter, Frontal activation during a verbal-fluency task as measured by near-infrared spectroscopy, *Brain Res. Bull.* **61**(1), 51-56 (2003).
3. F. Okada, Y. Tokumitsu, Y. Hoshi, and M. Tamura, Impaired interhemispheric integration in brain oxygenation and hemodynamics in schizophrenia, *Eur. Arch. Psychiatry Clin. Neurosci.* **244**(1), 17-25 (1994).
4. A. J. Fallgatter, and W. K. Strik, Reduced frontal functional asymmetry in schizophrenia during a cued continuous performance test assessed with near-infrared spectroscopy, *Schizophr. Bull.* **26**(4), 913-919 (2000).
5. T. Shinba, M. Nagano, N. Kariya, K. Ogawa, T. Shinozaki, S. Shimosato, and Y. Hoshi, Near-infrared spectroscopy analysis of frontal lobe dysfunction in schizophrenia, *Biol. Psychiatry* **55**(2), 154-164 (2004).
6. T. Suto, M. Fukuda, M. Ito, T. Uehara, and M. Mikuni, Multichannel near-infrared spectroscopy in depression and schizophrenia: cognitive brain activation study, *Biol. Psychiatry* **55**(5), 501-511 (2004).
7. F. Okada, N. Takahashi, and Y. Tokumitsu, Dominance of the 'nondominant' hemisphere in depression, *J. Affect. Disord.* **37**(1), 13-21 (1996).
8. K. Matsuo, T. Kato, M. Fukuda, and N. Kato, Alteration of hemoglobin oxygenation in the frontal region in elderly depressed patients as measured by near-infrared spectroscopy. *J. Neuropsychiatry Clin. Neurosci.* **12**(4), 465-471 (2000).
9. G. W. Eschweiler, C. Wegerer, W. Schlotter, C. Spandl, A. Stevens, M. Bartels, and G. Buchkremer, Left prefrontal activation predicts therapeutic effects of repetitive transcranial magnetic stimulation (rTMS) in major depression, *Psychiatry Res.* **99**(3), 161-172 (2000).
10. K. Matsuo, N. Kato, and T. Kato, Decreased cerebral haemodynamic response to cognitive and physiological tasks in mood disorders as shown by near-infrared spectroscopy, *Psycol. Med.* **32**(6), 1029-1037 (2002).
11. M. J. Herrmann, A. C. Ehlis, and A. J. Fallgatter, Bilaterally reduced frontal activation during a verbal fluency task in depressed patients as measured by near-infrared spectroscopy, *Neuropsychiatry Clin. Neurosci.* **16**(2), 170-175 (2004).
12. K. Matsuo, K. Taneichi, A. Matsumoto, T. Ohtani, H. Yamasue, Y. Sakano, T. Sasaki, M. Sadamatsu, K. Kasai, A. Iwanami, N. Asukai, N. Kato, and T. Kato, Hypoactivation of the prefrontal cortex during verbal fluency test in PTSD: a near-infrared spectroscopy study, *Psychiatry Res.* **124**(1), 1-10 (2003).
13. K. Matsuo, T. Kato, K. Taneichi, A. Matsumoto, T. Ohtani, T. Hamamoto, H. Yamasue, Y. Sakano, T. Sasaki, M. Sadamatsu, A. Iwanami, N. Asukai, and N. Kato, Activation of the prefrontal cortex to trauma-related stimuli measured by near-infrared spectroscopy in posttraumatic stress disorder due to terrorism, *Psychophysiology* **40**(4), 492-500 (2003).
14. J. Akiyoshi, K. Hieda, Y. Aoki, and H. Nagayama, Frontal brain hypoactivity as a biological substrate of anxiety in patients with panic disorders, *Neuropsychobiology* **47**(3), 165-170 (2003).
15. A. J. Fallgatter, D. Brandeis, and W. K. Strik, A robust assessment of the NoGo-anteriorisation of P300 microstates in a cued continuous performance test, *Brain Topography* **9**(4), 295-302 (1997).
16. C. Hock, K. Villringer, F. Muller-Spahn, R. Wenzel, H. Heekeren, S. Schuh-Hofer, M. Hofmann, S. Minoshima, M. Schwaiger, U. Dirnagl, and A. Villringer, Decrease in parietal cerebral hemoglobin oxygenation during performance of a verbal fluency task in patients with Alzheimer's disease monitored by means of near-infrared spectroscopy (NIRS)-correlation with simultaneous rCBF-PET measurements, *Brain Res.* **755**(2), 293-303 (1997).
17. K. Matsuo, A. Watanabe, Y. Onodera, N. Kato, and T. Kato, Prefrontal hemodynamic response to verbal-fluency task and hyperventilation in bipolar disorder measured by multi-channel near-infrared spectroscopy, *J. Affect Disord.* **82**(1), 85-92 (2004).
18. A. Watanabe, and T. Kato, Cerebrovascular response to cognitive tasks in patients with schizophrenia measured by near-infrared spectroscopy, *Schizophr. Bull.* **30**(2), 435-444 (2004).

38

QUANTIFICATION OF ADULT CEREBRAL BLOOD VOLUME USING THE NIRS TISSUE OXYGENATION INDEX

Ilias Tachtsidis[§], Terence S. Leung[§], Caroline Oliver[□], Julian R. Henty[§], Holly Jones[□], Martin Smith[□], David T. Delpy[§], Clare E. Elwell[§]

1. INTRODUCTION

Near-infrared spectroscopy (NIRS) is increasingly used as a non-invasive technique for monitoring cerebral oxygenation and haemodynamics[1, 2]. Simple continuous-wave (CW) NIRS systems utilising differential spectroscopy can measure quantitative changes in oxy- and deoxy- haemoglobin ($\Delta[O_2Hb]$, $\Delta[HHb]$) but only from an arbitrary baseline. Numerous studies of changes in cerebral oxygenation and haemodynamics in adults have been published but only few absolute quantitative measurements have been reported. Recent advances in the NIRS technology have enabled quantitative assessment of haemoglobin concentration in tissue using near-infrared (NIR) phase and time resolved systems; and absolute measurements of tissue saturation using phase, time or spatially resolved spectroscopy (SRS) systems[3, 4, 5, 6].

This paper suggests a way to use a commercially available spectrometer, which has both CW and SRS capabilities in order to measure absolute tissue haemoglobin (Hb_{tc}) and hence cerebral blood volume (CBV). The methodology is based on that of Wyatt et al.[7] who developed a method for measuring absolute CBV, using NIRS measurements during controlled changes in inspired O_2 fraction. By using NIRS measured tissue $\Delta[O_2Hb]$ and comparing it to changes in arterial saturation (SaO_2) measured with a pulse oximeter it is possible to calculate absolute Hb_{tc} concentration. This is the so-called 'de-saturation method' or 'O_2 method' or 'SaO_2 method'[8, 9, 10, 11]. The purpose of the present study was to compare measurements of CBV made using the conventional 'SaO_2 method' with those using a new method employing the SRS derived absolute cerebral tissue oxygenation index (TOI), which will be called the 'TOI method'.

[§]Medical Physics and Bioengineering, University College London, Shropshire House, 11- 20 Capper Street, London WC1E 6JA; iliastac@medphys.ucl.ac.uk
[□]Department of Neuroanaesthesia & Neurocritical Care, The National Hospital for Neurology & Neurosurgery, London

2. THEORY

2.1 Hb_{tc} Calculation

The theory for absolute quantification of Hb_{tc} relies upon the induction of a small change in the inspired oxygen. During the manoeuvre the consequent change in cerebral $\Delta[O_2Hb]$ is equivalent to the product of the Hb_{tc} and the change in fractional tissue saturation. If CBV is constant then $\Delta[Hb_{diff}]$ (i.e. $\Delta[O_2Hb]$- $\Delta[HHb]$) can be used to derive Hb_{tc}. The 'SaO$_2$ method' uses a CW NIRS system to monitor the changes in $\Delta[O_2Hb]$ and $\Delta[HHb]$ and a pulse oximeter to measure SaO$_2$. If cerebral blood flow (CBF), CBV and O$_2$ consumption remain constant during the manoeuvre the ΔSaO$_2$ measurement can be related to the tissue saturation. Therefore absolute Hb_{tc} can be obtained using Eq. (1).

Instead of using ΔSaO$_2$ as an indicator of tissue saturation one can use a direct measurement of tissue saturation, which modern NIRS systems measure. One such system the Hamamatsu NIRO 300 utilises the SRS technique, in which multiple closely separated detectors measure the attenuation slope[12]. From these measurements it is possible to calculate scaled absolute haemoglobin concentrations and hence accurately obtain a tissue oxygenation index (TOI) as $k[O_2Hb]/[kO_2Hb+kHHb]*100\%$ where k is a scaling factor dependent upon the tissue scattering coefficient (μ_s). TOI is a measure of oxygen saturation in tissue; therefore one can use Eq. (2) to measure absolute Hb_{tc}.

$$Hb_{tc}(\mu moles/l) = \frac{\Delta[Hb_{diff}]}{2 \cdot \Delta SaO_2} \quad (1) \qquad Hb_{tc}(\mu moles/l) = \frac{\Delta[Hb_{diff}]}{2 \cdot \Delta TOI} \quad (2)$$

2.2 CBV Calculation

It is important to remember that NIRS measures change in chromophore concentrations in micromolar units. Estimates of blood volume are obtained from these measurements by converting the concentration data into the more conventional clinical units of millilitres/100 grams (see Eq. (3)). This conversion requires knowledge of the red blood cell concentration, which is measured from a venous sample.

$$CBV(ml/100g) = \frac{[Hb_{tc}] \cdot MW_{Hb} \cdot 10^{-4}}{d_t \cdot [Hb_t \cdot 10^{-2}] \cdot CLVHR} \quad (3)$$

where $MW_{Hb}=64500$ is the molecular weight of haemoglobin, $d_t=1.05g/ml$ is the cerebral tissue density, Hb_t (g/dl) is the haemoglobin concentration obtained from a venous sample, and CLVHR=0.69 is the cerebral to large vessel haematocrit ratio[13].

3. PARTICIPANTS AND PROCEDURE

3.1 Participants

Data were recorded during three consecutive graded arterial hypoxaemias in 12 healthy volunteers of mean $\pm$SD age 32 $\pm$4years (the local ethics committee approved the protocol for the study, and all subjects gave informed consent for participation).

3.2 Instrumentation

A continuous wave near-infrared spectrometer (NIRS), with a sampling rate of 6Hz (NIRO 300, Hamamatsu Photonics KK) was used to measure absolute cerebral TOI over the frontal cortex using the SRS technique[12], together with changes in $\Delta[O_2Hb]$) and $\Delta[HHb]$ by using the modified Beer-Lambert law[14]. The optodes were placed on the forehead (taking care to avoid the midline sinuses) and were shielded from ambient light by using an elastic bandage and a black cloth. An optode spacing of 4cm was used and optical attenuators were used where necessary to optimise the signal to noise ratio. The differential pathlength factor (DPF) applied was 6.26[15].

A modified anaesthetic machine supplied a controlled mixture of nitrogen and oxygen to the subject via a face mask. Arterial saturation was monitored from the ear with a pulse oximeter (Novametrix 500) modified to measure in a beat-to-beat mode. Inspired oxygen (FiO₂) was monitored continuously via the Merlin modular monitor (Hewlett Packard). A Portapres® system (TNO Institute of Applied Physics), was used to continuously and non-invasively measure blood pressure from the finger.

3.3 Procedure

All measurements were made with the volunteers sitting comfortably in an armchair. The subject initially breathed normal air for 3-5min. The FiO₂ in the circuit was then gradually reduced to 10-15% by mixing air and nitrogen, until a baseline SaO₂ of 90% was achieved. At that point the subject was given 100% O₂ for five consecutive breaths. This was repeated a total of three times for each subject (see Fig 1).

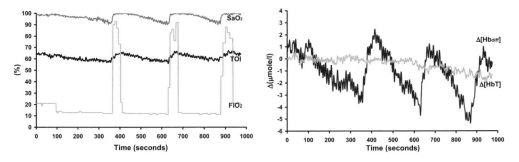

Figure 1. Representative data collected from one volunteer during the three hypoxic episodes. For illustrative purposes data has been filtered to remove heart rate oscillations.

3.4 CBV Measurement

Software has been written in MatLab to calculate absolute CBV using each episode of hypoxaemia. We first set a baseline range for the changes in cerebral total haemoglobin ($\Delta[HbT]=\Delta[O_2Hb]+\Delta[HHb]$). Within this range, the software identifies the corresponding $\Delta[Hb_{diff}]$ and TOI and calculates the absolute cerebral Hb_{tc} from the slope of the plot of these variables using linear regression. With a known haemoglobin concentration (g/dl) from a venous blood sample, the absolute cerebral Hb_{tc} is converted into absolute CBV.

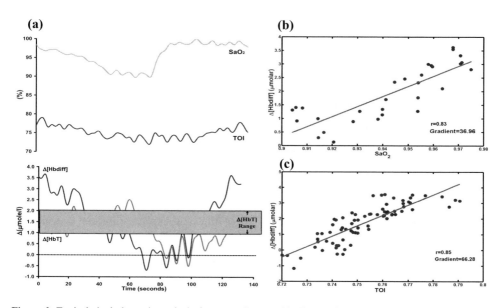

Figure 2. Typical single hypoxia analysis for one volunteer. (a) Changes in SaO$_2$, TOI, Hb$_{diff}$ and HbT after smoothing to remove cardiac oscillations, a baseline range of the changes in HbT is set, within this range, the software identifies the corresponding [Hb$_{diff}$], SaO$_2$ and TOI data points. (b) For the 'SaO$_2$ method' using the pulse oximeter measured SaO$_2$ a correction has to be made for any temporal differences and only the desaturation period is consider for analysis, the software calculates the absolute cerebral Hb$_{tc}$ from the slope of the plot of Δ[Hb$_{diff}$] against Δ[SaO$_2$] using linear regression. (c) For TOI analysis all the hypoxic swing data is considered and the software calculates the absolute cerebral Hb$_{tc}$ from the slope of the plot of Δ[Hb$_{diff}$] against Δ[TOI] using linear regression.

3. RESULTS

From a total of 36 hypoxic swings only 21 measurements (9 volunteers) proved suitable for calculation of CBV using the 'SaO$_2$ method' and 27 measurements (10 volunteers) proved suitable for calculation of CBV using the 'TOI method'. A criterion for rejection was the linearity correlation coefficient r. If r was less than 0.8, that hypoxic swing was excluded from the analysis.

The mean ±SD CBV was 2.23±1.06ml/100g using the 'SaO$_2$ method' and 2.62±0.61ml/100g using the 'TOI method'. For all the 36 hypoxic swings the intra-subject coefficient of variation using the 'SaO$_2$ method' ranged from 12 to 83% (mean 35%) and using the 'TOI method' ranged from 3 to 21% (mean 11%).

Table 1. Summary of results (mean values ±SD).

	CBV±SD (ml/100g)	Hb$_{tc}$±SD (μmoles/l)	Inter-subject Coefficient of variation
'SaO$_2$ method' (n=21)	2.23±1.06	38.4±18.6	48%
'TOI method' (n=27)	2.62±0.61	45.0±10.8	23%

4. DISCUSSION

This study has described an alternative method for measuring absolute Hb_{tc} through the use of a single NIRS system capable of both differential and SRS spectroscopy measurements and we have compared this methodology with the previously described 'SaO$_2$ method'. The absolute Hb_{tc} and CBV measurements calculated here are comparable with previously published data[8, 16]. Most importantly we have shown that using the 'TOI method', the data is more robust with lower intra-subject and inter-subject coefficients of variation.

The limitations of the 'SaO$_2$ method' have been described elsewhere[17] the most important being the low sampling rate and instrumental noise of the pulse oximeter. Limitations of NIRS include the inter-subject variability in DPF which has been shown to be on the order of 12-17%[15]. However, in this study, DPF acts purely as an equal scaling factor on the $\Delta[Hb_{diff}]$ and hence Hb_{tc} data to convert the units from micromoles/centimeter to micromoles and as such will not contribute to the discrepancy between the two measurements. There is also criticism regarding the tissue specificity of the NIRS instrumentation. For single optode measurements the influence of the extracranial compartment is inevitable. In a recent study Choi and colleagues[6] reported significant differences in the absolute haemoglobin measurements of the different layers of the adult head (scalp/skull and brain) using an NIRS multi-distance frequency domain method. NIR interrogates the whole head including scalp, skull and brain resulting in an averaged blood volume, which may be lower than in the brain itself. We are currently investigating this effect using a multi-layer model.

The estimated mean CBV is smaller than that obtained from positron emission tomography (PET)[18] or single-photon emission-computed tomography (SPECT)[18] studies. A further possible error in the calculation of CBV is the CLVHR where there are some differences in the literature values. Lammertsma et al.[13] measured the cerebral-to-large vessel haematocrit ratio as 0.69, whereas Sakai et al.[19] found a value of 0.76. The value measured depends on the distribution of vessel sizes considered. Sakai et al.[19] also found that the haematocrit may change with blood volume.

The ideal method of measuring CBV would be accurate over a broad range of haemodynamic conditions, reproducible, sensitive to changes in physiologic or pathologic states, non-invasive, and readily incorporated into current clinical protocols. Unfortunately none of the current methods for measuring CBV are satisfactory. CBV may be measured in human subjects by using PET or SPECT. These methods require the use of radioisotopes and are not bedside tests. Magnetic resonance imaging (MRI) can also be used, however the long scan times of equilibrium T1-weighted MRI imposes a major limitation and typical clinical scans using T2-weighted, gradient echo MRI yield only relative CBV changes[20]. NIRS, which was first described in 1977[21] has been used to monitor changes in $\Delta[HbT]$ on an arbitrary scale, but recent NIRS developments enable absolute measurements of tissue saturation and Hb_{tc} to be made. These methods avoid the use of radioisotopes and X-rays and may be performed at the bedside.

In the current study we have demonstrated an alternative method, which we call the 'TOI method' to calculate absolute Hb_{tc} and CBV using a single NIRS instrument during small hypoxic swings. We conclude that the 'TOI method' is better than the previous used 'SaO$_2$ method' with the major advantage being the use of a single system.

5. ACKNOWLEDGMENTS

The authors would like to thank all the volunteers who participated in this study and Hamamatsu Photonics KK for providing the NIRO 300 spectrometer. This work was supported by the EPSRC/MRC, grant No GR/N14248/01 (IT) and the Wellcome Trust grant No GR/062558 (TSL).

6. REFERENCES

1. M. Ferrari, L. Mottola, V. Quaresima, Principles, Techniques and Limitations of Near Infrared Spectroscopy. *Can J Appl Physiol* **29**(4), 463-487, (2004).
2. H. Obrig, , A. Villringer, Beyond the visible--imaging the human brain with light, *J Cereb.Blood Flow Metab* **23**(1), 1-18, (2003).
3. S.J. Matcher, P. Kirkpatrick, K. Nahid, M. Cope, D.T. Delpy, Absolute quantification methods in tissue near infrared spectroscopy, *Proc.SPIE* **2389**, 486-495 (1995).
4. S.J. Matcher, K. Nahid, M.Cope, D.T. Delpy, Absolute SO2 measurements in layered media, *OSA TOPS on Advantages in Optical Imaging and Photon Migration* **3**, 83-91 (1996).
5. D. T. Delpy, M. Cope, Quantification in tissue near-infrared spectroscopy, *Philosophical Transactions of the Royal Society of London Series B-Biological Sciences* **352**(1354), 649-659 (1997).
6. J. Choi, M. Wolf, V. Toronov, U. Wolf, C. Polzonetti, D. Hueber, LP Safonova, R. Gupta, A. Michalos, W. Mantulin, E. Gratton, Noninvasive determination of the optical properties of adult brain: near-infrared spectroscopy approach, *Journal of Biomedical Optics* **9**(1), 221-229 (2004).
7. J.S. Wyatt, M. Cope, D.T. Delpy, C.E. Richardson, A.D. Edwards, S. Wray, E.O. Reynolds, Quantitation of cerebral blood volume in human infants by near-infrared spectroscopy, *Journal of Applied Physiology* **68**, 1086-1091 (1990).
8. C.E. Elwell, M. Cope, A.D. Edwards, J.S. Wyatt, D.T. Delpy, E.O. Reynolds, Quantification of adult cerebral hemodynamics by near-infrared spectroscopy, *Journal of Applied Physiology* **77**, 2753-2760 (1994).
9. H. Owen-Reece, C.E. Elwell, J.S. Wyatt, D.T. Delpy, The effect of scalp ischaemia on measurement of cerebral blood volume by near-infrared spectroscopy, *Physiol Meas* **17**, 279-286 (1996).
10. A.K Gupta, D.K. Menon, M. Czosnyka, P. Smielewski, P.J. Kirkpatrick, J.G. Jones, Non-invasive measurement of cerebral blood volume in volunteers, *British Journal of Anaesthesia* **78**, 39-43 (1997).
11. M.J.T. Van de Ven, W.N.J.M. Colier, M.C. van der Sluijs, D. Walraven, B. Oeseburg, H. Folgering Can cerebral blood volume be measured reproducibly with an improved near infrared spectroscopy system? *J Cereb.Blood Flow Metab* **21**, 110-113 (2001).
12. S. Suzuki, S. Takasaki, T. Ozaki, Y. Kobayashi, A tissue oxygenation monitor using NIR spatially resolved spectroscopy, *Proc.SPIE* **3597**, 582-592 (1999).
13. A.A. Lammertsma, D.J. Brooks, R.P. Beaney, D.R. Turton, M.J. Kensett, J.D. Heather, J. Marshall, T. Jones, In vivo measurement of regional cerebral haematocrit using positron emission tomography, *J. Cereb. Blood Flow Metab.* **4**, 317-322 (1984).
14. D.T. Delpy, M. Cope, P. van der Zee, S. Arridge, S. Wray, J.S. Wyatt Estimation of optical pathlength through tissue from direct time of flight measurement, *Physics in Medicine and Biology* **33**(12), 1433-1442 (1988)
15. A.Duncan, J.H. Meek, M. Clemence, C.E. Elwell, P. Fallon, L. Tyszczuk, M. Cope, D.T. Delpy, Measurement of cranial optical path length as a function of age using phase resolved near infrared spectroscopy, *Pediatric Research* **39**(5), 889-894 (1996).
16. T.S. Leung, C.E. Elwell, I. Tachtsidis, J.R. Henty, D.T. Delpy, Measurement of the optical properties of the adult human head with spatially resolved spectroscopy and changes of posture, *Adv Exp Med & Biol* **540**, 13-18 (2004).
17. M. Firbank, C.E.Elwell, C.E Cooper, D.T. Delpy, Experimental and theoretical comparison of NIR spectroscopy measurements of cerebral hemoglobin changes, *Journal of Applied Physiology* **85**, 1915-1921 (1998).
18. E. Rostrup, I. Law, F. Pott, K. Ide, G.M. Knudsen, Cerebral haemodynamics measured with simultaneous PET and near-infrared spectroscopy in humans, *Brain Research* **954**, 183-193 (2002).
19. F. Sakai, K. Nakazawa, Y. Tazaki, I. Katsumi, H. Hino, H. Igarashi, T. Kanda, Regional cerebral blood volume and haematocrit measured in normal human volunteers by single emission computed tomography, *J. Cereb. Blood Flow Metab.* **5**, 207-213 (1985).

20. G.C. Newman, E. Delucia-Deranja, A. Tudorica, F.E. Hospod, C.S. Patlak, Cerebral blood volume measurements by T2-weighted MRI and contrast infusion, *Magnetic Resonance in Medicine* **50**, 844-855 (2003).
21. F.F. Jöbsis, Noninvasive infrared monitoring of cerebral and myocardial oxygen sufficiency and circulatory parameters, *Science* **198**, 1264-1267 (1977).

39.

DO SLOW AND SMALL OXYGEN CHANGES AFFECT THE CEREBRAL CYTOCHROME OXIDASE REDOX STATE MEASURED BY NEAR-INFRARED SPECTROSCOPY?

Martin Wolf, Matthias Keel, Vera Dietz, Kurt von Siebenthal, Hans-Ulrich Bucher, Oskar Baenziger[*]

1. INTRODUCTION

Cytochrome oxidase ($CytO_2$), an enzyme of the mitochondrial electron transport chain, is responsible for the majority of cellular oxygen consumption. The change of the redox state of the CuA center of the $CytO_2$ can be detected by near infrared spectroscopy (NIRS). NIRS techniques make it possible to non-invasively measure quantitative changes in $CytO_2$ concentration[1,2]. As cellular oxygen levels are well above the saturating concentration, it might be expected that $CytO_2$, in vivo, would have a very low level of reduction and would be insensitive to small changes in oxygen supply during normoxia. From isolated mitochondria Km is 0.01 kPa[3] and the critical PaO_2 0.9kPa[4], therefore changes in $CytO_2$ redox state are only expected to occur in severe hypoxia. However, in vivo studies with optical spectrophotometry have not confirmed these hypotheses. In adult subjects the $CytO_2$ redox state is altered by changing the arterial oxygen saturation (SaO_2)[5], which may be explained by the concept that some cells are always at the verge of compromised oxygenation, because cells have different distances to the nearest blood vessel. In contrast, neonatal studies on the influence of changes in SaO_2 did not reveal any reproducible and systematic correlation of SaO_2 and $CytO_2$ redox state[6]. Therefore the importance of measurements of $CytO_2$ is very controversial and direct validations of the data have not yet been performed[7].

The aim of this study was to evaluate changes of cerebral $CytO_2$ redox state, induced by slow and small changes in SaO_2, and to investigate the influence of different algorithms to calculate changes in $CytO_2$ concentration.

[*] Clinic for Neonatology, University Hospital, Frauenklinikstr. 10, CH-8091 Zurich
email: martin.wolf@alumni.ethz.ch

2. METHODS

During the first two days of life, in 22 mechanically ventilated preterm neonates (mean gestational age 30.4±3.4weeks, birth weight of 1539±706g) a minimum of 6 slow O_2 changes ($\Delta SaO_2 > 1.5\%$) was achieved by altering the inspired O_2 fraction (Fig. 1). The infants were asleep. SaO_2 was measured by pulse oxymetry (Hellige SMK132 or Nellcor N-200) on the right hand and kept between 85 and 95%. The pCO_2, measured by continuous transcutaneous CO_2-monitoring (Hellige, Germany), was constant during the whole procedure. Arterial blood pressure was monitored (Hellige SMK 132, Germany) via an intra-aortic catheter. Cerebral parameters were measured by a Critikon 2020 Cerebral RedOx Monitor at four wavelengths at 776.5, 819.0, 871.4 and 908.7nm. Neonatal blood contains fetal hemoglobin with a higher O_2 affinity than adult hemoglobin. This leads to a lower pO_2 in neonatal tissue.

Critikon algorithm: The goal of the Critikon algorithm is to determine cerebral concentrations of hemoglobin without the influence of the superficial layers, which is achieved using a multi-distance approach and a coupling compensation system. The algorithm is described in detail in the literature[8]. The absorption spectra were taken from the literature: hemoglobin[9], $CytO_2$[10].

UCL algorithm: If only the signal from detector at 37mm is analyzed, the Critikon instrument is technically equivalent to systems such as the Hamamatsu (Japan) NIRO 500 (wavelengths: 775, 810, 870, 904nm), and NIRO 300 (wavelengths: 775, 807, 850, 913nm) or the Critikon 2001 (UK). The algorithm corresponds to the UCL4 algorithm[11] and implements the absorption spectra supplied by Hamamatsu.

The data were averaged over 10s. The beginning and end of an O_2 change was identified by looking at the SaO_2 trace only.

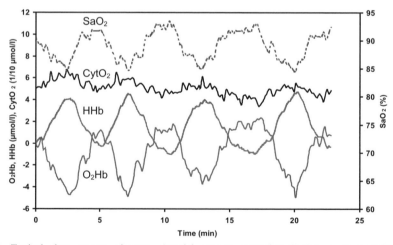

Figure 1. Typical slow oxygen changes. Arterial oxygen saturation (SaO_2), oxyhemoglobin (O_2Hb), deoxyhemoglobin (HHb) and cytochrome ($CytO_2$).

A program was used to calculate the difference between the beginning and end, and the mean for SaO_2, O_2Hb, HHb, $tHb = O_2Hb+HHb$, $CytO_2$, heart rate, blood pressure, and pCO_2. Variables were examined using bivariate linear regression analysis (SPSS, NJ,

USA). The study was approved by the Ethics Committee of the University Children's Hospital of Zurich.

3. RESULTS

11% of the O_2 changes were eliminated due to insufficient magnitude ($\Delta SaO_2 < 1.5\%$) or movement artifacts (sudden steps in SaO_2). 298 increases or decreases in SaO_2 were analyzed. The median change in SaO_2 was 5.9% (range 1.5 - 12%). The median pCO_2 was 5.9kPa (4.4 - 8.3kPa) and heart rate was 146bpm (117 - 186bpm). Changes in tHb, O_2Hb, HHb, O_2Hb- HHb and $CytO_2$ for the two applied algorithms are given in table 1.

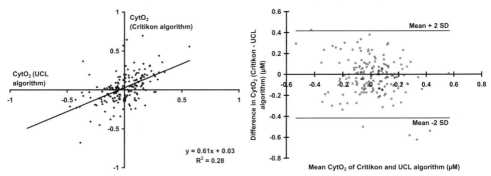

Figure 2. Left: Relation between the changes of the cytochrome oxidase concentration ($CytO_2$) for every oxygen change calculated by the two algorithms. Right: Bland-Altman plot to compare the agreement of the two algorithms.

Table 1. Median changes of hemoglobin and cytochrome concentration in μM per % change in SaO_2 induced by slow and small SaO_2 changes and calculated with the Critikon and UCL algorithm

Algorithm	$\Delta tHb/\Delta SaO_2$	$\Delta O_2Hb/\Delta SaO_2$	$\Delta HHb/\Delta SaO_2$	$\Delta O_2Hb-\Delta HHb/\Delta SaO_2$	$\Delta CytO_2/\Delta SaO_2$
Critikon	0.114	0.941	-0.768	1.645	-0.014
UCL	0.043	0.574	-0.535	1.116	-0.016

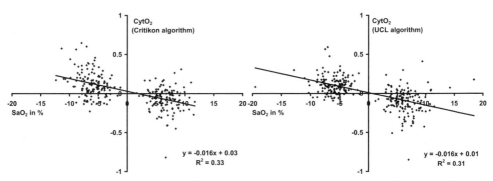

Figure 3. Relation between the changes of the cytochrome oxidase concentration ($CytO_2$) and arterial oxygen saturation (SaO_2) for every oxygen change (Left: Critikon algorithm; right: UCL algorithm).

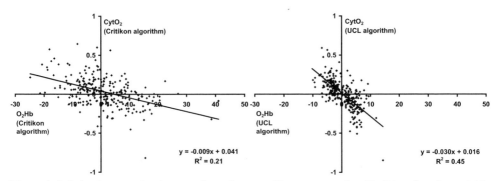

Figure 4. Relation between the changes of cytochrome oxidase concentration (CytO$_2$) and oxyhemoglobin concentration (O$_2$Hb) for every oxygen change (Left: Critikon algorithm; right: UCL algorithm).

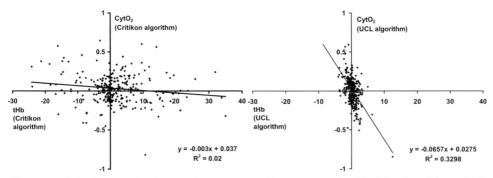

Figure 5. Relation between the changes of cytochrome oxidase concentration (CytO$_2$) and total hemoglobin concentration (tHb) for every oxygen change (Left: Critikon algorithm; right: UCL algorithm).

The correlation coefficient (R^2) for the two algorithms used for O$_2$Hb was 0.79 ($p<0.005$) and for CytO$_2$ (Fig. 2) was 0.28 ($p<0.01$). Further results are displayed in Fig. 3, 4 and 5. The R^2 between CytO$_2$ and O$_2$Hb-HHb was 0.36 ($y=-0.015x+0.011$, $p<0.01$) for the UCL, or 0.20 ($y=-0.007x + 0.037$, $p>0.05$) for the Critikon algorithm.

4. DISCUSSION

In the mitochondrial respiratory chain CytO$_2$ is the terminal electron acceptor, that catalyses the O$_2$ consumption, and is needed to generate cellular ATP. One of the four redox active metal centers of the CytO$_2$ can be detected with in vivo NIRS (CuA). The factors that control the steady-state redox level of CuA in CytO$_2$ are controversial[12].

There are concerns about the reliability of the algorithms used to determine the redox state of CytO$_2$ by NIRS[11], their interchangeability from one biological system to another (i.e. muscle, adult and neonatal brain)[6,11]. The concentration of CuA is less than 10% of that of hemoglobin and is therefore difficult to distinguish from hemoglobin[12]. Thus small errors in the algorithm or measurement, can potentially lead to significant errors in the estimation of CytO$_2$.

In a hemoglobin free fluorocarbon animal model the reduction of $CytO_2$ was only observed for severely impaired O_2 delivery[13-15] leading to the conclusion that no substantial changes in cerebral $CytO_2$ redox state occur at physiological levels of O_2 delivery and that NIRS detects O_2-dependent changes only when O_2 delivery is extremely compromised[15]. In contrast, one study demonstrates a decrease in $CytO_2$ before the O_2 consumption of the brain of fluorocarbon perfused cats decreases[16].

Two algorithms of commercial NIRS instruments were compared. The correlation between O_2Hb values was reasonable, but the Critikon algorithm yielded higher values, probably because the Critikon algorithm is more sensitive to deeper tissue layers. For the $CytO_2$, the values obtained by the two algorithms were less correlated. The R^2 values are higher for the UCL- than for the Critikon-algorithm. The Critikon algorithm measures the whole optical signal, of which the chromophore concentration changes constitute a small percentage. The UCL algorithm considers changes in the optical signal. Thus the signal to noise ratio is better for the UCL algorithm, which is reflected in higher R^2 values.

To our surprise we found a negative correlation of $CytO_2$ redox state with changes in O_2Hb and SaO_2, which is particularly significant for the UCL algorithm. A physiological explanation for an oxidation of $CytO_2$ simultaneous to a decrease in O_2Hb, tHb and SaO_2 is difficult. During cardiopulmonary bypass operations of cyanotic congenital heart defects $CytO_2$ decreased while the cerebral (O_2Hb - HHb) increased[7], but these changes may be due to the induced hypothermia[17]. In piglets undergoing gradual hypoxia (FiO_2 of 12% and 8%), a small oxidation of $CytO_2$ occurred[18]. In adult rats in a hyperoxic state, the increase in O_2Hb continued but the concentration of $CytO_2$ in its oxidized form remained stable[19]. A dissociation of the responses of O_2Hb and $CytO_2$ was also observed after the injection of epinephrine under severely hypoxic conditions, i.e. $CytO_2$ was reoxidized with increasing blood pressure while hemoglobin oxygenation was not changed[19]. Thus, an increase in SaO_2 could lead to a decrease in CBF by reducing sympathetic vasoconstriction and temporarily decreasing O_2 delivery. This hypothesis could be true during marginal tissue oxygenation, which is not expected in our patients. Indeed it seems a conclusion from recent studies that O_2 dependent redox changes of $CytO_2$ occur only at extremely compromised O_2 delivery[20]. In contrast to our results a study on 8 ventilated preterm infants[6] failed to demonstrate a correlation of changes in SaO_2 (85-99%) and $CytO_2$, whereas an increase in pCO_2 from 4.3 to 9.6 kPa was accompanied by an increase in tHb and $CytO_2$. The authors argued, that due to the low cerebral metabolism in preterm infants, the O_2 gradient between blood and mitochondria is small and therefore small changes in SaO_2 do not affect $CytO_2$. A coherence between SaO_2 and $CytO_2$ was suggested primarily for rapidly metabolizing tissue[21]. Neonatal and adult $CytO_2$ may act differently to changes in SaO_2[5]. In animal studies it is described that the activity of Na+, K+-ATPase and of $CytO_2$ increase with gestational age and postnatal age[22,23]. This may be an explanation for a missing coherence between $CytO_2$ and SaO_2, but not for the negative correlation in our patients.

Thus, our $CytO_2$-signal is probably erroneous. An instrument that measures the wavelength dependence of the optical pathlength may provide more accurate results[11]. By inducing forearm ischemia an initial increase in $CytO_2$, with about half the size of the increase in our study, was observed[11]. The authors concluded that some algorithms including the UCL-algorithm produce questionable results for $CytO_2$ and found evidence for significant crosstalk between the hemoglobin signal and $CytO_2$ signals during the hyperemic phase. Our results extend this observation to the neonatal brain.

5. CONCLUSION

Changes in concentrations of O_2Hb, HHb and tHb, induced by small and slow O_2 changes, measured by NIRS over the neonatal brain, are inversely correlated to the measured changes in the oxidation state of $CytO_2$. For small O_2 changes the methodology $CytO_2$ measurement needs to be improved.

6. REFERENCES

1. D.W. Lubbers, in: *Oxygen Supply: Theoretical and Practical Aspects of Supply and Microcirculation in Tissue*, edited by M. Kessler, et al. (University Park, Baltimore,1973), 151-155.
2. H. Beinert, R.W. Shaw, R.E. Hansen, and C.R. Hartzell, Studies on the origin of the near infrared (800-900 nm) absorption of cytochrome c oxidase. *Biochim. Biophys. Acta.* 591, 458-470 (1980).
3. H.F. Bienfait, J.M. Jacobs, and E.C. Slater, Mitochondrial oxygen affinity as a function of redox and phosphate potentials, *Biochim. Biophys. Acta.* **376**, 446-57 (1975).
4. F.F. Jobsis, Oxidative metabolism at low pO_2, *Fed. Proc.* **31**, 1404-13 (1972).
5. N.B. Hampson, E.M. Camporesi, B.W. Stolp, R.E. Moon, J.A. Griebel, and C.A. Piantadosi, Cerebral oxygen availability by NIR spectroscopy during transient hypoxia in humans, *J. Appl. Physiol.* **69**, 907-13 (1990).
6. A.D. Edwards, G.C. Brown, M. Cope, D.T. Delpy, and E.O. Reynolds, Quantification of concentration changes in neonatal human cerebral oxidized cytochrome oxidase. *J. Appl. Physiol.* **71**, 1907-13 (1991).
7. L. Skov and G. Greisen, Apparent cerebral cytochrome aa3 reduction during cardiopulmonary bypass in hypoxaemic children with congenital heart disease, *Physiol. Meas.* **15**, 447-57 (1994).
8. M. Wolf, P. Evans, and H.U. Bucher, The Measurement of Absolute Cerebral Haemoglobin Concentration in Adults and Neonates, *Adv. Exp. Med. Biol.* **428**, 219-227 (1997).
9. S. Wray, M. Cope, D.T. Delpy, J.S. Wyatt, and E.O. Reynolds, Characterization of the near infrared absorption spectra of cytochrome aa3 and haemoglobin for the non-invasive monitoring of cerebral oxygenation. *Biochim Biophys Acta.* **933**, 184-92 (1988).
10. M. Brunori, E. Antonini, and M.T. Wilson, In: *Metal Ions in Biological Systems XXIII*, edited by H. Siegel (Marcel Dekker, New York) 187-228.
11. S.J. Matcher, C.E. Elwell, C.E. Cooper, M. Cope, and D.T. Delpy, Performance comparison of several published tissue near-infrared spectroscopy algorithms, *Anal. Biochem.* **227**, 54-68 (1995).
12. C.E. Cooper and R. Springett, Measurement of cytochrome oxidase and mitochondrial energetics by near-infrared spectroscopy, *Philos. Trans. R. Soc. Lond. B. Biol. Sci.* **352**, 669-676 (1997).
13. M. Ferrari, D.F. Hanley, D.A. Wilson, and R.J. Traystman, Redox changes in cat brain cytochrome-c oxidase after blood-fluorocarbon exchange. *Am. J. Physiol.* **258**, H1706-13 (1990).
14. A.L. Sylvia and C.A. Piantadosi, O_2 dependence of in vivo brain cytochrome redox responses and energy metabolism in blood less rats, *J. Cereb. Blood. Flow. Metab.* **8**, 163-172 (1988).
15. M. Ferrari, M.A. R.J. Traystman, and D.F. Hanley, Cat brain cytochrome-c oxidase redox changes induced by hypoxia after blood-fluorocarbon exchange transfusion, *Am. J. Physiol.* **269**, H417-24 (1995).
16. R. Stingele, B. Wagner, R.J. Traystman, and D.F. Hanley, Reduction of cytochrome-c oxidase copper precedes failing cerebral O_2 utilization in fluorcarbon-perfused cats, *Am. J. Physiol.* **271**, H579-87 (1996).
17. A.J. du Plessis, J. Newburger, H. Naruse, M. Tsuji, A. Walsh, G. Walter, D. Wypij, and J.J. Volpe, Cerebral oxygen supply and utilization during infant cardiac surgery, *Ann. Neurol.* **37**, 488-97 (1995).
18. M. Tsuji, H. Naruse, J. Volpe, and D. Holtzman, Reduction of cytochrome aa3 measured by near-infrared spectroscopy predicts cerebral energy loss in hypoxic piglets, *Pediatr. Res.* **37**, 253-9 (1995).
19. Y. Hoshi, O. Hazeki, Y. Kakihana, and M. Tamura, Redox behavior of cytochrome oxidase in the rat brain measured by near-infrared spectroscopy, *J. Appl. Physiol.* **83**, 1842-8 (1997).
20. C.E. Cooper, S.J. Matcher, J.S. Wyatt, M. Cope, E.M. Nemoto, and D.T. Delpy, Near-infrared sepctroscopy of the brain: relevance to cytochrome oxidase bioenergetics. *Biochem. Soc. Trans.* **22**, 974-980 (1994).
21. B. Chance, Metabolic heterogeneities in rapidly metabolizing tissues, *L. Appl. Cardiol.* **4**, 207-221 (1989).
22. C. Cooper, M. Sharpe, C. Elwell, R. Springett, J. Penrice, L. Tyszczuk, P. Amess, J. Wyatt, V. Quaresima, and D. Delpy, The cytochrome oxidase redox state in vivo, *Adv. Exp. Med. Biol.* **428**, 449-56 (1997).
23. J.M. Cimadevilla, L.M. Garcia Moreno, M.C. Zahonero, and J.L. Arias, Glial and neuronal cell numbers and cytochrome oxidase activity in CA1 and CA3 during postnatal development and aging of the rat, *Mech. Ageing. Dev.* **99**, 49-60 (1997).

40.

NIRS MEASUREMENT OF VENOUS OXYGEN SATURATION IN THE ADULT HUMAN HEAD

Derek W. Brown[1], Daniel Haensse[1,2], Andrea Bauschatz[1], Martin Wolf[1,2]

1. INTRODUCTION

Knowledge of the venous oxygen saturation (SvO_2) is of significant interest for both general diagnostic purposes and in measurement of the cerebral metabolic rate of oxygen. While measurement of the tissue mixed oxygen saturation (StO_2), a weighted average of arterial, capillary, and venous oxygen saturations provides useful diagnostic information, it tells us little about the state of oxygen utilization in the brain.

In the case of hypoxic-ischemic injury, changes in blood flow and oxygen extraction are likely present before clinical manifestation of the injury[1-4]. However, as increases in blood flow serve to maintain oxygen delivery to the brain during mild hypoxic stress, measurement of StO_2 alone does not provide sufficient information to predict impending cerebral injury. With knowledge of both arterial oxygen saturation (SaO_2) and SvO_2 it is possible to determine cerebral oxygen utilization, or the oxygen extraction fraction (OEF), a measurement that may provide the ability to proactively assess the risk of cerebral hypoxic-ischemic injury[5].

Unfortunately, while non-invasive SaO_2 measurements are now commonplace, techniques allowing accurate non-invasive measurement of cerebral SvO_2 are lacking. The most promising technique, spiroximetry, first proposed by Wolf, has been used to non-invasively measure SvO_2 in the human and pig thigh[6], and in ventilated newborn infants[7], but has not yet been applied on the adult human head.

In the present paper we investigate the use of fast near infrared spectroscopy (NIRS) spiroximetry for the non-invasive measurement of SvO_2 in the adult human head. Furthermore, we determine both the oxygen extraction fraction and the distribution of cerebral blood volume into arterial and venous compartments using tissue, arterial, and venous oxygen saturations measured with spatially resolved NIRS.

---

[1] Clinic for Neonatology, University Hospital, Frauenklinikstr. 10, 8091 Zurich, Switzerland email: derek.brown@usz.ch
[2] Biomedical Engineering Laboratory, Swiss Federal Institute of Technology, Zurich, Switzerland

2. METHODOLOGY

2.1. Near-infrared Spectrophotometry (NIRS)

The NIRS system, MCP II, consists of 4 sets of light emitting diodes at 730 and 830 nm and four silicon PIN photodiode detectors arranged in a paired alternating rectangular pattern encased in medical grade silicon with source-detector separations of 3.75 and 2.5 cm. The system is capable of providing intensity measurements across all pairs with a temporal resolution of 100 Hz. Changes in attenuation are used to calculate changes in concentration of oxy and deoxyhemoglobin using standard NIRS techniques[8].

2.2. Tissue Mixed Oxygen Saturation (StO$_2$)

The combination of source-detector pairs at the outermost corners of the MCP II probe provide sufficiently different source-detector separation distances (3.75 and 2.5 cm) to allow for implementation of the multi-distance tissue oxygenation measurement[9, 10]. The MCP II enables calculation of StO$_2$ with a temporal resolution equal to the measurement frequency (100 Hz), providing one value per measurement. Presented StO$_2$ values are averages over the 30 second measurement period. SaO$_2$ was measured using a standard pulse oximeter.

2.3. Venous Oxygen Saturation (SvO$_2$)

The venous oxygen saturation calculation is based on a similar assumption to that used for the standard calculation of SaO$_2$. Here using the assumption that changes in blood volume at the respiratory frequency result primarily from venous blood.

Briefly, changes in attenuation at the breathing frequency, due to changes in venous blood volume, can be isolated from raw attenuation signals using an FFT approach[7]. Once isolated, these changes are used to calculate changes in concentrations of oxy and deoxyhemoglobin. Respiration induced changes in oxy and deoxyhemoglobin are used to calculate venous oxygen saturation as follows:

$$SvO_2 = \Delta O_2Hb_{resp} / (\Delta HHb_{resp} + \Delta O_2Hb_{resp})$$

where ΔO_2Hb_{resp}, and ΔHHb_{resp} are respiration induced changes in oxy and deoxyhemoglobin, respectively.

2.4. Compartmental Distribution and Oxygen Extraction Fraction (OEF)

Calculated StO$_2$ (multi-distance approach), SvO$_2$ (changes in blood volume at the breathing frequency), and SaO$_2$ (pulse oximeter) were used to determine the percentage of cerebral blood volume in venous and arterial compartments as follows:

$$StO_2 = x \cdot SvO_2 + (1-x)) \cdot SaO_2$$

where x represents the fractional distribution of cerebral blood volume in the venous compartment. OEF was determined by subtraction of SvO$_2$ from SaO$_2$.

2.5. Subjects and Studies

Measurements were performed on 10 healthy human adults. In three of the subjects it was not possible to identify changes in attenuation at the breathing frequency. Data from those subjects were not included in the study. Table 1 presents age, height, and weight of the 7 remaining subjects. The NIRS probe was placed on the right side of the forehead, 2 cm above the right temple. Measurements consisted of a single 30 second period.

3. RESULTS

3.1. Tissue Mixed, Arterial, and Venous Oxygen Saturation

Figure 1 a) presents a representative frequency spectrum showing clearly visible peaks at 0.3 and 1.1 Hz, the breathing and heart rate frequencies, respectively. Figure 1 b) presents calculated tissue mixed, arterial, and venous oxygen saturations for individual subjects. Average arterial, tissue mixed, and venous oxygen saturations were $97.1 \pm 0.6\%$, $71.0 \pm 2.9\%$, and $60.6 \pm 6.1\%$, respectively. Figure 1 b) also presents OEF for individual subjects (dashed line). Average OEF was $31.0 \pm 7.9\%$.

3.2. Compartmental Distribution

Figure 1 b) also presents the percent of cerebral blood volume in the venous compartment, x as described in section 2.4, determined from $SaO2$, $StO2$, and $SvO2$ measurements (solid line). Average distribution of cerebral blood volume in the arterial and venous compartments was $27.1 \pm 12\%$ (range $9 - 41\%$) and $72.9 \pm 12\%$ (range $59 - 91\%$), respectively.

Table 1. Subject height, weight, age, SaO_2, SvO_2, StO_2, x, and OEF

Subject	Height (m)	Weight (kg)	Age (years)	SaO_2 (%)	SvO_2 (%)	StO_2 (%)	x (%)	OEF (%)
1	1.68	58	36	97.0	55.7	69.5	66.5	39.4
2	1.63	60	25	97.0	62.1	66.1	88.6	30.9
3	1.82	75	26	97.5	55.9	72.7	59.5	35.9
4	1.73	72	45	97.0	72.7	75.0	90.6	18.9
5	1.83	85	25	97.0	56.5	71.7	62.5	36.5
6	1.55	61	59	98.0	58.6	69.4	72.6	33.7
7	1.80	95	31	96.0	62.6	72.6	70.0	21.4

a)

b)

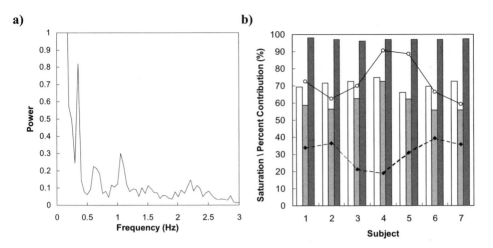

Figure 1. a) Representative frequency spectrum showing peaks at the breathing frequency, 0.3 Hz. **b)** Tissue mixed (white bar), venous (light gray bar), and arterial (dark gray bar) oxygen saturations, percent distribution of cerebral blood volume in the venous compartment (x from section 2.4) (solid line), and OEF (dashed line).

4. DISCUSSION

The method presented in this paper has previously been employed in the non-invasive measurement of both cerebral SvO_2 in ventilated preterm infants and peripheral SvO_2 in the human adult thigh. Wolf et al. report a mean cerebral SvO_2 of $73.0 \pm 8.9\%$ in the ventilated preterm infant[7] while Franceschini et al. report a mean baseline peripheral SvO_2 of approximately 76.5% in the adult human thigh[6]. We report a mean cerebral SvO_2 of $60.6 \pm 6.1\%$ in the adult human head. The apparent discrepancy between these results likely stem from differences in age and measurement location. Watzman et al. measured a mean SvO_2 of approximately 69% using a jugular bulb catheter in infants less than 8 years of age[11], while Pollard et al. report jugular bulb SvO_2 measurements of approximately 53% in healthy human adults[12]. Our findings lie between these two reports.

Elwell et al., using a NIRO 500 NIR spectrometer with a temporal resolution of 2 Hz, report data that suggest changes in cerebral concentrations of O_2Hb and HHb at the breathing frequency result primarily from changes in cerebral blood flow[13]. These conclusions, however, are derived from measurements that do not distinguish between arterial and venous contributions. One possible explanation for this is the use of a NIR spectrometer with comparatively low temporal resolution. Our instrument provides attenuation measurements with a temporal resolution of 100 Hz. Increased data acquisition over individual respiration cycles improves the resolving power of the FFT method and enables isolation of changes in attenuation at the breathing frequency alone, thereby allowing the detection of changes in blood volume at the breathing frequency.

Our approach does, however, assume that O_2Hb and HHb change in phase; an assumption that may not always be valid. Given that out of phase changes in O_2Hb and HHb would severely skew the calculation of venous saturation and that we report values that are within the expected saturation range, it is highly unlikely that our data are affected by this phenomenon.

It should be noted that the FFT method requires that the subject maintain a constant breathing frequency. In the current study, 3 of 10 measurements were not usable, likely due to a lack of constant breathing frequency. For this reason, the technique is best suited for use in ventilated subjects. However, the use of a strain gauge apparatus to monitor breathing frequency may provide sufficient information to isolate changes in attenuation due to changes in venous blood volume in subjects that do not maintain constant breathing frequency.

Under normal physiological conditions the cerebral blood volume is generally considered to be distributed into arterial, capillary, and venous compartments in the ratio 20, 10, 70[14]. Our presented compartmental distribution data are in agreement with this ratio. As some cerebro-vascular diseases alter the distribution of cerebral blood volume, the ability to measure this distribution may prove useful for disease monitoring.

Likely, however, the most interesting use of the presented technique will be found in the combination of SaO_2, SvO_2, and cerebral blood flow measurement. Combined measurement of these three parameters would enable calculation of not only the oxygen extraction fraction but also the cerebral metabolic rate of oxygen, one of the most useful indicators of normal brain function. In the neonate, these measures may provide, for the first time, a clinically usable, affordable means of proactively assessing the risk of, and classifying damage caused by, cerebral hypoxic-ischemic injury.

5. CONCLUSION

Provided that both the breathing frequency remains constant and that the temporal resolution of the instrument is sufficiently high, NIRS spiroximetry enables measurement of cerebral SvO_2 in healthy human adults. Furthermore, simultaneous measurements of StO_2, SaO_2, and SvO_2 enable calculation of both OEF and the compartmental distribution of cerebral blood volume.

6. ACKNOWLEDGMENTS

The authors thank volunteers for their participation.

7. REFERENCES

1. A. L. Carney and E. M. Anderson, The system approach to brain blood flow, Adv Neurol. **30** 1-30 (1981).
2. T. Hauge, Catheter vertebral angiography, Acta Radiol. **109** 1-219 (1954).
3. N. A. Lasse and M. S. Christiansen, Physiology of cerebral blood flow, Br. J. Anaesthesiol. **48** 719-734 (1976).
4. R. Roski, R. F. Spetzler, M. Owen, K. Chandar, J. G. Sholl, and F. E. Nulsen, Reversal of seven-year old visual field defect with extracranial-intracranial arterial anastomosis, Surg Neurol. **10**(4), 267-268 (1978).
5. P. Frykholm, J. L. Andersson, J. Valtysson, H. C. Silander, L. Hillered, L. Persson, Y. Olsson, W. R. Yu, G. Westerberg, Y. Watanabe, B. Langstrom, and P. Enblad, A metabolic threshold of irreversible ischemia demonstrated by PET in a middle cerebral artery occlusion-reperfusion primate model, Acta Neurol Scand. **102**(1), 18-26 (2000).
6. M. A. Franceschini, D. A. Boas, A. Zourabian, S. G. Diamond, S. Nadgir, D. W. Lin, J. B. Moore, and S. Fantini, Near-infrared spiroximetry: Noninvasive measurements of venous saturation in piglets and human subjects, J Appl Physiol. **92**(1), 372-384 (2002).

7.   M. Wolf, G. Duc, M. Keel, P. Niederer, K. von Siebenthal, and H. U. Bucher, Continuous noninvasive measurement of cerebral arterial and venous oxygen saturation at the bedside in mechanically ventilated neonates, Crit Care Med. **25**(9), 1579-1582 (1997).
8.   M. Cope, D. T. Delpy, E. O. Reynolds, S. Wray, J. Wyatt, and P. van der Zee, Methods of quantitating cerebral near infrared spectroscopy data, Adv Exp Med Biol. **222** 183-189 (1988).
9.   D. M. Hueber, M. A. Franceschini, H. Y. Ma, Q. Zhang, J. R. Ballesteros, S. Fantini, D. Wallace, V. Ntziachristos, and B. Chance, Non-invasive and quantitative near-infrared haemoglobin spectrometry in the piglet brain during hypoxic stress, using a frequency-domain multidistance instrument, Phys Med Biol. **46**(1), 41-62 (2001).
10. J. Matcher, P. Kirkpatrick, K. Nahid, M. Cope, and D. T. Delpy, Absolute quantification methods in tissue near-infrared spectroscopy, Proc SPIE. **2389** 486-495 (1995).
11. H. M. Watzman, C. D. Kurth, L. M. Montenegro, J. Rome, J. M. Steven, and S. C. Nicolson, Arterial and venous contributions to near-infrared cerebral oximetry, Anesthesiology. **93**(4), 947-953 (2000).
12. V. Pollard, D. S. Prough, A. E. DeMelo, D. J. Deyo, T. Uchida, and H. F. Stoddart, Validation in volunteers of a near-infrared spectroscope for monitoring brain oxygenation in vivo, Anesth Analg. **82**(2), 269-277 (1996).
13. C. E. Elwell, H. Owen-Reece, M. Cope, A. D. Edwards, J. S. Wyatt, E. O. Reynolds, and D. T. Delpy, Measurement of changes in cerebral haemodynamics during inspiration and expiration using near infrared spectroscopy, Adv Exp Med Biol. **345** 619-626 (1994).
14. M. A. Mintun, M. E. Raichle, W. R. Martin, and P. Herscovitch, Brain oxygen utilization measured with O-15 radiotracers and positron emission tomography, J Nucl Med. 25(2), 177-187 (1984).

41.

WHAT IS TISSUE ENGINEERING?
WHAT IS ISOTT'S ROLE?

Duane F. Bruley[*]

1. INTRODUCTION

Tissue Engineering is still a developing field with its definition being "in the mind of the researcher." Presently there are a plethora of explications for this cutting edge medical technology. While serving at the National Science Foundation (NSF) in the late 1980s we attempted to define Tissue Engineering by sponsoring a national meeting of researchers in La Jolla, CA for this purpose. At that time I proposed that the first real tissue engineers were Dr. August Krogh and Mr. Erlang in 1919. Engineering typically involves system quantification *via* mathematics, which they accomplished with the definition of the Krogh Capillary Tissue Cylinder and the first mathematical model representing the tissue oxygen distribution.

ISOTT objectives include the investigation of tissue and organ function, repair, renewal and replacement *in vitro* and *in vivo*. For example, angiogenesis and blood hemostatis are essential considerations for oxygen and nutrient supply and metabolic byproduct removal from tissue. Oxygen tension measurement and prediction are necessary tools to determine the health of individual cells and organs in the body. In this paper, the "Bruley Template" will be used to illustrate the interaction of Bioprocess Engineering and Biomedical Engineering and the interface with clinical practice for Tissue Engineering applications. Other scientific and engineering challenges will be discussed including various problems that might be encountered in the commercialization of engineered tissue.

2. HISTORICAL BACKGROUND

Tissue Engineering is evolving as an important discipline with potential ubiquitous clinical impact that some researchers estimate could involve as many as thirty-two

[*] Duane F. Bruley, Ph.D., P.E., College of Engineering, University of Maryland Baltimore County, Baltimore, MD 21250.

million procedures per year with a one-hundred billion dollar market for engineered tissues. Some of these tissues might include kidney, urethra, pancreas, liver, blood vessels, bone and cartilage, etc. Various investigators have used different names that include, in my view, a narrow definition of tissue engineering. The terms *regenerative medicine* and *reparative medicine* are used in some circles.

It is my opinion that tissue engineering was first defined through the work of Dr. August Krogh and the mathematician Mr. Agner K. Erlang (Figure 1, Equation 1). Engineering involves the quantification of systems *via* mathematical modeling. This was accomplished in the early 1900s when Dr. Krogh conceptualized the capillary tissue system and Mr. Erlang derived a mathematical model describing oxygen transport in the tissue. This linear model led to an analytical solution for which in part Dr. Krogh received the Nobel Prize in 1919.[1,2,3,4] In reality, vascular beds and the microcirculation are found to be a tortuous network that defies a single capillary solution. However, much valuable insight and information can be gained from simplified models of the complex structure.

Many investigators advanced this initial study by increasing the complexity of the mathematical models. Two of the most significant early works in the analytical modeling arena were accomplished by the work of Opitz and Schneider[5] and Gerard Thews.[6] For conservation of space, I will now use my own work to represent studies that I feel should be included in the definition of "Tissue Engineering." Certainly many other ISOTT investigators have contributed through theoretical and experimental research. In 1962, I met Dr. Melvin H. Knisely in the anatomy department of the medical school of South

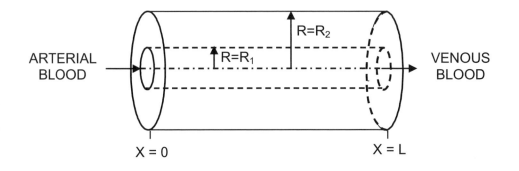

Figure 1. The Krogh Tissue Cylinder

$$\alpha_t \frac{\partial P_t}{\partial t} = k \left[\frac{1}{r} \frac{\partial}{\partial r} r \frac{\partial P_t}{\partial r} + \frac{\partial^2 P_t}{\partial z^2} \right] - M$$

Equation 1. The approximate Krogh-Erlang Tissue Model

Carolina. Since his Quartz Rod Crystal technique determined that over one hundred disease states led to blood agglutination (blood sludging), and since he studied under Dr. Krogh, he was keenly interested in predicting cell survival under many pathologic states. His concern involved the plugging of capillaries which prevented the transport of oxygen and nutrients to the cells as well as the removal of toxic metabolic byproducts *via* the capillaries.

My interest with Knisely was to examine more realistically the functioning of the microcirculation, both mathematically and experimentally, and to search for anticoagulants[7,8] that could be used to prevent cell death. Being experienced in finite difference calculus and computer simulation (analog, hybrid, and digital), I developed the Bruley model (Equations 2, 3, and 4). This model included time dependent, convective, diffusion, and metabolic capillary tissue interaction, and the inclusion of nonlinearities such as the sigmoidal shape of the oxygen dissociation curve in the capillary and the Machaelis-Menten kinetics in the tissue. The model was solved using the Crank-Nicholson and Euler finite difference strategies for a large number of case studies.[9] This was the first computerized model for the microcirculation.

More intricate capillary tissue systems were investigated looking separately at the red blood cell, the plasma phase, and the tissue.[10] This system was solved using simultaneous, time dependent, partial differential equations for the components of interest. These included oxygen, carbon dioxide, lactic acid and glucose.

In an attempt to generate a more realistic picture of oxygen transport in the microcirculation, stochastic models were developed and compared with the deterministic model solutions. Our stochastic model allowed quantification of oxygen levels in and around a single erythrocyte as it flowed from the arterial to the venous end of the

$$\frac{\partial P}{\partial \theta} = \frac{D_1}{\left[1 + \dfrac{N}{C_1}\dfrac{d\psi}{dP}\right]} \left[\frac{\partial^2 P}{\partial r^2} + \frac{1}{r}\frac{\partial P}{\partial r} + \frac{\partial^2 P}{\partial z^2}\right] - V\frac{\partial P}{\partial z} a$$

$$\text{where, } \psi = \frac{KM^m}{1 + KP^m}$$

Equation 2. Bruley Model, The First Computer Solution: Capillary Equation

$$P_{i_{capillary}} = P_{i_{tissue}}$$

$$D_1 C_1 \frac{\partial P}{\partial r}\bigg|_{capillary} = D_2 C_2 \frac{\partial P}{\partial r}\bigg|_{tissue}$$

Equation 3. Bruley Model, The First Computer Solution: Interface Conditions

$$\frac{\partial P}{\partial \theta} = D_2 \left[\frac{\partial^2 P}{\partial r^2} + \frac{1}{r} \frac{\partial P}{\partial r} + \frac{\partial^2 P}{\partial z^2} \right] - \frac{A}{C_2}$$

Equation 4. Bruley Model, The First Computer Solution: Tissue Equation

capillary. These discrete models provided insight into diffusion resistances at the surface of the erythrocyte as well as through the endothelial lining of the capillaries.[11] To predict oxygen tensions throughout heterogeneous capillary tissue systems, a probabilistic/deterministic technique, The Bruley, Williford, Kang (BWK) method[12] was developed. This is a powerful computer strategy that allows gross solutions throughout a complex capillary tissue system. This calculation scheme allows solutions for three-dimensional, time dependent, convection, diffusion/conduction nonlinear simulations. It is extremely fast as compared with finite difference and finite element techniques.

In 1967 Dr. Haim Bicher joined our group to develop polarographic microelectrodes for experimental measurements in the cerebral grey matter of curarized, anesthetized cats, and to compare the results with the computer simulations. The advantage of distributed parameter analysis over lumped parameter analysis was the ability to simulate gradients of oxygen in tissues. In this same area of interest only microelectrodes can be used to approximate or evaluate point values of oxygen and thus measure tissue oxygen gradients from the capillary to the neurons or cells in the tissue.

3. WHAT IS TISSUE ENGINEERING?

The limited contemporary definition of tissue engineering is the synthesis of artificial implants, laboratory grown cells, cells, or molecules to replace or support the function of defective or injured parts of the body. A broader definition of tissue engineering can be illustrated by the Bruley Template (Figure 2).

Problems that limit the successful manufacturing and implantation of tissues might include stem cell biology, tissue scaffold design, angiogenesis, blood hemostasis, and immunogenesis. Some of the areas where progress is being made would include skin, cartilage, bone, kidney, pancreas, cardiac, and liver tissues. To achieve the goal for tissue development, analysis and replacement, it requires the input of many experts. Therefore a team approach is recommended with guidelines for the research as specified by a strategy such as Total Quality Management (TQM). The customer requirements, the patient, determines the composition of the team. This allows the manufacture of the desired tissue product with ultimate safety and efficacy.

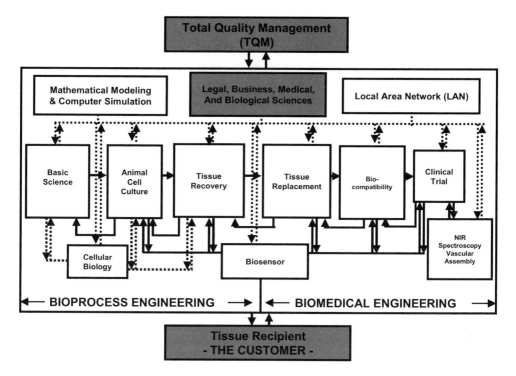

Figure 2. The Bruley Template – Tissue Engineering Design

4. WHAT IS ISOTT'S ROLE?

ISOTT is full of particular examples of research that are integral to the success of modern tissue engineering. However, due to space limitations and a concern regarding offending someone by omission, I have not focused on any of the extremely important current research projects in our society.

This lecture was developed to initiate thinking regarding the long-term role of ISOTT in tissue engineering. This would include expanding the contemporary definition to consider oxygen and other metabolite and anabolite measurements and predictions *via* mathematics, along with such topics as angiogenesis and blood hemostasis for optimum tissue growth and health. Considering this idea the question arises, "Should ISOTT try to recruit more members who are researchers in the new tissue engineering?" Also, "Are there existing subgroups of ISOTT that would be interested in pursuing and interacting with outside groups who are using bioprocessing to develop *in vitro* tissues for implantation?"

5. ACKNOWLEDGEMENTS

The author wishes to thank Dr. William Wagner, Dr. Francis Wang, Dr. Kyung Kang, and Dr. Huiping Wu for material used in the presentation and to Ms. Eileen Thiessen for her help in developing Power Point slides, organizing my talk, typing the manuscript and for feedback on the material.

6. REFERENCES

1. Krogh, A., "The Rate of Diffusion of Gases Through Animal Tissue with Some Remarks on the Coefficient of Invasion," *J. Physiol.*, **52**:391 (1918-1919).
2. Krogh, A., "The Number and Distribution of Capillaries in Muscle with Calculations of the Oxygen Pressure Head Necessary for Supplying the Tissue," *J. Phyiol.*, **52**:409 (1918-1919).
3. Krogh, A., "The Supply of Oxygen to the Tissues and the Regulations of the Capillary Circulation," *J. Physiol.*, **52**:457 (1918-1919).
4. Krogh, A., *The Anatomy and Physiology of Capillaries*, Yale University Press, New Haven, Conn., 1st ed., 1922.
5. Opitz, E. and M. Schneider, "The Oxygen Supply of the Brain and the Mechanism of Deficiency Effects," *Ergebnisse der Physiologie, biologischem Chemie, und experimentallen Pharmakologie*, **46**: 126 (1950).
6. Thews, G., "Oxygen Diffusion in the Brain. A Contribution to the Question of the Oxygen Supply of the Organs," *Pflugers Archiv.*, **271**:197 (1960).
7. Bicher, H. I., D. F. Bruley, and M. H. Knisely, "Anti-Adhesive Drugs and Tissue Oxygenation," *Advances in Experimental Medicine and Biology: International Society on Oxygen Transport to Tissue (ISOTT)*. Plenum Press, **37B**:657 (1973).
8. Bruley, D. F., "Protein C – The Ultimate Anticoagulant/Antithrombotic?" *Anticoagulant, Antithrombotic and Thrombolytic Therapeutics II*, ed. C. A. Thibeault, IBC Library Series Publication, Boston, 129 (1998).
9. Reneau, D. D., Jr., D. F. Bruley, and N. H. Knisely, "A Mathematical Simulation of Oxygen Release, Diffusion, and Consumption in the Capillaries and Tissue of Human Brain," *Chemical Engineering in Medicine and Biology*, D. Hershey, ed., Plenum Press, New York, 135 (1967).
10. Artigue, R. S. and D. F. Bruley, "The Transport of Oxygen, Glucose, Carbon Dioxide and Lactic Acid in the Human Brain: Mathematical Models," *Advances in Experimental Medicine and Biology Series: Oxygen Transport to Tissue Vol. IV*, Plenum Press, **159** (1983).
11. Groome, L. J., D. F. Bruley, and M. H. Knisely, "A Stochastic Model for the Transport of Oxyen to the Brain," *Advances in Experimental Medicine and Biology: Oxygen Transport to Tissue II*, Plenum Publishing Corporation, **75**:267 (1976).
12. Kang, K. A. and D. F. Bruley, "The B-W-K Technique Applied to Problems of Hyperthermia," *AIChE Symposium Series*, ed. M. S. El-Genk, New York, **93**(314):174 (1997).

42

OXYGENATION OF CULTURED PANCREATIC ISLETS

Richard Olsson[1] and Per-Ola Carlsson[2]

1. INTRODUCTION

The rich vascularization of the pancreatic islets was already noted in early studies performed by Langerhans [1]. The dense glomerular-like islet angioarchitecture assures that no portion of an islet is more than one cell away from arterial blood [2,3]. The blood perfusion of the pancreatic islets is also very high and constitutes 7-10% of the whole pancreatic blood flow, despite the islets contributing only 1-2% to the pancreatic volume [4,5]. This high blood perfusion of pancreatic islets results in a tissue oxygen tension of 40 mmHg [6], which is approximately twice as high as in the kidney cortex [7]. This may reflect the need for adequate glucose sensing of the pancreatic β-cells and may facilitate disposal of secreted islet hormones. Pancreatic islets also normally have a very high demand on metabolic activity and oxygen consumption to meet the varying needs for insulin secretion [8,9].

Prior to transplantation, islets are isolated and cultured for a couple of days. Transplantation of cultured islets is preferred compared to transplantation of freshly isolated islets, since culture reduces islet immunogenicity [10] and culture also supplies time to find the most suitable recipient. In order to optimize the condition of the islet graft, it is very important to limit the dysfunction and necrosis/apoptosis of islet cells during the pre-transplantation culture. Endogenous islets are used to a high level of oxygenation *in vivo*, and one must therefore supply enough oxygen *in vitro* to protect the cultured islets from hypoxia. It is well-known that culture of especially large islets results in a central necrosis due to limitations in oxygen diffusion *in vitro*. Studies of islet oxygenation *in vitro* have been limited by a lack of available experimental techniques and in former studies only mathematical calculations of islet cell oxygenation have been performed [11].

[1] Richard Olsson, Department of Medical Cell Biology, Uppsala University, Box 571, 751 23 Uppsala, Sweden. Phone: +46 18 471 40 33. Telefax: +46 18 471 40 59.
 e-mail: richard.olsson@medcellbiol.uu.se
[2] Per-Ola Carlsson, Departments of Medical Cell Biology and Medical Sciences, Uppsala University, Sweden.

Low oxygen tension in cells and tissue may be detected by the use of biochemical markers such as 2-nitroimidazoles [12]. Under low oxygen conditions (pO_2 less than 10 mmHg), bioreductive metabolism of 2-nitroimidazoles proceeds inversely proportional to the oxygen tension, eventually generating reactive adducts that bind to macromolecules in the cells [13]. The 2-nitroimidazole-derived side chains can then be recognized by antibodies and used to assess tissue areas/cells low in oxygen tension. In this study we evaluated the oxygenation of cultured islets with the 2-nitroimidazole pimonidazole, which is known to accumulate in tissues with an oxygen tension less than 10 mmHg.

2. METHODS

2.1. Animals

Adult male C57BL/6 mice (~30 g) and Wistar-Furth rats (~300 g) were purchased from B&K, Sollentuna, Sweden. The animals had free access to water and pelleted food throughout the course of the study. All experiments were approved by the local animal ethics committee of Uppsala University.

2.2. Islet Isolation and Culture

Pancreatic islets from mice and rats were prepared by collagenase digestion and cultured free-floating in groups of 150 islets in 5 ml RPMI 1640 medium supplemented with 11.1 mmol/l glucose, 10% (vol/vol) fetal calf serum, 0.17 mmol/l sodium benzylpenicillinate and 0.17 mmol/l streptomycin at 95%air/5%CO_2 [14]. The medium was changed every second day.

Human islets were kindly provided by Professor Olle Korsgren, Department of Clinical Immunology, Uppsala University Hospital, Sweden. Approximately 10 000 human islets were cultured in 10 ml RPMI 1640 medium supplemented with 5.6 mmol/l glucose, 10% (vol/vol) human serum, 0.17 mmol/l sodium benzylpenicillinate and 0.17 mmol/l streptomycin at 95%air/5%CO_2. The medium was changed every second day.

2.3. Evaluation of Oxygen Dependent Accumulation of Pimonidazole

Rat islets were cultured for 1, 5 or 10 days. The islets were split in two experimental groups, where one group was incubated with a normal amount of medium (5 ml) and the second group was incubated with an increased amount of medium (15 ml). After 24h of culture, all culture dishes were incubated, in the presence of pimonidazole (200 µmol/l), at 95%air/5%CO_2 for 2 hours. The rationale for this experimental set-up was that excess non-stirred culture medium markedly increases the diffusion distance of oxygen from air to the pancreatic islets and, thus, decrease the oxygenation of isolated pancreatic islets with resulting increased frequency of central islet necrosis [15].

In order to evaluate if oxygen dependent accumulation of pimonidazole occur in islets *in vivo*, rats were anesthetized with thiobutabarbital sodium (120 mg/kg body weight i.p.; Inactin; Research Biochemicals International, Natick, MA), placed on an operating table maintained at body temperature (37°C) and tracheostomized. Polyethylene catheters were inserted into the right femoral artery and vein. The arterial catheter was pre-heparinized to avoid blood clot formation and was connected to a blood pressure transducer (PDCR 75; Groby, UK) to monitor blood pressure. The abdominal cavity of the animals was opened by a midline incision and the pancreas was exposed and its surface islets visualized, as previously described [16].

In brief, the pancreas was immobilized over a cylindrical plastic block attached to the operating table and superfused with mineral oil to prevent excessive evaporation. Sterile-filtered neutral red (0.8 ml, 2% wt/vol, Kebo Grave, Stockholm, Sweden) was administered intravenously to selectively stain the islets within the pancreas [16]. We have previously evaluated the use of neutral red without noticing any adverse effects on islet function, blood flow or oxygen tension in the endogenous pancreas [6,16]. Tissue oxygen tension in surface pancreatic islets was then measured with a polarographic technique [6]. Modified Clark microelectrodes with an outer tip diameter of 2-6 μm (Unisense, Aarhus, Denmark) [17] were polarized at -0.8V, which gave a linear response between the oxygen tension and the electrode current. The latter was measured by a picoamperemeter (University of Aarhus, Aarhus, Denmark). Two-point calibration of the electrodes was performed in water saturated with $Na_2S_2O_5$ or water saturated with air at 37°C, respectively. The electrodes were inserted into pancreatic islets by the use of a micromanipulator under a stereomicroscope. Islet oxygen tension and femoral artery blood pressure were then continuously recorded during the application of a graded aortic vascular clamp superior to the celiac artery. The vascular clamp was positioned to allow a femoral arterial blood pressure of ~20 mmHg and an islet oxygen tension of 5-10 mmHg for 2 hours, meanwhile pimonidazole (60 mg/kg), injected intravenously at the start of the vascular clamp, were allowed to bind to low oxygenated regions. Blood pressure, islet oxygen tension and body temperature were continuously recorded during the experiment with a MacLab Instrument (AD instruments, Hastings, UK) connected to a Macintosh Power-PC 6100 (Apple Computers, Cupertino, CA).

In control experiments, pimonidazole (60 mg/kg) was injected intravenously to awake rats. After 2 hours the animals were anesthetized, sacrificed by cervical dislocation and the pancreata were removed for immunohistological evaluation of pimonidazole accumulation.

2.4. Immunohistochemical Staining for Pimonidazole

Isolated islets were fixed in 4% paraformaldehyde overnight, dehydrated in ethanol and embedded in paraffin. Pancreata and livers were fixed in 10% (vol/vol) neutral buffered formalin overnight, washed in water, dehydrated in ethanol and embedded in paraffin. Sections, 5 μm thick, were mounted on L-polylysine-treated glass slides. The slides were deparaffinized, washed and stained with antibody for pimonidazole according to manufacturer's instructions (Chemicon International, Temecula, CA).

An automatic staining procedure (Dako Autostainer: DakoCytomation, Älvsjö, Sweden) was applied to facilitate reproducible staining. The slides were incubated with Dako blocking solution (DakoCytomation) for 30 min and incubated with primary antibody against pimonidazole (Chemicon International), diluted 1:25 in TBS with 0.02% (vol/vol) Brij 35, for 40 min at room temperature (RT). The slides were washed in TBS and incubated with a biotinylated secondary goat anti-mouse antibody (Southern Biotechnology Inc., Birmingham, AL), diluted 1:500 in TBS with 0.02% (vol/vol) Brij 35 and 0.035% (vol/vol) Dako blocking solution for 30 min at RT. The slides were again washed and incubated with Alkaline Phosphatase Standard ABC kit (Vector Laboratories, Burlingame, CA) for 30 min at RT. The slides were washed and incubated with Vector Red Alkaline Phosphatase substrate kit (Vector Laboratories) for 25 min. The slides were counterstained with hematoxylin. Liver tissue was used as positive control slides, whereas omission of primary antibody was used as negative control.

3. RESULTS

All animals were normoglycemic (non-fasting blood glucose < 9 mmol/l). The staining pattern for pimonidazole was similar in mouse, rat and human islets. A majority of isolated pancreatic islets incubated at normal culture conditions (95%air/5%CO₂, 5 ml culture medium) stained negative for pimonidazole, irrespective of preceding culture period. In islets which stained positive for pimonidazole, an obvious gradient with pronounced central staining and unstained periphery of the islets was often discerned (Figure 1). In some larger islets, especially when investigated without preceding culture, central necrosis surrounded by pimonidazole-staining was seen. Incubation of islets with excessive culture medium (15 ml) increased the fraction of pimonidazole positive islets. Isolated large pancreatic islets incubated with excessive culture medium regularly developed central necrosis, which was surrounded by cells staining positive for pimonidazole. The accumulation of pimonidazole in the islets was also increased by increased density of islet tissue in the culture dish. Moreover, cluster formation of islets resulted in increased staining for pimonidazole and islets in the centre of the cluster were occasionally completely stained for pimonidazole.

A   **B**

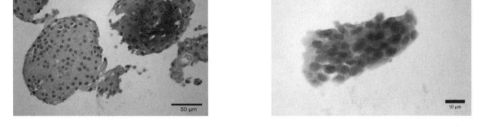

Figure 1. Mouse (A) and human (B) islets exposed to pimonidazole (200 μmol/l) for 2 hours during standard culture conditions and thereafter stained with antibody for adducts of pimonidazole incorporated intracellularly.

All pancreatic islets in the animals subjected to a graded aortic vascular clamping applied cranially to the celiac artery stained strongly positive for pimonidazole and had pronounced dilatation of islet capillaries. The surrounding exocrine pancreatic parenchyma stained more weakly for pimonidazole, but for the peripheral margins of the pancreatic lobuli. In control rats the exocrine pancreatic parenchyma in all cases remained negative for pimonidazole and almost all islets were unstained.

4. DISCUSSION

For clinical transplantation islets are cultured prior to implantation. During culture an adequate oxygenation of the islets is important to prevent cell dysfunction and death, which may reduce the success rate of the transplantations. In the present study, we aimed to evaluate the oxygenation of pancreatic islets *in vitro*. For this purpose we used pimonidazole, which is a 2-nitrimidazole that accumulates in cells with an oxygen tension less than 10 mmHg. Pimonidazole has, to our knowledge, not been used in any former studies of islet oxygenation. We report that the intercellular incorporation of pimonidazole in pancreatic islets cells is increased by experimentally induced hypoxia both *in vitro* and *in vivo*. The fraction of pimonidazole-positive islets is increased by increased density of islets in the culture dish. Moreover, cluster formation of islets in the culture dish results in an increased accumulation of pimonidazole, most likely due to increased oxygen consumption and decreased oxygen diffusion in that particular spot with a high density of islet tissue.

In contrast to islets *in vivo*, no isolated islets stained homogenously with pimonidazole. Instead, an obvious gradient with pronounced central staining and unstained periphery of the islets was often observed, which argues against oxygen-insensitive nitroreduction as cause of pimonidazole-binding to pancreatic islet cells. In some larger islets investigated *in vitro*, central necrosis surrounded by pimonidazole-staining was seen. The necrotic islet core did not stain, because accumulation of 2-nitroimidazoles requires functional nitroreductase activity, which does not occur in dead cells [18].

Islet cells exposed to oxygen tension levels less than 12 mmHg in vitro have in perifusion experiments been observed to have disturbed glucose-stimulated insulin secretion [19]. Moreover, experiments in an insulinoma cell line, βTC3 cells, have reported that oxygen tension levels below 25 mmHg gradually shift these cells from aerobic to anaerobic metabolism with a concomitant increased lactate production [20,21]. This suggests that pimonidazole-positive islet cells, which have an oxygen tension less than 10 mmHg, are hypoxic and dysfunctional.

Since human islets are cultured more densely than rodent islets, they are less oxygenated (cf. above). Good culture conditions are crucial to prepare the human islets for transplantation and the oxygenation during culture most be optimized to avoid islet cell dysfunction and death. Those islets that despite of the hypoxia survive the culture are most likely depleted of ATP and this will probably make them more vulnerable in the hypoxic environment of the implantation organ in the early post-transplantation period [22].

In conclusion, the present study suggests that 2-nitroimidazoles are suitable for measurements of islet oxygenation both *in vitro* and *in vivo*. During culture islet oxygenation is impaired by high densities of islet tissue. Less dense cultures of human islets will increase the oxygenation of the islets and may improve islet quality at transplantation.

5. ACKNOWLEDGEMENTS

The skilled technical assistance of Eva Törnelius is gratefully acknowledged. This study was supported by grants from The Swedish Research Council (72XD-15043), The Juvenile Diabetes Research Foundation, The Swedish Society of Medicine, The Swedish Society for Medical Research, The Novo Nordic Fund, The Swedish Juvenile Diabetes Fund, The Magnus Bergvall Foundation, The Åke Wiberg Foundation, The Thuring Foundation, The Lars Hierta Memorial Fund, The Siblings Svensson Foundation, The Anér Foundation, The Clas Groschinsky Memorial Fund and The Family Ernfors Fund.

6. REFERENCES

1. P. Langerhans, Beiträge zur mikroskropischen Anatomie der Bauchspeiseldrüse, *Inaugural dissertation* (Gustave Lange, Berlin, 1869).
2. S. Bonner-Weir, The microvasculature of the pancreas, with special emphasis on that of the islets of Langerhans. Anatomy and functional implications, in: *The Pancreas. Biology, pathobiology and*

disease. 2nd ed., edited by V.L.W. Go, E.P. DiMagno, J.D. Gardner, E. Lebenthal, H.A. Reber, and G.A. Scheele (Raven Press, New York, 1993), pp. 759-768.

3. F.C. Brunicardi, J. Stagner, S. Bonner-Weir, H. Wayland, R. Kleinman, E. Livingston, P. Guth, M. Menger, R. McCuskey, M. Intaglietta, A. Charles, S. Ashley, A. Cheung, E. Ipp, S. Gilman, T. Howard, E. Passaro Jr, Microcirculation of the islets of Langerhans, *Diabetes* 45, 385-392 (1996).

4. N. Lifson, K.G. Kramlinger, R.R. Mayrand, E.J. Lender, Blood flow to the rabbit pancreas with special reference to the islets of Langerhans, *Gastroenterology* 79, 466-473 (1980).

5. L. Jansson, The regulation of pancreatic islet blood flow, *Diabetes Metab. Rev.* 10, 407-416 (1994).

6. P-O Carlsson, P. Liss, A. Andersson, L. Jansson, Measurements of oxygen tension in native and transplanted rat pancreatic islets, *Diabetes* 47, 1027-1032 (1998).

7. P-O Carlsson, F. Palm, A. Andersson, P. Liss, Markedly decreased oxygen tension in transplanted rat pancreatic islets irrespective of the implantation site, *Diabetes* 50(3), 489-495 (2001).

8. C. Hellerström, Effects of carbohydrates on the oxygen consumption of isolated pancreatic islets of mice, *Endocrinology* 81, 105-114 (1967).

9. J.C. Hutton, W.J. Malaisse, Dynamics of O_2 consumption in rat pancreatic islets, *Diabetologia* 18, 395-405 (1980).

10. M. Kedinger, K. Haffen, J. Grenier, R. Eloy, In vitro culture reduces immunogenicity of pancreatic endocrine islets, *Nature* 270, 736-738 (1977).

11. C.K. Colton, Implantable biohybrid artificial organs, *Cell Transplant.* 4(4), 415-436 (1995).

12. R.J. Hodgkiss, Use of 2-nitroimidazoles as bioreductive markers for tumour hypoxia, *Anti-cancer drug design* 13, 687-702 (1998).

13. A.J. Varghese, G.F. Whitmore, Modification of guanine derivatives by reduced 2-nitroimidazoles, *Cancer Res.* 43, 78-82 (1983).

14. A. Andersson, Isolated mouse pancreatic islets in culture: effects of serum and different culture media on the insulin production of the islets, *Diabetologia* 14, 397-404 (1978).

15. A. Andersson, Tissue culture of isolated pancreatic islets, *Acta Endocrinol. Suppl. (Copenh)* 205, 283-294 (1976).

16. P-O Carlsson, L. Jansson, C-G Östenson, Ö. Källskog, Islet capillary blood pressure increase mediated by hyperglycemia in NIDDM GK rats, *Diabetes* 46, 947-952 (1997).

17. N.P. Revsbech, An oxygen microsensor with a guard cathode, *Limnol. Oceanogr.* 34, 474-478 (1989).

18. J.S. Rasey, Z. Grunbaum, S. Magee, N.J. Nelson, P.L. Olive, R.E. Durand, K.A. Krohn, Characterization of radiolabelled fluromisonidazole as a probe for hypoxic cells, *Radiat. Res.* 111, 292-304 (1987).

19. K.E. Dionne, C.K. Colton, M.L. Yarmush, Effect of oxygen on isolated pancreatic tissue, *ASAIO Transactions* 35, 739-741 (1989).

20. K.K. Papas, R.C. Long Jr, I. Constantinidis, A. Sambanis, Effects of oxygen on metabolic and secretory activities of βTC3 cells, *Biochimica et Biophysica Acta* 1291, 163-166 (1996).

21. K.K. Papas, R.C. Long Jr, A. Sambanis, I. Constantinidis, Development of a bioartificial pancreas: II. Effects of oxygen on long-term entrapped βTC3 cell cultures, *Biotechnol. Bioeng.* 66, 231-237 (1999).

22. A.M. Davalli, L. Scaglia, D.H. Zangen, J. Hollister, S. Bonner-Weir, G.C. Weir, Vulnerability of islets in the immediate posttransplantation period. Dynamic changes in structure and function, *Diabetes* 45(9), 1161-1167 (1996).

43.

ALTERATION OF BRAIN OXYGENATION DURING "PIGGY BACK" LIVER TRANSPLANTATION

Piercarmine Panzera[**], Luigi Greco[*], Giuseppe Carravetta[*], Antonella Gentile[*], Giorgio Catalano[**], Giuseppe Cicco[**], Vincenzo Memeo[**1]

1. INTRODUCTION

Cerebral blood flow (CBF), thanks to dilatation or constriction of the cerebral resistance vessels, is kept reasonably constant within a wide range of mean arterial blood pressure (MAP) values. This is commonly called CBF autoregulation. It works if the MAP is between the lower and higher limits of 60 to 150 mmHg[1, 2]. The systemic circulation of patients with liver failure is characterized by a low vascular resistance and a compensatory increased cardiac output (CO)[3]. Moreover, it has been demonstrated that because these patients show functional loss of CBF autoregulation, they are vulnerable to sudden relevant changes of MAP. For the same reason, paradoxically, these patients could be considered less suitable for major surgery with a considerable blood loss and fluid shifts, such as Orthotopic Liver Transplantation (OLT). Arterial hypertensive episodes during OLT could provoke cerebral hyperperfusion with an increase in cerebral vascular volume and subsequent elevation of intracranial pressure that could lead to aggravation of cerebral edema and even cerebral hemorrhage[4, 5]. On the other hand, hypotensive episodes could lead to ischemic lesions. Therefore, the present study indirectly investigated CBF by monitoring brain oxygenation (BO) during "Piggy Back" OLT using the noninvasive technique of near infrared spectrophotometry. The aim of the study was to evaluate if there are significant changes in cerebral oxygenation and, if so, to identify the surgical maneuvre mainly causing such changes.

2. PATIENTS AND METHODS

2.1 Patients

The last 15 patients undergoing OLT in our institute, 8 men and 7 women, aged 38 to 60 years, were enrolled for the study (patients with acute liver failure were not

---

[1] * Department of General Surgery and Liver Transplantation, Faculty of Medicine , University of Bari
**CEMOT Centre of research in Hemorheology, Microcirculation and Oxygen Transport, University of Bari

admitted). The protocol was approved by the institutional ethics committee. All the patients or their next of kin gave written informed consent for the study.

Seven patients had hepatitis C virus (HCV) and related cirrhosis (HCV$^+$C), 4 had hepatitis B virus (HBV) and related cirrhosis (HBV$^+$C), 1 had alcohol liver disease (ALD), 1 had ALD+ HBV$^+$, and 2 had hepatocellular carcinoma (HCC) and HBV$^+$. Four patients had encephalopathy on admission to hospital. For all patients the Child-Pugh score and the model for end stage liver disease (MELD) score were calculated to stratify the severity of their liver failure. Patient details are listed in table 1.

Table 1. Data on the study population

Patient no.	Sex	Age	Diagnosis	Encephalopathy	Child-Pugh Score	MELD Score
1	F	58	HCV$^+$C	no	B/8	13
2	M	60	HCV$^+$C	no	A/6	13
3	F	49	HBV$^+$C	yes	C/11	27
4	M	38	HBV$^+$C	yes	C/12	23
5	M	58	HCV$^+$C	no	B/9	18
6	M	47	HCC+HBV$^+$	no	A/6	12
7	M	51	ALD+HBV$^+$	yes	C/11	26
8	M	55	HCC+HBV$^+$	no	B/8	15
9	M	56	HCC+HCV$^+$	no	A/7	12
10	F	56	HBV$^+$C	no	B/8	14
11	F	51	HCV$^+$C	no	B/8	14
12	M	55	ALD	yes	C/12	27
13	F	48	HCV$^+$C	no	A/7	12
14	M	46	HCV$^+$C	no	B/8	14
15	M	55	HBV$^+$C	no	B/9	17

2.2. Surgery

All patients underwent OLT with the "Piggy Back" technique. After laparotomy and liver mobilization the first procedure (T1) was dissection of the liver hilus: first the ligature of the biliary duct then of the hepatic artery (T2) and then of the portal vein (T3). After dissection of the liver from the vena cavae, left in situ, the right suprahepatic vein was ligatured (T4) and the common ostium of the medium and left suprahepatic vein was clamped (T5) at its origin from the vena cavae and the native liver was removed. Then the donor liver was implanted by the following anastomoses (in order): the donor retrohepatic vena cavae with the recipient ostium of the suprahepatic veins, portal vein, hepatic artery and biliary duct. After performance of the portal vein anastomosis the graft was flushed out (T6) with the recipient's blood by temporary declamping of the portal vein and, after reperfusion of the organ (T7), by declamping of the portal vein and the suprahepatic common ostium. After the arterial anastomosis was completed the implanted liver also received arterial perfusion (T8). The main advantage of the "Piggy Back" technique is that caval blood flow is not interrupted during OLT, even without the use of extracorporeal perfusion.

2.3. Detection of Cerebral Oxygenation

Monitoring of cerebral oxygenation was performed throughout surgery with near-infrared spectroscopy[6,7] using the NIRO300 (Hamamatstu). This instrument can detect concentration changes in oxygenated hemoglobin (ΔHbO_2), deoxygenated hemoglobin (ΔHHb), total volume of hemoglobin (ΔHbT) and oxidized cytochrome aa$_3$ ($\Delta CytOx$). The cromophore concentration change in micromole per litre of tissue is acquired at a sampling time of two seconds. The NIRO300 also calculates the tissue oxygenation index (TOI), that is the result of HbO_2/HbT expressed as percentage, and the tissual hemoglobin index (THI) that results from $KHbO_2 + KHHb$ (K=constant scattering contribution) expressed in arbitrary units (au). The optodes were placed on the central forehead just below the hairline. To keep the optodes at a recommended distance of 4 cm from each other and to shield them from the background light we used a dedicated black-rubber pad. The NIRO300 calculates cerebral oxygenation change in optical densities (OD) by an algorithm based on a modified Lambert-Beer's law : $A = \alpha \cdot c \cdot d \cdot B + G$, where A is the measured attenuation in OD, α is the specific extinction coefficient of the adsorbing compound in μM cm^{-1}, c is the concentration of the absorbing compound measured in μM, d is the distance between the optodes, B is the differential pathlength factor, and G is a factor introduced to account for scattering of light in the tissue and this factor was assumed to remain constant. For the pathlength factor a fixed 6.26 factor was used[8].

2.4. Hemodynamic Parameters

Particular attention was paid to the following systemic hemodynamic parameters: cardiac output (CO), mean arterial pressure (MAP), central venous pressure (CVP), and the O_2 Delivery ($D_A O_2$) calculated by the formula: $D_A O_2 = (0.003$ x Pa $O_2 +$ Hb x SpO_2 x 1.34) x CO. The CO was measured using the Baxter Vigilance, a continuous CO monitoring instrument based on the fluid thermodilution method[9].

2.5. Statistics

Results are expressed as mean ± standard error (SE). Statistical comparison was done with the Mann-Whitney non parametric test and by one-way analysis of variance (ANOVA) tests to compare values at the different time points.

3. RESULTS

3.1. Patients

All 15 transplant recipients survived the surgery. All patients were extubated in the operating room. The mean blood loss during OLT was 4.6± 4.2 L. One patient died 22 days after OLT because of primary non-functioning of the graft. One of the patients with encephalopathy (no.12) did not make a complete mental recovery after transplantation.

3.2. Cerebral Perfusion

ΔHbO_2: A significant change of the oxygenated hemoglobin was found at the portal ligature (T3) (-6.9±2.1 μmol l^{-1}; p<0.05), at the wash out (T6)(-13±1.7 μmol l^{-1}; p<0.05) and at portal reperfusion (T7)(+0.8±2.4 μmol l^{-1}; p<0.01) Figure 1.

ΔHHb: A significant change of desoxygenated hemoglobin was found only at T6 (+3.87±2.1 µmol l⁻¹; p<0.05) Figure 1.

ΔHbT: Significant changes in total volume of hemoglobin were recorded at the following surgical times: at T3 (-6.7±2.4 µmol l⁻¹; p<0.05) and at T7 (1.8±3 µmol l⁻¹; p<0.01) Figure 1.

ΔCytOx : No significant change of the oxidized cytochrome aa₃ was recorded.

TOI: The cerebral Tissue Oxygenation Index showed large changes during OLT at T6 (54.75±3.6%; p<0.05) a decrease of more than 15% of the T5 value, and at T7 (72.4±1.2 %; p<0.01) an increase of more than 33% of the T6 value.

Patients with encephalopathy had a higher a basal mean TOI than patients without encephalopathy 74.5±2.16 vs 65.9±2.23% (p<0.01). This difference is no longer significant after reperfusion. (Figure 2)

THI : No significant changes were found for the Tissue Hemoglobin Index during OLT.

Patients with encephalopathy had a higher basal mean THI than patients without encephalopathy 54.8±1.68 a.u. vs 34.6±4.03 a.u. (p<0.01). This difference remained significant throughout OLT. (Figure 2)

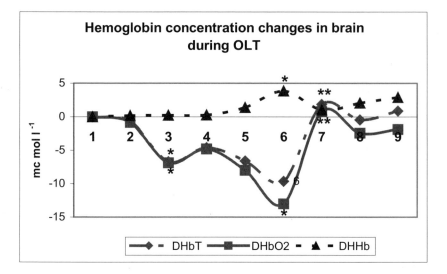

Figure 1. ΔHbO₂ (oxygenated hemoglobin), ΔHHb (desoxygenated hemoglobin) and ΔHbT (total volume of hemoglobin) changes during OLT (µmol l⁻¹) : 1) pre-operative; 2) hepatic artery ligature; 3) portal vein ligature; 4) right suprahepatic vein ligature; 5) clamping of the common suprahepatic vein ostium; 6) wash out; 7) portal reperfusion; 8) artery reperfusion; 9) end of surgery. (*=p<0.05; **=p<0.01)

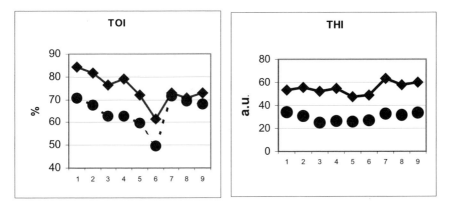

Figure 2. TOI and THI in patients with (♦) or without (•) encephalopathy: 1) pre-operative; 2) hepatic artery ligature; 3) portal vein ligature; 4) right suprahepatic vein ligature; 5) clamping of the common suprahepatic vein ostium.

3.3. Hemodynamic Parameters

At start of surgery the mean CO was 7.96 ± 0.34 ml/min and the mean $D_A O_2$ was 1080 ± 139 ml/min. Both these parameters progressively decreased during the anhepatic phase (CO 6.07 ± 0.46 ml/min; $D_A O_2$ 798 ± 54) and increased at reperfusion (CO 6.77 ± 0.88; $D_A O_2$ 879 ± 130). (Figure 3)

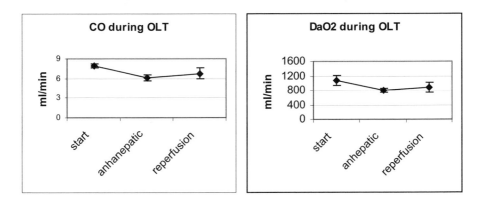

Figure 3 Cardiac output (CO) and oxygen delivery ($D_A O_2$) during liver transplantation.

4. DISCUSSION

Continuous monitoring of cerebral oxygenation by the NIRO300 throughout OLT provides new functional data about the efficacy of CBF. It can integrate data from other techniques previously used, such as Transcranial Doppler (TCD). TCD provides data on the velocity of flow but not of the flow itself because changes in the artery size are unknown[10].

Moreover, previous studies on cerebral perfusion during OLT used the artery-jugular difference in oxygen content that required taking venous samples at specific times previously decided to be the most significant[11, 12]. However, use of the NIRO allows non-invasive continuous monitoring of cerebral oxygenation, without the need to previously select the surgery time, a point considered relevant. Thus, in this study it was possible to establish from the surgical data which surgical maneuvres were specifically related to changes in cerebral oxygenation.

Thus it was so possible to show that the first significant changes in cerebral oxygenation occur with portal ligature, with a relevant decrease of HbO_2 and HbT.

This is the first time that this kind of measurement has been performed during a "Piggy Back" OLT, which is now the most widely used technique. This type of surgery avoids the hemodynamic problems associated with the two previous techniques: i.e. caval flow interruption caused by cross clamping of the vena cavae, and the blood shift of the venous-venous by-pass. The anhepatic phase during the "Piggy Back" OLT unexpectedly showed the same decreases of cerebral oxygenation found during OLT performed with the older techniques[11, 12]. Particularly significant is the decrease of TOI at clamping of the common ostium of the suprahepatic veins (T6), i.e. the maneuvre at the start of the anheptic phase during "Piggy Back" OLT.

The NIRO continuous monitoring showed that the wash out (T6), fortunately a brief surgical period, causes the most important drop in brain oxygenation. At this time point the lowest levels of TOI and HbO_2 were recorded, whereas no significant decrease of HbT and THI was found, probably because of the simultaneous increase of HHb. This might be due to a short relative stasis of CBF, perhaps caused by the decrease in CO at this time point. The possible cause of this sudden oxygenation change may be the rapid loss of the blood used to wash out the transplanted liver, associated with the maneuvres on the cavae vein at the declamping of the caval suprahepatic common ostium. Even if a reduction in CO was found immediately before the reperfusion it was not possible to confirm its relationship with the wash out maneuvre and thus its correlation with the changes of cerebral oxygenation at this time point. Because the instrument used to detect the CO (Vigilance) takes about 3 minutes between each measurement, it was not possible to exactly specify in time the fast changes in CO recorded at the washout-reperfusion time.

Immediately afterwards, at reperfusion, the highest peak of HbO_2, TOI and HbT was found. In a short period, HbO_2, HbT, THI and TOI changed from their lowest levels (recorded at the wash out) to their highest levels (recorded immediately after) at reperfusion. It remains unclear whether these sudden changes are related to changes of CO or to the release of vasoactive substances from the ischemic liver[13].

This sudden increase in CBF may lead to risks related to cerebral hypertensive episodes; e.g. it has been demonstrated that patients with acute liver failure that presented a marked increased CBF died of cerebral vascular diseases[14]. It was interesting to note that our patient with the largest washout-reperfusion peak had mental disorders after OLT.

Patients with encephalopathy had higher levels of cerebral TOI and THI. Whereas THI remained elevated throughout OLT, the difference in the TOI level seems to disappear immediately at liver reperfusion.

5. CONCLUSION

Relevant changes in cerebral circulation occur during "Piggy Back" liver transplantation. Particularly at the washout-reperfusion time the cerebral perfusion suddenly changes from its lowest to its highest values. Further investigation is required to evaluate whether patients with the greatest change in cerebral oxygenation at this time point will suffer neurological complications after transplantation.

It is remarkable that TOI and THI levels are higher in the most compromised patients, perhaps because of an increased CO and, mainly, because of a decreased vascular resistance.

6. REFERENCES

1. NA Lassen. Autoregulation of Cerebral blood flow. Circ. Res. 1992; 15:201-204
2. OB Paulson, S Strandgaard, L Edvinsson. Cerebral autoregulation. Cerebrovasc Brain Metab Rev 1990;2:161-192
3. FS Larsen, GM Knudsen, OB Paulson, H Vilstrup. Cerebral blood flow autoregulation is absent in rats with thioacetamide induced hepatic failure. J Hepatology 1994;21:465-468
4. FS Larsen, E Ejlersen, B Hansen, GM Kundsen, N Tygstrup, NH Secher. Functional loss of cerebral blood flow autoregulation in patients with fulminant hepatic failure. J Hepatology 1995;23:212-217
5. SJ Munoz. Difficult management problems in fulminant hepatic failure. Semin Liver Dis 1994;13:395-413
6. PL Madsen, NH Secher. Near infrared oxymetry of the brain. Prog Neurobiol 1999;58:541-560
7. DM Mancini, L Bolinger, H Li, K Kendrik, B. Chance. Validation of near-infrared spectroscopy. J Appl Physiol 1994;77:2740-2747
8. A Duncan, J H Meek, M Clemence, C E Elwell, L Tyszczuck, M Copy, D T Delpy 1995 optical pathlength measurements on adult head, calf and forearm and the head of newborn infant using phase resolved optical spectroscopy. Phys. Med. Biol. 40 295- 304
9. T Mihaljevic, L von Segesser, M Tonz, B Leskosek, B Seifert, R Jenni, M Turina. Continuous versus bolus thermodilution cardiac output measurements- a comparative study. Crit Care Med 1995; 23:944-949
10. FS Larsen, KS Olsen, BA Hansen, OB Paulson, GM Kundsen. Transcranial Doppler is valid for determination of the lower limit of cerebral blood flow autoregulation. Stroke 1994;25:1985-1988
11. BJ Philips, IR Armstrong, A Pollock, A Lee. Cerebral blood flow and metabolism in patients with chronic liver disease undergoing liver transplantation. Hepatology 1998;27(2):369-376
12. P Pere, K Köckerstedt, H Isoniemi, L Lindgren. Cerebral blood flow and oxygenation in liver transplantation for acute or chronic hepatic disease without venovenous bypass. Liver Transplantation 2000;6(4):471-479
13. DD Doblar,YC Lim, L Frenette. The effect of acute hypocapnia on middle cerebral artery transcranial Doppler velocity during orthotopic liver transplantation: changes at reperfusion. Anesth Analg 1995;80:1194-8
14. FS Larsen, F Pott, BA Hansen et al. Transcranial Doppler sonography may predict brain death in patients with fulminant hepatic failure. Transplant Proc 1995;6:3510-11

44.

EFFECTS OF PRESERVATION SOLUTIONS ON BLOOD

Piercarmine Panzera**,Luigi Greco*, Maria T. Rotelli*,Vito Lavolpe**, Anna M. Salerno**, Antonella Gentile*, Giorgio Catalano**, Giuseppe Cicco**, Vincenzo Memeo** [1]

1. INTRODUCTION

Preservation solutions in organ transplantation are designed to satisfy three principles: they have to rapidly wash out the blood and cool the organ, prevent cell swelling and interstitial oedema formation, and prevent excessive cellular acidosis [1,2].

University of Wisconsin (UW) solution has been considered the gold standard for liver graft preservation. A new preserving solution, Celsior solution (CS)[3], is now available. This is an extracellular type of solution, initially used successfully in heart transplantation[4], and more recently shown to be an effective alternative to UW solution also in kidney[5], pancreas[6], lung[7], small intestine[8] and liver[9] transplantation.

Although clinical and experimental studies show a similar effectiveness in organ preservation, the Celsior solution formula theoretically promises some advantages. Its low potassium content should avoid glomerular capillary contraction due to calcium-associated contracture during kidney transplantation or endothelial impairment and biliary tract damage, that seems to occur in liver transplantation after a lengthy cold ischemia time using UW solution[10,11].

CS contains oxygen radical scavengers, with reduced glutathione and mannitol and histidine, that might prevent oxidative injury caused by post-reperfusion free radicals, while the rapidly oxidized glutathione in shelf-stored UW solution definitely lacks this important property[11].

[1]
  * Department of General Surgery and Liver Transplantation
    Faculty of Medicine , University of Bari
  ** CEMOT
    Centre of research in Hemorheology, Microcirculation and Oxygen Transport
    University of Bari

Another advantage of the Celsior formula is that it does not contain macromolecules. The presence of a fairly high concentration of macromolecules in the suspension medium is known to be associated with red blood cell (RBC) aggregation[12].

It has been demonstrated that the Hydroxyethyl Starch (HES) contained in UW solution (50g/L) causes erythrocyte aggregation[13, 14]. The formation of erythrocyte rouleaux during organ perfusion might affect the blood flush-out in the microcirculation, ultimately affecting organ preservation and increasing the risk of primary dysfunction after transplantation[15].

The effect of UW solution on RBC deformability is still unclear. Some authors have shown a stiffening effect of UW solution on RBC[16], but this has not been confirmed by others [17].

The aim of this study is to investigate whether the CS solution presents any hemorheological advantage compared with UW solution.

2. MATERIALS AND METHODS

2.1. Preparation of Blood Samples

Venous blood samples were obtained from 17 healthy individuals aged 25 to 31 years by vein puncture from an antecubital vein and anticoagulated with EDTA. RBC were separated from the blood by centrifugation at 1400 g for 10 min and then resuspended in autologous plasma at a hematocrit of 38%. Then the blood was admixtured in vitro at 4°C with the UW or the CS at the following mixing ratios: blood:UW=3:1; 6:1; blood:CS=3:1; 6:1. Pure blood and blood diluted with Saline solution (SS) (blood:SS=3:1; 6:1) were used as controls. All the admixtures were prepared immediately before use. For the RBC deformability determination, 200 μl of the different admixtures were diluted in 5 ml of a polyvinylpyrovidone solution. The experiments were done at 22°C.

2.2. Determination of RBC Aggregation

RBC aggregation was quantified using a laser-assisted optical rotation red cell analyser (LORCA RR Mechatronics, Hoorn, The Netherlands)[17, 18]. This instrument consists of a laser light, a rotating thermostated cup and a video camera connected to an ellipse-fit computer program. Using the ektacytometric principle it quantifies the aggregation. The blood was brought under a shear rate of 500sec^{-1}, after which the shear was stopped at t=0. The backscattered intensity from the blood layer was measured for 120 sec after shear stop. The intensity decreased with erythrocyte aggregation. We considered the following parameters: the Aggregation Index (AI) and the time necessary to obtain 50% of the final aggregation ($t_{1/2}$).

2.3. Determination of Erythrocyte Deformability

To investigate RBC deformability, a dedicated software applied to LORCA was used, measuring the diffraction pattern of the RBC under various shear stresses in the range of 0.3 to 30 Pa. The Elongation Index (EI) was the parameter used to express RBC deformability at several shear stresses. An increased EI indicates greater cell deformability.

2.4. Images

The morphological aspect of erythrocytes and of the erythrocyte rouleaux was detected by light microscopy (LEINTZ) using x240 magnification. Images were then acquired by a JVC camera and elaborated by an image-processing program.

2.5. Statistical Analysis

Results are expressed as mean ± standard error (S.E.). Statistical comparison between groups was done by Student's paired t test. P values <0.05 were taken as statistically significant.

3. RESULTS

3.1. Aggregation

3.1.1. Aggregation Velocity

$t_{1/2}$ is an index of the velocity of the RBC aggregation process. This is the time the solution takes to reach 50% of the final RBC aggregation.

Pure blood control samples have a mean $t_{1/2}$ of 10.816±0.477 sec. The presence of UW causes faster aggregation of erythocytes (more than 25% quicker than controls); the difference in increase at the higher UW concentration (1:3) is statistically significant (P<0.01). On the contrary, in the presence of CS aggregation takes a much longer time , this difference is statistically significant even at the low concentration (P= <0.01) and increases to three times the control at the higher concentration (P= <0.01)(Fig.1). CS samples showed a longer $t_{1/2}$ even when compared with the SS samples, although this difference was small and significant only at the higher concentration (P<0.05).

3.1.2. Level of RBC aggregation

AI (aggregation index) indicates the percentage of blood aggregation.

The mean AI measured in pure blood control samples is 28.00±0.73%. UW solution causes a concentration-dependent increase in the percentage of aggregated RBC: in fact, the AI is significantly higher in samples with a blood:UW ratio of 3:1 (P= <0.01), as much as 30% higher than controls. On the contrary, the presence of CS at a higher dilution reduces the aggregation index. Already at the CS:blood=1:6 ratio the AI is significantly reduced compared with the pure blood controls (P= <0.01) (Fig 2). When the CS is present at a 1:3 ratio the AI is less than half the control value (P= <0.01). Even compared to the SS samples, the CS samples showed a slightly lower AI , this difference was significant only for the 3:1 ratio (P<0.05).

3.2 Deformability

The use of UW or CS does not result in any statistically significant change in RBC deformability.

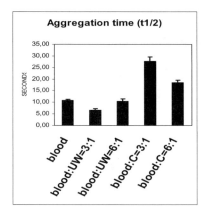

Figure 1. Aggregation time

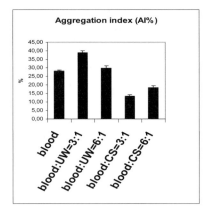

Figure 2. Aggregation index

Figure 3. Light microscope image from control samples

3.3. Imaging

Imaging does not show any appreciable aggregation in control samples (Fig.3) and those with CS solution (Fig 4). On the contrary, in samples with UW solution we always find erythrocyte clusters and the size of the aggregates seems to be proportional to the concentration of UW. Samples with the higher concentration of UW (UW:blood=1:3) have larger clusters (Fig.5).

4. DISCUSSION

This study developed from two starting points: the theory that relates RBC aggregation to the presence of macromolecules in the medium[12] and the knowledge that the hydroxyethyl starch present in UW solution causes RBC aggregation[13].
The importance of hydroxyethyl starch, together with other components such as raffinose, in the UW formula lies in the increased colloid pressure that should prevent interstitial oedema in the preserved organs.

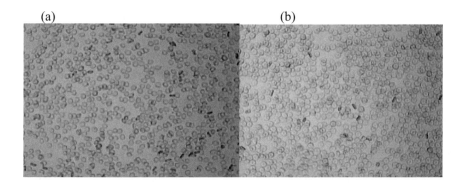

Figure 4. Light microscope images :(a) blood:CS=6:1; (b) blood:CS=3:1

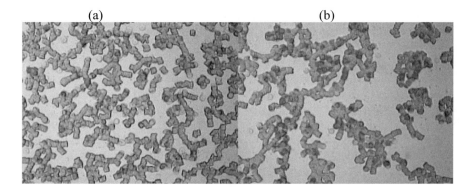

Figure 5. Light microscope image (a) blood:UW=6:1; (b)blood:UW=3:1

Soon after UW solution became available some authors showed it to be better for organ preservation when compared to the solution previously used, but to have drawbacks for organ flush-out because of the HES content and high viscosity [21].

CS solution achieves the same efficacy in preventing oedema formation using molecules such as mannitol, lactobionate and histidine instead of HES. The absence of macromolecules in this formula should avoid the formation of RBC rouleaux.

In this study we directly compared the effect of UW and CS solutions in vitro, at the same dilutions, on erythrocyte aggregability.

UW solution causes massive, immediate RBC aggregation, the higher the percentage of UW in the solution the faster the RBC aggregation; on the contrary, CS solution reduces RBC aggregability even when compared with controls (lower AI and longer t1/2). The stronger the CS solution the less the erythrocyte aggregation; even when compared with SS diluted controls, CS samples showed slightly reduced and slower RBC aggregation, although this difference was statistically significant only at the higher CS/blood ratio. Thus, the inhibiting effect of CS, like that of SS, seems to be correlated to a dilution effect. Nevertheless, there may also be some RBC-antiaggregating property of the CS formula itself.

Both dilutions were shown to preserve RBC deformability. In this study RBC deformability was investigated immediately after preparation of the samples because we were particularly interested in the perfusion solution effect during graft perfusion (that takes place within a short time).

Other authors have shown that UW solution induces RBC stiffness after a longer (1/2 hour) incubation time[16], which may be important for organ preservation. Instead, RBC deformability reduction at reperfusion is mostly to be correlated to superoxide anions formation[19].

The formation of RBC rouleaux during perfusion in organ procurement may seriously affect correct blood flush-out. It is difficult to believe that RBC aggregates can pass through the organs microcirculation and reach the venous discharge. So we have to assume that using UW solution in the microcirculation of the organ does not result in complete substitution of solution for blood. This means that the peripheral tissue does not receive the benefits of the cold perfusion. What is more, the presence of RBC trapped in aggregates in the organ microcirculation may also affect the microcirculation flow at organ reperfusion. This could be one of the causes of primary non-functioning of the transplanted organ[14].

Although clinical studies[9, 20] show similar effects of the two solutions, this study reveals some haemorheological advantages of the CS solution.

5. CONCLUSION

Celsior solution does not aggregate erythrocytes, whereas Wisconsin solution massively does. This feature could make CS preferable to UW during organ procurement.

6. REFERENCES

1. RJ Ploeg, D Goossens, LF McAnuly et al. Successful 72-hour storage of dog kidneys with UWsolutions. Transplantation 1988;46:191
2. NA 't Hart, HGD Leuvenik HGD, RJ Ploeg. New solutions in organ preservation. Transplantation Reviews 2002;16:131-141.
3. JB Cofer, GB Clintmalm, TK Howard , CV Morris, BS Husberg, RM Goldstein, TA Gonwa. A comparison of UW with Eurocollins preservation solution in liver transplantation. Transplantation 1990;49:1088-1093.
4. P Menaschè, JL Termigon, F Pradier, C Grousset, C Mouas, G Alberici et al. Experimental evaluation of Celsior®,a new heart preservation solution. Eur J Cardiothorac Surg 1994;8:207-213.
5. N Baldan, M Toffano, R Cadrobbi, R Cordello,Calabrese F,Bacelle L,Rigotti P.Kidney preservation solution. Transplant Proc 1997 ;29 :3539-3540.
6. N Baldan, P Rigotti, L Furian, ML Valente, F Calabrese, L Di Filippo et al. Pancreas preservation with Celsior solution in a pig autotransplantation model :Comparative study with University of Winsconsin solution. Transplant Proc 2001;33:873-875.
7. RF Roberts, GP Nishanian, JN Carey, Y Sakamaki, VA Starnes, ML Barr. A comparison of the new preservation solution Celsior to Eurocollins and University of Wisconsin in lung reperfusion injury. Transplantation 1999;67:152-155.
8. T Minor, T Vollmar, MD Meger,W Isselhard. Cold preservation of the small intestine with the new Celsior solution. Tranpl Int 1998;11:32-37.
9. A Cavallari, U Cillo, B Nardo, F Filipponi, E Gingeri, R Montalti, F Vistoli et al. A multicenter pilot multicenter study comparing Celsior and University of Wisconsin preserving solutions for use in liver transplantation. Liver Tranpl2003;9(8):814-821.
10. A Cavallari, B Nardo, G Pasquinelli, L Badiali de Giorgi, R Bellusci, F Ferlaino et al. Subcellular lesion of the biliary tract in human liver transplant incurred during preservation. Tranplant Proc 1992;24:1979-1980.
11. OI Pisarenko,ES Solomatina, VE Ivanov, IM Studneva, VI Kapelko, VN Smirnov. On the Mecchanism of enhanced ATP formatin in hypoxic myocardium caused by glutamic acid. Basic Res Cardiol 1985;80:126-134.

12. DE Brooks. Current models of red blood cell aggregation. Biorheology 1995;32: 103.
13. AM Morariu, A Plaats, W Oeveren,N t'Hart, H Leuvenink, R Graaf, RJ Ploeg, G Rakhorst. Hyperaggregating effect of hydroxyethyl starch components and University of Wisconsin solution on human red blood cells: a risk of impaired graft perfusion in organ procurement? Transplantation 2003;76:37-43.
14. A Plaats,A Nils, N t'Hart, A Morariu, G Verkerke, H Leuvenink et al. Effect of University of Winsconsin organ-preservation solution on haemorheology Transplant Int 2004;
15. RJ Ploeg, AM D'Alessandro, SJ Knechtle,et al.Risk factors for primary dysfunction after liver transplantation:A multivariate analysis. Transplantation 1993;55:807-813.
16. B Chimiel, L Cierpka. Organ preservation solutions impair deformability of erythrocytes in vitro. Transplantation Proceedings 2003;35:2163-2164.
17. MR Hardeman, PT Goedhart, KP Lettinga. Laser-assisted optical rotational cell analyser (L.O.R.C.A.). A new instrument for measurement of various structural hemorheological parameters. Clin Hemorheol 1994;14:605-618.
18. K Osterloh, P Gaehtgens, AR Pries. Determination of microvascular flow pattern formation in vivo. Am J Physiol Heart Circ Physiol 2000;278:H1142-H1152.
19. OK Baskurt, A Temiz, HJ Meiselman. Effect of superoxide anions on red blood cell rheologic properties. Free Rad Biology1998;24:102-110.
20. A De Roover, L de Leval, J Gilmaire, O Detry, J Boniver, P Honorè, M Meurisse. A new model of human intestinal preservation :comparison of University of Wisconsin and Celsior preservation solutions. Transplant Proc 2004;36:270-272.
21. J Jacobson, J Wahlberg, L Frodin, B Odlind, G Tufveson. Organ flush out solutions and cold storage preservation solutions: effect on organ cooling and post ischemic erythrocyte trapping in kidney grafts. An experimental study in the rat. Scand J Urol Nephrol. 1989;23(3):219-22.

45

POST ISCHEMIC NO-REFLOW AFTER 60 MINUTES HEPATIC WARM ISCHEMIA IN PIGS

Luigi Greco[1], Antonella Gentile[1], Piercarmine Panzera[1], Giorgio Catalano[1], Giuseppe Cicco[2], Vincenzo Memeo[1] *

1. INTRODUCTION

Prolonged liver ischemia is a frequently required condition in liver surgery[1]. During major resection for cancer or in cases of severe trauma, it may be necessary to interrupt blood inflow or outflow, by clamping the pedicle of Glisson's capsule (Pringle's maneuver), or even resorting to total vascular exclusion, by clamping the supra and subhepatic inferior vena cava, to limit intraoperative blood loss[2,3].

Although it is universally acknowledged that the organ can tolerate prolonged ischemia for a period of up to 60 minutes, the anatomical and functional variations of the liver parenchyma inevitably induced by the procedure are receiving ever greater attention[4,5]. Ischemia-reperfusion damage is also an essential factor during liver transplant and the degree of damage suffered will affect the functional efficiency of the graft[6].

In the present experimental study, the alterations of the liver microcirculation induced by warm ischemia lasting 60 minutes were studied by continuous measurement of the microcirculation with an intraparenchymal probe (Periflux).

2. MATERIAL AND METHODS

The study was conducted on 8 female pigs with a mean weight of 30 kg($\pm$5). Under general anesthesia, a median laparotomy approximately 15 cm was made and the optical

* [1] Department of General Surgery and Liver Transplantation, Faculty of Medicine , University of Bari
[2] CEMOT: Centre of research in Haemorheology, Microcirculation and Oxygen Transport,University of Bari

fiber flowmeter probe, inserted in a spinal needle, was inserted in the liver parenchyma of the left lobe of the liver at a depth of about 5 cm.

The fiber was connected to the probe and protracted measurements lasting 30 minutes were started (T0). At this stage the left branches of the hepatic artery and portal vein were dissected and isolated and clamped for 60 minutes (T1). Flow measurement continued throughout this period. It was judged preferable to localize the ischemia in the left lobe only, in order to maintain normal portal blood flow through the right lobe and hence avoid prolonged bowel stasis. This is because the latter condition is ill tolerated in the pig and leads to massive release of toxic substances at the moment of declamping, which can themselves cause alterations in the circulation during reperfusion.

The clamps were removed after 60 minutes and the blood flow was recorded for a further 60 minutes (T2).

Great care was taken during all the surgical maneuvers to avoid disturbing the position of the optical fiber inserted in the liver parenchyma.

2.1. Technical Periflux Data

The data of the laser Doppler flowmeter are expressed in arbitrary units (PU). The device was calibrated in a suspension of 2-micron microspheres of latex[7,8].

3. RESULTS

The mean baseline value, expressed in PU, was 264($\pm$15). An abrupt reduction in blood flow was recorded immediately after clamping (PU 79$\pm$10), the percentage reduction being by 70% ($\pm$10%).

After reperfusion, the T2 value rose quite quickly (PU 184$\pm$18) but still remained constantly below the T0 value, by a mean of 30% (5$\pm$%) (Fig.1). In this phase the blood flow remained virtually unchanged for the first 30 minutes, and then showed a mild positive trend in the second half of the measurement time. However, in no case did the T2 values return to the T0 values by the end of the experiment.

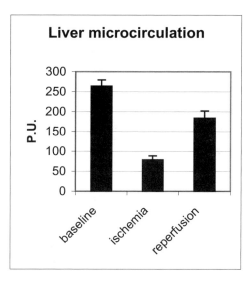

Figure 1. Laser Doppler perfusion values before and 60 minutes after hepatic artery and portal vein clamping, and 60 minutes after release of the clamps.

4. DISCUSSION

The liver damage caused by prolonged warm ischemia depends on a series of factors that are interdependent and cumulative, and have negative effects on the liver microcirculation during the reperfusion phase[9].

Periods of ischemia as brief as 30 to 40 minutes followed by reperfusion with oxygenated blood usually result in some degree of injury to the organ.

The extent of injury may be still reversible, but prolonged ischemia may cause cell death and organ failure.

Although several hypotheses have been proposed to explain no-reflow injury, at the present time only certain steps of this process have been clearly demonstrated.

The energy deficit rapidly leads to edema and swelling of the endothelial cells[10,11]. This phenomenon seems to be triggered by a local alteration of the balance between endothelin, which increases the permeability of the hepatocyte membrane, and nitric oxide, which by contrast induces cell contraction [12,13].

There is contemporary activation of the Kuppfer cells, which generates a series of inflammatory type responses with overproduction of mediators such as the cytokines and oxygen-derived free radicals. The cytokines mainly involved are tumor necrosis factor (TNF alpha) and interleukin (L). In addition to their direct action, they trigger the production of other inflammatory mediators such as the icosanoids and cytokines[14,15]. Electron microscope studies have clearly demonstrated that the Kuppfer cells are activated during reperfusion, and that this phenomenon is associated with evident damage to the liver parenchyma[16,17].

However, this inflammatory response to ischemia can also have a positive value in surgery. In fact, in liver tumors, radiofrequency thermoablation using Pringle's maneuver has been shown to have a greater oncologic efficacy[18]. During the ischemia-reperfusion phase ODFR endothelial cells and protease enzymes are released, which boost the chemotactic action and induce an accumulation of leukocytes in the sinusoid spaces. In turn, the neutrophils are responsible for releasing other analogous substances that strengthen this effect[19].

At the same time, there is formation of platelet agglomerations due to a local increase of the platelet activating factor. These reduce the sinusoid spaces and hence pose a mechanical obstacle to the liver microcirculation[20,21].

In conclusion, during liver transplantation, ischemia reperfusion damage has a primary clinical importance. The mechanisms involved are those described above, but it must be emphasized that the entity of the syndrome is governed by other factors such as the donor's hemodynamic equilibrium, the cold ischemia time, the duration of the operation and any periods of intra-operative hypertension, as well as the host's overall condition[22,23].

In the present study we focused on alterations in the hepatic microcirculation of the pig induced by 60 minutes of warm ischemia obtained by sectorial clamping of the left Glisson's pedicle. Measurements of the blood flow were made with a laser Doppler flowmeter probe positioned directly inside the parenchyma. The tool used, the Periflux, seems to be very reliable for this kind of research[24,25].

Analysis of the results showed that blood flow was reduced by 30% after ischemia (T2), compared to the baseline value (T0). These data clearly demonstrate altered microcirculation conditions and can be interpreted as a direct expression of the entity of ischemia-reperfusion injury. Further studies are necessary to confirm these results and above all to assess the progress of the damage over time and the degree of reversibility.

5. REFERENCES

1. B. Vollmar., J. Glasz, R. Leiderer .et al. Hepatic microcirculatory perfusion failure is a determinant for liver dysfunction in warm ischemia-reperfusion. Am.J.Pathol. 1994:145, 141-3.
2. C. Huguet, P. Addario Chicco, A. Gavelli et al.Techique of hepatic vascular exclusion for extensive liver resection.Am.J.Surg. 1992:163:602-605
3. N. Habib, G. Zografos., L. Greco, A. Bean Liver resection with total vascular exclusion for malignant tumours. Brit.J.Surg. 1994,81,1181-1184.
4. M.E. Moussa, C.E. Sarral, S. Uemoto al. Effect of total hepatic vascular exclusion during liver resection on hepatic ultrastrupture. Liver Transplant Surg. 1996: 2, 461-467.
5. J.M. Badia, L.C. Ayton, T.J. Evans et al. Systemic cytokine response to hepatic resection under total vascular exclusion. Eur. J.Surg. 1998:164, 185-190.
6. J.M. Henderson Liver transplantation and rejection: an overview. Ann.Surg. 1989: 46(suppl. 2); 1482-1484.
7. N.E. Almond,A.M. Wheatley. Measurement of hepatic perfusion in rats by laser-Doppler flowmetry. Am.J.Physiol. 1992: 262; G203-209
8. M.N. Tawadrous, Yi Zhang Xing, A.M. Wheatley. Microvascular origin of Laser- Doppler Flux signal from the surface of normal and Injured liver of the rat. Microvascular Research 2001; 62, 355-365.
9. F.S. Inglott, N. Habib, R.T. Mathie Hepatic ischemia-reperfusion injury. Am.J of Surg. 2001: 181, 161-166.
10. I. Marzi, Y. Takei, Ruker et all. Endothelin-1 is involved in hepatic sinusoidal vasoconstriction after ischemia and reperfusion. Transplant Int.: 7, 503-506.
11. N. Kawada, T Tran Thi , H Klein, K. Decker. The contraction of hepatic stellate cells stimulated with vasoactive substances. Possible involvement of endothelin-1 and nitric oxide in the regulation of the sinusoidal tonus. Fur.J. Biochem. 1993: 213, 815-823.
12. E. Kawamura, N. Yamanaka, F. Okamoto et al. Response of plasma and tissue endothelin –1 to liver ischemia and its implications in ischemia-reperfusion injury. Hepatology 1995: 21, 1138-1143.
13. B.H.J. Pannen, M. Bauer, G.F.E. Noldge-schomburg et al. Regulation of hepatic blood flow during resuscitation from hemorrhagic shock: role of NO and endothelins. Am.J.Physiol 1997: 272, 2736-45.

14. Y. Shiratori, H. Kiriyana, Y. Fusushi et al. Modulation of ischemia-reperfusion induced hepatic injury by Kuppfer cells. Dig. Dis. Sci. 1994: 39, 1263-72.
15. H. Jaeschke, A.P. Bautista, Z. Spolaries, J.J. Spitzer. Superoxide generation by neutrophils amd Kuppfer cells during in vivo reperfusion after ischemia in rats. J. Leukoc. Biol. 1992: 52, 377-382.
16. S. Kobayashi, M.G. Clemens Kuppfer cell exacerbation of hepatocite hypoxia/ reoxygenation injury: Circ. Shock 1992: 37, 245-252.
17. H. Jaeschle, A. Farhood Neutrophil and Kuppfer cell-induced oxidant stress and ischemia- reperfusion injury in rat liver. Am.J.Physiol. 1991: 260, 355-362.
18. L.R. Jiao, P.D. Hansen, R. Havlik et al. Clinical short- term results of radiofrequency ablation in primary and secondary liver tumours. Am.J.Surg. 1999: 177, 303-306.
19. H. Jaeschle, A. Farhood, C.W. Smith Neutrophils contribute to ischemia/ reperfusion injury in rat liver in vivo. FASEB J. 1990: 4, 3355-3359.
20. A. Khandoga, P. Biberthaler, G. Enders, S. Axmann, J. Hutter. Platelet adhesion mediated by fibrinogen-intercellular adhesion molecule-1 binding induces tissue injury in the post ischemic liver in vivo. Transplantation 2002: 74, 681-688.
21. G. Martinez Mier, L.H. Toledo Pereira, P.A. Ward. Adhesion molecules in liver ischemia and reperfusion. J. Surg. Research 2000: 94, 185-194.
22. S. Massberg, K. Messner. The nature of ischemia/ reperfusion injury Transplantation Proceedings 1998: 30, 4217-4223.
23. A.M. Seifalian, S.V. Mallet, K. Rolles, B.R. Davidson Hepatic microcirculation during human orthotopic liver transpantation. Brit.J.Surg. 1997: 84, 1391-1395.
24. D. Arvidssson, H. Svensson, U. Haglund Laser-Doppler flowmetry for estimating liver blood flow Am.J.Physiol. 1988: 254, 471-476.
25. A.M. Wheatley, N.E. Almond, E.T. Stuart, D. Zhao. Interpretation of the laser-Doppler signal from the liver of the rat. Microvasc.Research 1995: 45, 290-301.

46.

APOLIPOPROTEIN E GENOTYPE AND CBF IN TRAUMATIC BRAIN INJURED PATIENTS

Mary E. Kerr, M. Ilyas Kamboh, Yuan Kong, Sheila Alexander, and Howard Yonas[1]

1. INTRODUCTION

Prior to the discovery in 1993 of the link between apolipoprotein (ApoE) and Alzheimer disease[1], ApoE was primarily studied for its role in cholesterol transport and cardiovascular pathology [2]. Since that time, ApoE has been increasingly recognized for its function in hypercholesterolemia, vascular elasticity[3, 4], compliance[5] and endothelial dysfunction[6]. These studies suggest that vascular mechanisms of ApoE may contribute to neurodegeneration [7]. The mechanism for the ApoE role in cerebrovascular pathology is unknown but theories include increased production of $O(2)(-)$ and inactivation of nitric oxide (NO) synthase enzyme activity [6, 8]. From this perspective, we hypothesize that ApoE 4 allele carriers have differential cerebral blood flow (CBF) immediately after injury that may affect outcomes in the traumatic brain injured (TBI) population.

2. SUBJECTS AND METHODS

This prospective study recruited 120 patients with severe TBI [Glasgow Coma Score (GCS) ≤ 8] admitted to a NeuroTrauma ICU at a University Hospital with a Level I Trauma Service from Aug. 06, 1994 to Feb. 02, 2001. This analysis includes those patients with a XeCT CBF test within the first 24 hours after injury (n=54). The Xe/CT CBF tests were ordered by the attending neurosurgeon as part of routine care. The medical records were reviewed to obtain information regarding demographics, symptoms, and treatmentj. Admission GCS were verified by the resident and on-call neurosurgeon.

[1] Mary E. Kerr, Yuan Kong, &. Sheila Alexander, University of Pittsburgh School of Nursing. Howard Yonas, Department of Neurological Surgery, University of Pittsburgh Medical Center. M. Ilyas Kamboh, Graduate School of Public Health, University of Pittsburgh, Pittsburgh, PA.

3. APOE GENOTYPING

Cerebrospinal fluid (CSF) samples were collected and stored in a -80°C freezer for future genotyping in batch. Genetic determinations were performed in Dr. Kamboh's laboratory. Genomic DNA was isolated from CSF samples by the Qiagen kit (Qiagen, Chatswirth, CA). ApoE genotyping was performed as described by Kamboh et al. (1995).

4. Xe CT CBF

A baseline 3 or 4 level stable xenon-enhanced CT (Xe/CT) CBF test was conducted and 20 2-cm circular regions of interest (ROI) were measured per level. The CBF software calculated the mean CBF within each of the 20 ROIs. The 20 ROIs were averaged per level and the 3 or 4 levels were averaged to determine hemispheric (right, left, ipsilateral, contralateral and difference between ipsilateral and contralateral) and global ROI CBF. XeCT CBF tests with poor confidence images were excluded. The baseline CT scan was reviewed independently by investigators blinded to the clinical data. The primary side of injury was coded as the ipsilateral hemisphere when the injury was greater than the contralateral side. If the primary side of injury could not be identified (e.g. diffuse axonal injury, diffuse punctate hemorrhage, bilateral injury), it was excluded from the analysis comparing ipsilateral to contralateral or "control" CBF (n=43).

5. STATISTICAL ANALYSES

SPSS (version 11.0, SPSS Inc,) and SAS (version 8.2, SAS Institute Inc., Cary, NC) were used for all analyses. An analysis of variance was used to compare the mean CBFs among patients with ApoE genotypes of 3|2, 3|3 and 3|4 (n=51). All patients were dichotomized based on the presence or absence of the ApoE 4 allele (n=54) and a t-test was used to compare mean differences in hemispheric and global ROI CBF values. A regression analysis was used to predict the effect of ApoE 4 allele on CBF while controlling for potential co-variates (PaCO2, severity of injury, race, and gender).

6. RESULTS

Out of 54 patients, the ApoE genotype distribution was 2|2 (n=0; 0%) 2|4 (n=1; 1.9%), 3|2 (n=11; 20.7%), 3|3 (n=30, 55.6%), 3/4 (n=10, 18.5%), and 4|4 (n=2, 3.7%). After classification by the presence or absence of the ApoE 4 allele, 13 (24.1%) subjects were positive for the ApoE 4 allele and 41 (75.9%) subjects were negative. Table 1 shows the means and standard deviations of the demographic characteristics and time of baseline XeCT for the total sample. There was no statistically significant difference in the median GCS, but a greater proportion of the patients that were positive for the ApoE 4 allele had a GCS of 5 through 8 indicating that they had less injury on admission.

Table 1. Demographic characteristics and time of baseline XeCT by ApoE4 group

	ApoE4 (-) n=41		ApoE4 (+) n=13		Total N=54		Statistics	
	Mean	**SD**	**M**	**SD**	**M**	**SD**	**Test**	**P-value**
Age	32	14.6	26	10.0	31	14.4	T=1.47	0.14
TFI (hours)	17	4.7	20	2.5	17	4.5	T=-3.23	*0.0025*
	n	**%**	**n**	**%**	**N**	**%**	**Test**	**P-value**
Female	9	22.0	2	15.4	11	20.4	X2=0.26	0.61
Caucasian	35	85.4	13	100.0	48	88.9	X2=2.14	0.14
GCS (=5)	11	26.8	2	15.4	13	24.1	X2=6.59	0.25
GCS (5-8)	33	19.5	8	53.9	39	72.2	X2=5.80	*0.016*

7. GLOBAL AND HEMISPHERIC CBF

7.1 Comparison by ApoE Genotype

The overall average of the global, right hemispheric and left hemispheric CBF ROIs within the first 24 hours was 44.9 (17.2), 43.6 (16.5) and 44.2 (16.4) mL/100g/minute. Table 2 compared the right hemisphere, left hemispheres and global average CBF ROIs in patients by ApoE genotype. Post hoc analysis using Fisher's Least Significant Difference (LSD) indicated that patients with the ApoE 3|4 genotype had higher ROI flows than those with the ApoE 2|3 and 3|3 genotypes (See Figure 1). In a subset of subjects comparing the side of injury (ipsilateral versus contralateral, patients with the ApoE 3|4 genotype showed higher flows on the ipsilateral hemisphere (p=.04), but not on the contralateral hemisphere. There was no statistical significance in the side-to-side differences between the ipsilateral and contralateral hemispheres by ApoE genotype (p=.20). There also was no significant difference in the $PaCO_2$, however the ApoE 3|4 genotype group tended to have lower $PaCO_2$ levels (Refer to Table 2).

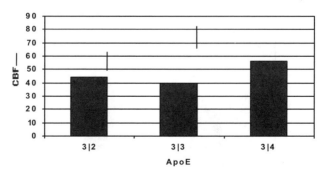

Figure 1. Right hemisphere CBF by ApoE genotype (p=.006)

Table 2. Comparison of right and left hemisphere ROI and global ROI CBF in patients with ApoE genotypes 2|3, 3|3 and 3|4

	Apoe (3|2) n=11 (20.4%)		ApoE (3|3) n=30 (55.6%)		ApoE (3|4) n=10 (18.5%)		Statistic	
	Mean	Std.	M	Std.	M	Std.	F	P-value
Paco2	35.2	3.6	38.2	5.7	34.9	6.7	1.88	0.16
Right Hemisphere	42.8	12.1	40.4	14.5	59.4	21.4	5.69	*0.006*
Left Hemisphere	46.4	9.6	38.8	14.5	53.8	22.8	3.72	*0.03*
Global	44.6	10.2	39.6	14.2	56.6	21.3	4.76	*0.013*
TFI (hours)	16.9	4.4	16.6	5.0	20.2	2.6	2.50	0.09
	n=8(19.5%)		n=25 (61.0%)		n=8 (19.5%)		F	P-value
Ipsi. CBF	41.3	9.4	39.4	15.0	56.8	24.8	3.45	*0.04*
Contr. CBF	42.0	7.0	40.0	15.1	51.8	24.2	1.61	0.21
Contr – Ipsi	0.64	7.06	0.64	4.94	-5.05	14.2	1.68	0.20

Table 3. Comparison of CBF in patients by ApoE4 allele presence

	ApoE4 (-) N=41 (75.9%)		ApoE4 (+) n=13 (24.1%)		Statistic	
	Mean	SD	M	SD	Test	P-value
Paco2	37.4	5.4	36.5	6.8	.48	.636
R hemis CBF	41.0	13.8	57.2	21.2	-2.59	*.020*
L hemis CBF	40.8	13.7	52.2	21.9	-1.76	.098
Global CBF	40.9	13.3	54.7	20.9	-2.24	*.041*
TFI (hours)	16.6	4.7	19.9	2.5	-3.23	*.0025*
	N=33 (76.7%)		n=10 (23.3%)		Test	P-value
Ipsilateral CBF	39.9	13.7	53.6	24.2	-1.71	.12
Contralateral CBF	40.5	13.5	48.8	23.2	-1.34	.30
Contralateral – ipsilateral CBF	.53	4.90	-3.24	11.41	1.16	.22

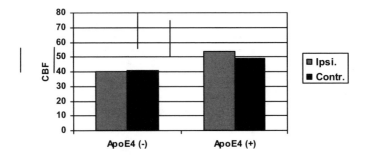

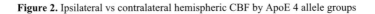

Figure 2. Ipsilateral vs contralateral hemispheric CBF by ApoE 4 allele groups

Table 4. Regression analysis results

Predictors		R hemis CBF (n=51)	L hemis CBF (n=51)	Global CBF (n=51)	Ipsilateral CBF (n=41)	Contra-lateral CBF (n=41)	Contralateral— Ipsilateral CBF (n=41)
					Outcomes		
ApoE4	B	11.4	5.9	8.6	9.2	4.3	-5.1
	P-value	*.02****	*.27*	*.08**	*.07**	*.43*	*.06**
GCS	B	3.7	2.9	3.3	6.2	1.52	-4.7
(3,4 VS. 5-8)	*P-value*	*.41*	*.55*	*.46*	*.77*	*.77*	*.07**
Paco2	B	1.1	.62	.87	1.2	.70	-.48
	P-value	*.0019****	*.10**	*.015****	*.0056****	*.11*	*.03****
TFI	B	1.6	1.7	1.6	2.1	2.0	N/A
	P-value	*.0018****	*.0027****	*.0016****	*.0004****	*.0013****	N/A

*p-value < 0.1;   ** p-value <0.05

7.2 ApoE 4 Allele

Table 3 shows the right hemisphere, left hemisphere and global mean CBF ROIs in all patients dichotomized by the presence or absence of the ApoE 4 allele. The Student's t-test analysis indicates that patients with the ApoE 4 alelle had significantly higher flows in the right hemisphere (p=.02) and global CBF ROIs (p=.04). Patients with the ApoE 4 allele had trends toward higher flows in the ipsilateral hemisphere (See Figure 2). There were no differences in the contralateral hemisphere or side-to side hemispheric differences. A stepwise regression analysis was conducted to control for age, gender, race, severity of injury [as indicated by the (GCS) on admission], and the level of $PaCO_2$ during the CBF test. Results showed that the significant predictors of CBF were the presence of the ApoE 4 allele, level of $PaCO_2$, GCS and time of conducting the test from injury (Refer to table 4).

8. DISCUSSION

This study shows that patients with the ApoE 4 allele have higher CBF within the first 24 hours after a severe TBI than those without the ApoE 4 allele. This difference was reflected in higher CBF within the ipsilateral hemisphere and after controlling for level of PaCO2, severity of injury, gender, race and age. One reason for higher CBF may be a genetically based differential flux of ApoE to the region of injury.

Regional differences in CBF have been reported in healthy individuals. Scarmeas et al found that the resting rCBF in carriers of the ApoE 4 allele was lower in the left and right inferior temporal gyri and higher in the left insula, right supramarginal gyrus, and the inferior occipital gyrus in college age young adults [9]. Cohen et al. did not find regional CBF differences but did find that global grey matter CBF was higher in elderly individuals with the ApoE 4 allele [10]. Orihara & Nakasono (2002) detected ApoE protein within the neurons of the traumatized cortical hemisphere, particularly around the capillaries and neuropil within the hemisphere. They did not detect ApoE within the neurons of the contralateral cortical hemisphere. In animal models, a lack of apolipoprotein E has been associated with decreased elasticity and increased vascular stiffness [3, 4], altered vascular resistance and compliance, and elevated cardiac outflow velocities[5].

This analysis suggests there are higher levels of global and hemispheric cortical flow in carriers of the ApoE 4 allele that appears independent of severity of injury or other personal characteristics. There is some evidence to suggest that ApoE genotype may influence the overexpression of the brain's metabolic processes as part of the acute phase response following brain injury. In a smaller study, we found limited evidence to show clinical changes in CBF and metabolism [11] with trends toward higher global CBF measured using bedside inhaled Xe133 in ApoE 4 allele carriers during the acute phase after injury. In addition, we found that levels of excitatory amino acids (EAA) in the cerebrospinal fluid may be associated with ApoE 4 allele differences during the acute phase (first five days) following a brain injury [12]. These trends in allele specific differences in CBF may increase risk for poorer outcomes in patients and may be related to impaired astrocyte-neuronal connectivity.

A developing and compelling body of research suggests that one's genetic profile may predispose patients to variations in response to disease or injury. The majority of epidemiologic, experimental, and clinical studies on genetics and brain injury have focused on Alzheimer's disease or systemic disorders (Hademenos, 2001). This project demonstrates that an allele specific differential in CBF within the first 24 hours after TBI supports the need for further research focused on allele specific CBF - metabolic coupling to changing metabolic needs, as one may experience after a TBI. Intervention that improves cardiovascular function via an ApoE pathway may increase brain perfusion and improve long term outcomes in the TBI population.

9. ACKNOWLEDGMENT

This project was supported in part by NIH, NINR (R01NR04801) and Brain Trauma Research Center (5P50NS30318-13). We thank Ryan L. Minnster for his excellent assistance in DNA isolation and ApoE genotyping.

10. REFERENCES

1. Strittmatter WJ. Apolipoprotein-E—high-avidity binding to beta-amyloid and increased frequency of type-4 allele in late-onset familial Alzheimer-disease. *PNAS.* March 1993;90(5):1977-1981.
2. Solyom A, Bradford RH, Furman RH. Apolipoprotein and lipid distribution in canine serum lipoproteins. *Biochim Biophys Acta.* Feb 16 1971;229(2):471-483.
3. Wang HY, Zhang FC, Gao JJ, et al. Apolipoprotein E is a genetic risk factor for fetal iodine deficiency disorder in China. *Mol Psychiatry.* Jul 2000;5(4):363-368.
4. Tham DM, Martin-McNulty B, Wang YX, et al. Angiotensin II injures the arterial wall causing increased aortic stiffening in apolipoprotein E-deficient mice. *Am J Physiol Regul Integr Comp Physiol.* Dec 2002;283(6):R1442-1449.
5. Hartley CJ, Reddy AK, Madala S, et al. Hemodynamic changes in apolipoprotein E-knockout mice. *Am J Physiol Heart Circ Physiol.* Nov 2000;279(5):H2326-2334.
6. d'Uscio LV, Baker TA, Mantilla CB, et al. Mechanism of endothelial dysfunction in apolipoprotein E-deficient mice. *Arterioscler Thromb Vasc Biol.* Jun 2001;21(6):1017-1022.
7. Kalaria RN, Ballard CG, Ince PG, et al. Multiple substrates of late-onset dementia: implications for brain protection. *Novartis Found Symp.* 2001;235:49-60; discussion 60-45.
8. Barton M, Haudenschild CC, d'Uscio LV, Shaw S, Munter K, Luscher TF. Endothelin ETA receptor blockade restores NO-mediated endothelial function and inhibits atherosclerosis in apolipoprotein E-deficient mice. *Proc Natl Acad Sci U S A.* Nov 24 1998;95(24):14367-14372.
9. Scarmeas N, Habeck CG, Stern Y, Anderson KE. APOE genotype and cerebral blood flow in healthy young individuals. *Jama.* Sep 24 2003;290(12):1581-1582.
10. Cohen RM, Podruchny TA, Bokde AL, et al. Higher in vivo muscarinic-2 receptor distribution volumes in aging subjects with an apolipoprotein E-epsilon4 allele. *Synapse.* Sep 1 2003;49(3):150-156.
11. Kerr ME, Kraus M, Marion D, Kamboh I. Evaluation of apolipoprotein E genotypes on cerebral blood flow and metabolism following traumatic brain injury. *Adv Exp Med Biol.* 1999;471:117-124.
12. Kerr ME, Ilyas Kamboh M, Yookyung K, et al. Relationship between apoE4 allele and excitatory amino acid levels after traumatic brain injury. *Crit Care Med.* Sep 2003;31(9):2371-2379.

47.

IS CYCLOXYGENASE-2 (COX-2) A MAJOR COMPONENT OF THE MECHANISM RESPONSIBLE FOR MICROVASCULAR REMODELING IN THE BRAIN?

Joseph C. LaManna, Xiaoyan Sun, Andre D. Ivy, and Nicole L. Ward[*]

1. INTRODUCTION

The mammalian brain is exquisitely dependent on timely availability of both oxygen and glucose. Only minimal stores of either are present in the adult brain. Yet, mammals can exist and even flourish continuously at altitudes of up to 14,000 feet or more, and for brief periods at even higher elevations. The adaptive mechanisms that allow the brain to function over this wide range of environmental oxygen content reveal a complex metabolic and vascular control system, and an inherent structural and functional plasticity that appears to be driven by oxygen adequacy, i.e., the balance between delivery of oxygen and utilization of oxygen. These mechanisms are continuously active during normal physiological adaptation, and also play a significant role in the protective/restorative, as well as, the pathological responses to oxidative challenges.

Chronic adaptation to mild hypoxia includes a robust angiogenetic response[1]. The capillary density increases by almost double after 3 weeks of exposure to 0.5 ATM; the equivalent of an altitude of 5500m. The adaptive response is driven primarily by vascular endothelial growth factor (VEGF)[2,3] and angi`opoietin-2 (ang-2)[4]. VEGF is now well known to be upregulated by Hypoxia Inducible Factor-1 through activation of the hypoxic response element of the gene[5-7]. Upregulation of ang-2 has been less well understood until recently when it was shown that, in vitro, ang-2 is upregulated by prostaglandin E2 which comes from increased endothelial cyclooxygenase-2 (COX-2) activity[8]. Vascular remodeling depends on the balance between HIF-1 dependent and HIF-1 independent processes[9]. In this study we report preliminary evidence for in vivo upregulation of endothelial COX-2 in response to hypoxia.

[*] Department of Anatomy, Case Western Reserve University School of Medicine, Cleveland, Ohio 44106

2. METHODS

2.1 Altitude Adaptation Model

All animals received humane care according to the US Department of Agriculture (USDA), Office for Laboratory Animal Welfare (OLAW), Public Health Services (PHS), and American Association for Accreditation of Laboratory Animal Care (AAALAC) using the guidelines set forth in the Animal Welfare Act and Guide for the Care and Use of Laboratory Animals. Animals were kept on a standard light cycle and fed food and water ad libitum throughout the course of the experiments.

CD1 mice were kept for up to 3 weeks at 0.5 ATM in hypobaric chambers. Littermate controls were kept in the same room in a similar chamber under similar conditions except that they were kept at normobaric pressure for the 3 week duration of the experiment.

2.2 Immunohistochemistry

Mice were deeply anesthetized and transcardially perfused with ice-cold phosphate buffered saline (PBS, pH 7.2) followed by 4% paraformaldehyde in PBS. Brains were removed, post-fixed in the same fixative for 24 h at 4°C and embedded in paraffin. Coronal serial sections (6μm) were deparaffinized, hydrated and subjected to antigen retrieval at 90°C for 15 min using a Target Retrieval Solution (Dako). Antigens were detected through the use of a goat anti-rabbit secondary (1:150, Vector Laboratories), an avidin-biotin complex (Vectastain Elite ABC Kit, Vector Laboratories), and a diaminobenzidine peroxidase substrate (Vector Laboratories). Antibodies used included polyclonal rabbit anti-Glut1 (Dako) and rabbit anti-COX2 (Cayman Chemicals). Slides were dehydrated and coverslipped using Permount (Fisher).

2.3 Microvessel Density

Capillary diameters and densities were measured on tissues stained with antibodies against the glucose transporter, GLUT-1[10,11]. A photo montage spanning the full depth of the parietal cortex was created using a SPOT digital camera connected to a Nikon E600 Eclipse microscope with a 20X objective. A computer-aided interactive image analysis system (ImagePro Plus) was used to count marked antibody-positive capillaries between 4-25 μm in diameter and determine the number per unit area of brain tissue.

2.4 Western Analyses

Mice were deeply anesthetized using an intraperitoneal injection of sodium pentobarbital (65mg/kg). Control and hypoxic mouse cortical samples were dissected and homogenized in ice-cold lysis buffer (1x PBS, 1% Nonidet P-40, 0.5% sodium deoxycholate, 0.1% SDS, 100mM Na_3VO_4) or (50 mM Tris-HCl/1mM EDTA, pH=7.4) containing protease inhibitors (1 μg/ml leupeptin, 10 μg/ml aprotinin, 100 μg/ml PMSF, 1 μg/ml pepstatin). Homogenates were centrifuged at 20,000 g for 30 minutes at 4° C and the supernatants used for western blot analysis. Protein content in the supernatant was determined by a Bradford assay (Bio-Rad) with bovine serum albumin as a standard.

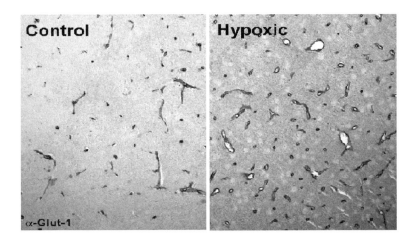

Figure 1. Exposure to chronic hypoxia causes increases in capillary density in the cortex of mice.

Control and hypoxic mouse cortical samples were processed for either Western blotting or immunoprecipitation followed by Western blotting. Equal amounts of whole cell lysate were immunoprecipitated with antibodies against Tie2, Ang1, or Ang2 (Santa Cruz) as described previously[4]. Western blotting was done as described previously[4]. Briefly, proteins from either whole cell lysate or following immunoprecipitation were separated using SDS gel electrophoresis and were transferred to PVDF membranes (Immobilon). Membranes were incubated in 5% skim milk in tris-buffered saline (TBS) for 1h to block nonspecific binding, and then incubated overnight in the same blocking solution with the primary antibody of interest. COX-2 (Cayman Chemicals), VEGF (A20, Santa Cruz), Tie2 (331, Pharmingen), Ang-1 and Ang-2 (Santa Cruz) proteins were detected using commercially available antibodies. Membranes were washed in TBS with triton, followed by incubation with the appropriate HRP-conjugated secondary (Bio-Rad). Immunoreactive protein bands were visualized using an enhanced chemiluminescence detection system (ECL kit, Amersham).

3. RESULTS

3.1 Increased Capillary Density

CD1 mice exposed to 0.5 ATM for three weeks had 36% more Glut-1 positive capillaries in the forebrain cortex than mice living under normobaric conditions (Figure 1). Control capillary density was about 230 capillaries/mm^2, whereas the hypoxic adapted group had about 310 capillaries/mm^2. This is very similar to the previous finding in Wistar rats[1]. This level of hypoxic exposure was tolerated by the mice very well.

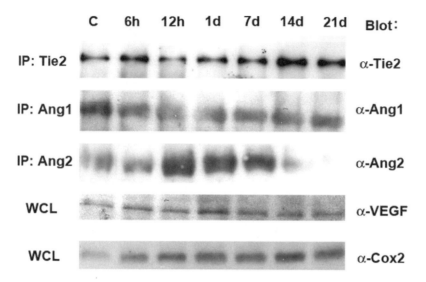

Figure 2. Ang2, Vegf, and COX2 protein are increased in mice exposed to hypoxia.

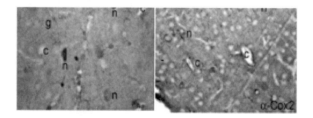

Figure 3. Hypoxic mice have increased Cox2 protein in endothelial cells of cortical capillaries.

3.2 Western Blot

Whole cell lysates from mouse cortex were immunoprecipitated (IP) or run as whole cell lysates and then blotted as indicated against Tie2, Ang1, Ang2, VEGF, or COX2 (Figure 2). Tie2 and Ang1 levels did not change from control normoxic cortex, following 6 hours (h), 12 hours, 1, 7, 14, or 21 days. Ang2, VEGF and Cox2 all increased as early as 6 hours following exposure to hypoxic conditions (WCL, whole cell lysate; IP, immunoprecipitate; h, hours; d, days).

3.3 Immunohistochemistry

CD1 mice exposed to chronic hypoxia compared with mice exposed to normoxic conditions show increased levels of COX-2 protein in endothelial cells of the capillaries (c) of the forebrain cortex. In the control brain, a few neurons (n) and a few astrocytic glial cells (g) show constitutive COX-2 protein.

4. DISCUSSION

In endothelial cells, Ang-2 expression is induced by hypoxic exposure independently of the HIF-1 transcription factor. Ang-2 expression is the result of hypoxic upregulation of cyclooxygenase-2 (COX-2) and subsequent metabolism of arachidonic acid to PGE2[12]. The physiological role of COX-2 is not well understood. There are 2 forms of cyclooxygenase, the constituitive COX-1 and the mitogen-inducible COX-2. The gross oversimplification that COX-1 is the "good" form and COX-2 the "bad" form,[13,14] is rapidly losing currency. Most of the literature concerns the negative effects of COX-2 and the protective role of COX-2 inhibitors. In the CNS, COX-2 is mostly associated with neurons[15,16] and has been linked to neurodegenerative diseases[17] such as Parkinson's[18] and Alzheimer's[19-21]. Inhibition of COX-2 is neuroprotective in cerebral ischemia[22-24]. But, more recently, COX-2 has been suggested to have physiological roles, for example, in protection against myocardial ischemic damage[14].

COX-2 is the key participant in inflammatory cytokine-induced angiogenesis[25] and has been suggested as a target for anti-angiogenesis therapies[26]. Hypoxic induction of COX-2 through interleukin-1β has been shown in human neuroblastoma cells[27] and microglia[28], and COX-2 expression was associated with early mechanisms leading to apoptosis in vitro and in vivo[29]. COX-2 has been implicated in NMDA-mediated neuronal cell death[30,31].

Although persistence of angiogenesis into the adult is usually linked to disease[26], there is clearly a role for physiological angiogenesis in the adult brain adaptation to hypoxia. The important distinction for the CNS might be that overexpression of COX-2 in neurons leads to degeneration, but that COX-2 expression in the microvasculature is required for capillary remodeling just as it is necessary for synaptic plasticity and LTP[16]. The paradoxical roles of COX-2 are important to understand, especially considering the rise in the use of prescription COX-2 selective inhibitors such as Vioxx® and Celebrex®.

5. CONCLUSION

We have used a relatively simple model of hypoxia that triggers adaptive structural changes in the cerebral microvasculature to study the process of physiological angiogenesis. This model can be used to obtain mechanistic data for the processes that probably underlie the dynamic structural changes that occur in learning and the control of oxygen availability to the neurovascular unit. These mechanisms are broadly involved in a wide variety of pathophysiological processes. This is the vascular component to CNS functional plasticity, supporting learning and adaptation. The angiogenic process may wane with age, contributing to the decreasing ability to survive metabolic stress and the diminution of neuronal plasticity.

6. ACKNOWLEDGEMENTS

This work was funded by USPHS grant NS038632.

7. REFERENCES

1. J. C. LaManna, L.M. Vendel, and R.M. Farrell: Brain adaptation to chronic hypobaric hypoxia in rats. *J.Appl.Physiol.* **72**:2238-2243 (1992).

2. N.-T. Kuo, D. Benhayon, R.J. Przybylski, R.J. Martin, and J.C. LaManna: Prolonged hypoxia increases vascular endothelial growth factor mRNA and protein in adult mouse brain. *J.Appl.Physiol.* **86**:260-264 (1999).

3. J. C. Chávez, F. Agani, P. Pichiule, and J.C. LaManna: Expression of hypoxic inducible factor 1α in the brain of rats during chronic hypoxia. *J.Appl.Physiol.* **89**:1937-1942 (2000).

4. P. Pichiule and J.C. LaManna: Angiopoietin-2 and rat brain capillary remodeling during adaptation and de-adaptation to prolonged mild hypoxia. *J.Appl.Physiol.* **93**:1131-1139 (2002).

5. M. Gassmann and R.H. Wenger: HIF-1, a mediator of the molecular response to hypoxia. *News Physiol Sci* **12**:214-218 (1997).

6. J. A. Forsythe, B.-H. Jiang, N.V. Iyer, F. Agani, S.W. Leung, R.D. Koos, and G.L. Semenza: Activation of vascular endothelial growth factor gene transcription by hypoxia-inducible factor 1. *Mol.Cell.Biol.* **16**:4604-4613 (1996).

7. H. F. Bunn and R.O. Poyton: Oxygen sensing and molecular adaptation to hypoxia. *Physiol.Rev.* **76**:839-885 (1996).

8. P. Pichiule, J.C. Chavez, and J.C. LaManna: Hypoxic regulation of angiopoietin-2 expression in endothelial cells. *J Biol.Chem.* **279**:12171-12180 (2004).

9. J. C. LaManna, J.C. Chavez, and P. Pichiule: Structural and functional adaptation to hypoxia in the rat brain. *J.Exp.Biol.* **207**:3163-3169 (2004).

10. S. I. Harik, R.N. Kalaria, L. Andersson, P. Lundahl, and G. Perry: Immunocytochemical localization of the erythroid glucose transporter: Abundance in tissues with barrier functions. *J.Neurosci.* **10**:3862-3872 (1990).

11. S. I. Harik, W.D. Lust, S.C. Jones, K.L. Lauro, S. Pundik, and J.C. LaManna: Brain glucose metabolism in hypobaric hypoxia. *J.Appl.Physiol.* **79**:136-140 (1995).

12. P. Pichiule, J.C. Chavez, and J.C. LaManna: Hypoxic regulation of angiopoietin-2 expression in endothelial cells. *J.Biol.Chem.* (in press)(2004).

13. C. D. Funk: Prostaglandins and leukotrienes: advances in eicosanoid biology. *Science* **294**:1871-1875 (2001).

14. L. Parente and M. Perretti: Advances in the pathophysiology of constitutive and inducible cyclooxygenases: two enzymes in the spotlight. *Biochem.Pharmacol.* **65**:153-159 (2003).

15. C. D. Breder, D. Dewitt, and R.P. Kraig: Characterization of inducible cyclooxygenase in rat brain. *J Comp Neurol.* **355**:296-315 (1995).

16. N. G. Bazan: Synaptic lipid signaling: significance of polyunsaturated fatty acids and platelet-activating factor. *J Lipid Res.* **44**:2221-2233 (2003).

17. T. Wyss-Coray and L. Mucke: Inflammation in neurodegenerative disease--a double-edged sword. *Neuron* **35**:419-432 (2002).

18. P. Teismann, M. Vila, D.K. Choi, K. Tieu, D.C. Wu, V. Jackson-Lewis, and S. Przedborski: COX-2 and neurodegeneration in Parkinson's disease. *Ann.N.Y.Acad.Sci.* **991**:272-277 (2003).

19. J. J. Hoozemans, R. Veerhuis, A.J. Rozemuller, and P. Eikelenboom: Non-steroidal anti-inflammatory drugs and cyclooxygenase in Alzheimer's disease. *Curr.Drug Targets.* **4**:461-468 (2003).

20. P. L. McGeer and E.G. McGeer: Inflammation, autotoxicity and Alzheimer disease. *Neurobiol.Aging* **22**:799-809 (2001).

21. Z. Xiang, L. Ho, J. Valdellon, D. Borchelt, K. Kelley, L. Spielman, P.S. Aisen, and G.M. Pasinetti: Cyclooxygenase (COX)-2 and cell cycle activity in a transgenic mouse model of Alzheimer's disease neuropathology. *Neurobiol.Aging* **23**:327-334 (2002).

22. M. Nagayama, K. Niwa, T. Nagayama, M.E. Ross, and C. Iadecola: The cyclooxygenase-2 inhibitor NS-398 ameliorates ischemic brain injury in wild-type mice but not in mice with deletion of the inducible nitric oxide synthase gene. *J Cereb.Blood Flow Metab* **19**:1213-1219 (1999).

23. S. Nogawa, F. Zhang, M.E. Ross, and C. Iadecola: Cyclo-oxygenase-2 gene expression in neurons contributes to ischemic brain damage. *J Neurosci.* **17**:2746-2755 (1997).

24. K. Hara, D.L. Kong, F.R. Sharp, and P.R. Weinstein: Effect of selective inhibition of cyclooxygenase 2 on temporary focal cerebral ischemia in rats. *Neurosci.Lett.* **256**:53-56 (1998).

25. T. Kuwano, S. Nakao, H. Yamamoto, M. Tsuneyoshi, T. Yamamoto, M. Kuwano, and M. Ono: Cyclooxygenase 2 is a key enzyme for inflammatory cytokine-induced angiogenesis. *FASEB J.* **18**:300-310 (2004).

26. M. A. Iniguez, A. Rodriguez, O.V. Volpert, M. Fresno, and J.M. Redondo: Cyclooxygenase-2: a therapeutic target in angiogenesis. *Trends Mol.Med.* **9**:73-78 (2003).

27. J. J. Hoozemans, R. Veerhuis, I. Janssen, A.J. Rozemuller, and P. Eikelenboom: Interleukin-1beta induced cyclooxygenase 2 expression and prostaglandin E2 secretion by human neuroblastoma cells: implications for Alzheimer's disease. *Exp.Gerontol.* **36**:559-570 (2001).

28. N. G. Kim, H. Lee, E. Son, O.Y. Kwon, J.Y. Park, J.H. Park, G.J. Cho, W.S. Choi, and K. Suk: Hypoxic induction of caspase-11/caspase-1/interleukin-1beta in brain microglia. *Brain Res.Mol.Brain Res.* **114**:107-114 (2003).

29. L. Ho, H. Osaka, P.S. Aisen, and G.M. Pasinetti: Induction of cyclooxygenase (COX)-2 but not COX-1 gene expression in apoptotic cell death. *J Neuroimmunol.* **89**:142-149 (1998).

30. C. Iadecola, K. Niwa, S. Nogawa, X. Zhao, M. Nagayama, E. Araki, S. Morham, and M.E. Ross: Reduced susceptibility to ischemic brain injury and N-methyl-D-aspartate-mediated neurotoxicity in cyclooxygenase-2-deficient mice. *Proc.Natl.Acad.Sci.U.S.A* **98**:1294-1299 (2001).

31. S. J. Hewett, T.F. Uliasz, A.S. Vidwans, and J.A. Hewett: Cyclooxygenase-2 contributes to N-methyl-D-aspartate-mediated neuronal cell death in primary cortical cell culture. *J Pharmacol.Exp.Ther.* **293**:417-425 (2000).

48.

RELATIONSHIP BETWEEN HANDGRIP SUSTAINED SUBMAXIMAL EXERCISE AND PREFRONTAL CORTEX OXYGENATION

Leonardo Mottola,[1] Stefano Crisostomi,[1] Marco Ferrari,[1] and Valentina Quaresima[1]

1. INTRODUCTION

Fatigue might be defined as an exercise-induced loss of power- or force-generating capacity.[1] It has not been fully clarified what the effect is of fatiguing skeletal muscle exercise on brain, and in particular on ipsi- and contralateral prefrontal cortex (PFC). A recent functional magnetic resonance imaging (fMRI) study[2] demonstrated that during sustained muscle contractions there is a progressive involvement of the ipsilateral and contralateral PFC, whose activation may be involved in processing fatigue-related feedback and/or adjusting the descending command for the ongoing task.[3]

Functional near infrared spectroscopy (fNIRS) has been used to monitor brain oxygenation changes over the motor cortex in a wide variety of exercise modalities: finger opposition,[4] finger tapping,[5] finger flexion/extension,[6] palm squeezing,[7] plantar flexion,[8] and walking.[9] Prefrontal cortex oxygenation response upon exhaustive cycling exercise was also investigated.[10-12]

The aim of this study was to assess bilateral PFC oxygenation during a sustained submaximal handgrip isometric exercise by two-channel fNIRS.

2. METHODS

Twelve right handed volunteers (age: 28.5±4.1 yrs) participated in this study. Subjects were physically active although none were engaged in daily, intensive or

[1] Department of Sciences and Biomedical Technologies, University of L'Aquila, I-67100 L'Aquila, Italy email: vale@univaq.it

specific training programs. All subjects gave their informed consent prior to participation after a full oral and written explanation of the experiments.

All subjects performed the motor task consisting of 4-min continuous isometric contraction at 30% of their maximal voluntary contraction (MVC) while their forehead was monitored by fNIRS (Fig. 1). The study was performed in a quiet room. For oxy- and deoxyhemoglobin (O_2Hb and HHb) measurements, a 2-channel NIRS oximeter (NIRO-300, Hamamatsu, Japan) was employed. The two sets of emission and detection NIRO-300 probes were attached bilaterally to the forehead of the subjects. In each set, the emission and detection probe were kept at a constant geometry and distance (5 cm apart) by a rubber shell that in turn was attached by a double-sided adhesive sheet. The detection probes were placed in correspondence of Fp1 and Fp2 of the 10-20 system for the EEG electrode placement with the emission probes being lateral by 5 cm on both sides approximately at F7 and F8, respectively. The anterior part of the prefrontal area, including Brodmann areas 9 and 10, may be the main area of contribution to the NIRS measurements. NIRS data were collected at the frequency of 6 Hz and transferred on-line to a computer for storage and subsequent analysis. The quantification of O_2Hb and HHb concentration changes, expressed in $\Delta\mu M$, was obtained by including an age-dependent constant differential pathlength factor $(4.99 + 0.067 * age^{0.814})$.[8]

Handgrip force was measured by a system consisting of a handgrip device and a digital handgrip analyzer (MIE, Medical Research, UK) (Fig. 1). The system in turn was connected to a computer for data acquisition and analysis (Mie Cas software). The pliers were fixed in vertical position to a rigid support (Fig.1). The grip span was 2 cm, and the exact point where the subject had to grasp was marked, in order to standardize the testing conditions. Subjects exerted handgrip contractions to match the output force to the target figure on a digital display. The sampling rate for force data was 33 Hz. MVC handgrip force was measured before the experiment itself (one day before) in each subject. The MVC value was calculated by averaging the measures over 5 trials. Heart rate (HR) was measured by a pulse oximeter (Nellcor N-200, USA).

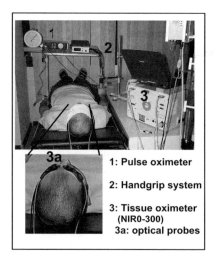

Figure 1. Experimental set-up

Data are reported as mean and standard deviation of O₂Hb, HHb, and HR and compared using repeated measures analysis of variance. Significant differences were identified using Turkey's honestly significant difference multiple comparison test. Each data point was calculated by averaging the last 5 s every 30 s of exercise. The paired *t* test was used to compare maximal changes in O₂Hb and HHb between left and right PFC. Significance level was set at P<0.05.

3. RESULTS

A typical example of force profile, heart rate and concurrent changes in O₂Hb and HHb in left and right PFC (subject #4) is reported in Fig. 2. The target force was maintained constant throughout the task performance. Heart rate increased progressively during the entire duration of the exercise period, and immediately decreased after its end. In all subjects PFC oxygenation of both hemispheres was modified according to the typical tracings shown in Fig. 2. A decrease in HHb accompanied by an increase in O₂Hb was observed in both hemispheres during the course of the task. Immediately after the end of the handgrip exercise, O₂Hb and HHb showed a tendency to return gradually to pre-exercise values.

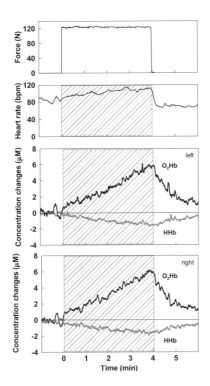

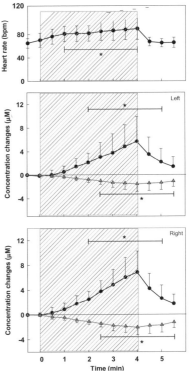

Figure 2. Time course of handgrip force, heart rate, and O₂Hb, HHb changes in left and right PFC (subject #4) before, during (▨) and after isometric contraction at 30% of his maximal voluntary contraction.

Figure 3. Mean time course of heart rate, and O₂Hb (-●-), HHb (-Δ-) changes in left and right PFC before, during (▨) and after isometric contraction at 30% of their maximal voluntary contraction (n=12). Each data point represents the average over the last 5 s every 30 s of exercise. *Significantly different from the baseline.

Figure 3 shows the time course of heart rate, O_2Hb and HHb changes in left and right PFC over the 12 subjects. Although the inter-subject variability in the amplitude of O_2Hb and HHb changes, O_2Hb significantly increased after the first 2 min of contraction, and HHb decreased after the first 2.5 min in both PFC sides (Fig. 3). The activation was not evident anymore after the first 60 s after the end of the task performance period.

The maximal concentration changes in O_2Hb and HHb found at the end of 4-min contraction are reported in Fig. 4. Changes in HHb were found significantly higher in the ipsilateral than in the contralateral PFC.

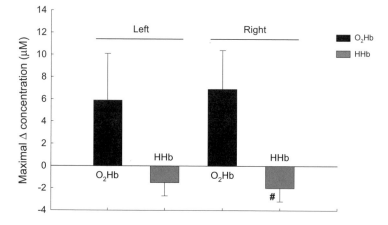

Figure 4. Maximal concentration changes at the end of 4-min continuous isometric contraction. [#] Significantly different from contralateral PFC.

4. DISCUSSION

Since 1993, more than 80 papers have reported frontal cortex activation, measured with fNIRS as changes in concentrations of O_2Hb and HHb in healthy subjects during specific tasks, such as arithmetic calculations, speech production, spatial/visual/object working memory, semantic processing, selected attention tests, anagrams, etc.

The typical fNIRS oxygenation response over an activated cortical area consists of a decrease in HHb accompanied with an increase in O_2Hb of 2–3 fold of magnitude. The physiological significance of this response has been extensively investigated in humans[13] and in the rat brain model.[14] O_2Hb is the most sensitive indicator of the increase in cerebral blood flow (neurovascular coupling) and the direction of the changes in HHb is determined by the oxygenation and volume of the venous blood.[14]

The prefrontal cortex is involved in many processes, some of them are related to motor activity and speech. Experimental data suggest that PFC activity occurs in relation to attention, short-term memory, and complex forms of motor behavior, i.e. anticipatory preparation, motor sequences, programming of speech, etc.

Notwithstanding fNIRS has been used to monitor brain oxygenation changes over the motor cortex during a wide variety of exercise modalities,[1-9] only a few fNIRS studies about the PFC oxygenation response during motor tasks are available.[10-12] Furthermore, substantial activation in the prefrontal region in addition to expected activations extending over the motor, premotor and supplementary motor areas was found during apple peeling by multichannel NIRS.[15]

To the best of our knowledge, this is the first time that fNIRS has been utilized to investigate PFC oxygenation during sustained submaximal handgrip exercise. In this study, the so-called cortical activation, was observed in both ipsi- and contralateral PFC sides. A potential minor contribution of skin blood flow increase to the changes in O_2Hb and HHb can be excluded during a short duration submaximal exercise, supporting the observed asymmetric HHb changes (Fig. 4). The results of this study confirm those obtained by fMRI[2] and suggest that PFC output signals drive right forearm muscles to maintain strength. The observed PFC changes are task related, as supported by the mismatched patterns of HR and O_2Hb changes (Fig. 3).

5. REFERENCES

1. S. C. Gandevia, Spinal and supraspinal factors in human muscle fatigue, *Physiol. Rev.* **81**(4), 1725-1789 (2001).
2. J. Z. Liu, Z. Y. Shan, L. D. Zhang, V. Sahgal, R. W. Brown, and G. H. Yue, Human brain activation during sustained and intermittent submaximal fatigue muscle contractions: an fMRI study, *J. Neurophysiol.* **90**(1), 300-312 (2003).
3. L. Nybo and N. H. Secher, Cerebral perturbations provoked by prolonged exercise, *Prog. Neurobiol.* **72**(4), 223-261 (2004).
4. M. A. Franceschini, S. Fantini, J. H. Thompson, J. P. Culver, and D. A. Boas, Hemodynamic evoked response of the sensorimotor cortex measured noninvasively with near-infrared optical imaging, *Psychophysiol.* **40**(4), 548-560 (2003).
5. D. J. Mehagnoul-Schipper, B. F. van der Kallen, W. N. Colier, M. C. van der Sluijs, L. J. van Erning, H. O. Thijssen, B. Oeseburg, W. H. Hoefnagels, and R. W. Jansen, Simultaneous measurements of cerebral oxygenation changes during brain activation by near-infrared spectroscopy and functional magnetic resonance imaging in healthy young and elderly subjects, *Hum. Brain Mapp.* **16**(1), 14-23 (2002).
6. G. Strangman, J. P. Culver, J. H. Thompson, and D. A. Boas, A quantitative comparison of simultaneous BOLD fMRI and NIRS recordings during functional brain activation, *NeuroImage* **17**(2), 719-731 (2002).
7. M. Wolf, U. Wolf, V. Toronov, A. Michalos, L. A. Paunescu, J. H. Choi, and E. Gratton, Different time evolution of oxyhemoglobin and deoxyhemoglobin concentration changes in the visual and motor cortices during functional stimulation: a near-infrared spectroscopy study, *NeuroImage* **16**(3 Pt 1), 704-712 (2002).
8. W. N. Colier, V. Quaresima, B. Oeseburg, and M. Ferrari, Human motor-cortex oxygenation changes induced by cyclic coupled movements of hand and foot, *Exp. Brain Res.* **129**(3), 457-461 (1999).
9. I. Miyai, H.C. Tanabe, I. Sase, H. Eda, I. Oda, I. Konishi, Y. Tsunazawa, T. Suzuki, T. Yanagida, and K. Kubota, Cortical mapping of gait in humans: a near-infrared spectroscopy tomography study, *NeuroImage* **14**(5), 1186-1192 (2001).
10. K. Shibuya, J. Tanaka, N. Kuboyama, S. Murai, and T. Ogaki, Cerebral cortex activity during supramaximal exhaustive exercise, *J. Sports Med. Phys. Fitness* **44**(2), 215-219 (2004).
11. K. Shibuya, J. Tanaka, N. Kuboyama, and T. Ogaki, Cerebral oxygenation during intermittent supramaximal exercise, *Respir. Physiol. Neurobiol.* **140**(2), 165-172 (2004).
12. K. Ide, A. Horn, and N. H. Secher, Cerebral metabolic response to submaximal exercise, *J. Appl. Physiol.* **87**(5), 1604-1608 (1999).
13. H. Obrig, and A. Villringer, Beyond the visible-imaging the human brain with light, *J. Cereb. Blood Flow Metab.* **23**(1), 1-18 (2003).
14. Hoshi, N. Kobayashi, and M. Tamura, Interpretation of near-infrared spectroscopy signals: a study with a newly developed perfused rat brain model, *J. Appl. Physiol.* **90**(5), 1657-1662 (2001).
15. M. Okamoto, H. Dan, K. Shimizu, K. Takeo, T. Amita, I. Oda, I. Konishi, K. Sakamoto, S. Isobe, T. Suzuki, K. Kohyama, and I. Dan, Multimodal assessment of cortical activation during apple peeling by NIRS and fMRI, *NeuroImage* **21**(4):1275-1288 (2004).

49.

EFFECT OF LOCAL COOLING (15° C FOR 24 HOURS) WITH THE CHILLERPAD™ AFTER TRAUMATIC BRAIN INJURY IN THE NONHUMAN PRIMATE

Edwin M. Nemoto, Gutti Rao, Timothy Robinson, Todd Saunders, John Kirkman, Denise Davis, Hiroto Kuwabara, C. Edward Dixon.[*]

1. INTRODUCTION

The efficacy of hypothermia is proven in rats [1] but unproven clinically in traumatic brain injury (TBI)[2]. Despite the failure of the NIH supported multicenter clinical trial to show improved outcome with hypothermia [3], optimism about the potential efficacy of hypothermia persists because of the delay of eight to twelve hours before the induction of hypothermia in the clinical trial. Furthermore, a subgroup analysis from that study showed that younger patients with TBI who received hypothermia and who were not rewarmed, had better outcome [2].

Our aim in this study was to evaluate the efficacy of hypothermia induced directly on the cerebral cortex using the ChillerPad™ to 15° C for 24 h beginning three hours after TBI in the Rhesus monkey (M. mulatta) with recovery for ten days after TBI.

2. METHODS

2.1 Surgical Procedures

Following a protocol approved by the Institutional Animal Care and Use Committee, six male monkeys (M. mulatta) weighing from 5.0 to 9.3 kg, (7.3 ± 5.1, mean ± SD) were intubated and mechanically ventilated on isoflurane/70% nitrous oxide/30% oxygen for anesthesia. A femoral artery catheter was inserted and the monkeys turned prone for head fixation in a stereotactic device. The dorsal calvarium was scrubbed with betadine for aseptic surgery. A circular craniotomy 22 mm in diameter was centered 3.5 cm from the orbital ridge. An intracranial pressure transducer (Camino 110-4G, Integra) was

---

[*] Edwin Nemoto, Gutti Rao, Edward Dixon, University of Pittsburgh, Pittsburgh, PA 15213; Timothy Robinson, John Kirkman, Seacoast Technologies, Inc., Portsmouth, NH, 03801; Hiroto Kuwabara, Johns Hopkins University, Baltimore, MD, 21287

inserted subdurally adjacent to the craniotomy through a 3-mm burr hole and secured in place with a cyanoacrylate glue-soaked sponge.

2.2   Controlled Cortical Impact (CCI)

A controlled cortical impact (CCI) was produced at a velocity of 3.5 m/sec and a depth of penetration of 7 mm from the cortical surface at the peak of inspiration as judged by the end-tidal CO_2 pressure wave. Immediately after the CCI, which generally caused a tear in the dura, the dura was sutured with 6-0 prolene, the bone flap replaced and secured in place by suturing the galeoperiosteal flap over the bone flap for the three hour delay before the placement of the ChillerPad™and the start of cooling the brain to 15° C.

2.3   Cooling and Rewarming

Beginning at 2.5 h after the CCI, the craniotomy was exposed and the ChillerPad™ was placed onto the dura with thermocouples in the center beneath the ChillerPad™. The ChillerPad™ was secured in place by suturing the galeoperiosteal flap over it and over that, the skin. Throughout the 24 h cooling period and 10 h of rewarming at 2.5° C/h, physiological variables were continuously monitored with intermittent monitoring of arterial blood gases and electrolytes. Electrolytes were maintained within normal limits by switching to plasmalyte with the addition of potassium if necessary. Fluid intake and urine output were monitored with adjustments in the fluid infusion rate to balance to a net positive uptake of about 300 ml over the 24 h of cooling and 10 h of rewarming.

After complete rewarming, the ChillerPad was removed and thecraniotomy was sealed with the bone flap which was cemented in place with methylmethacrylate cement. The skin was sutured subcutaneously and the monkey treated prophylactically with 250 mg Cefazolin, i.v. Anesthesia was switched from isoflurane/nitrous oxide/oxygen to ketamine and fentanyl, the latter by continuous infusion with continuous monitoring of noninvasive arterial blood pressure.

2.4 Magnetic Resonance Imaging

The monkey was transported to the MRI scanner for a 40 hr post TBI MRI scan with SPGR, proton and T2 weighted scanning and diffusion tensor scanning. The MRI scan was done using a 3.0T GE scanner.

2.5 Recovery

After completion of the MRI scan, the monkey was returned to the OR where it recovered from fentanyl anesthesia by washout rather than reversal with Narcan and placed into the recovery room. The monkey was observed continuously for the next 24 h of recovery before being placed in the normal holding facility. Analgesia was provided with Buprenorphine after the first 24 h of recovery to avoid depressing the monkey with sedation in the early postrecovery period.

Neurologic examinations were done daily over the 10 days of recovery. Repeat MRI scans were performed at 4, 7 and 10 days. In some cases, the day 4 MRI was not done because it was felt that because the monkeys had gone 3 days without food and after recovering for 24 hr again placed on NPO, would be too much of a stress. Therefore, two monkeys did not have an MRI on day four post TBI.

2.6 Necropsy

On day 10 after CCI, the monkeys were taken for their last MRI scan after which time they were taken to the necropsy room under fentanyl anesthesia and they were deeply anesthetized with sodium pentobarbital (10 mg/kg) while the fentanyl was still in effect and the brains perfused fixed transcardially with a liter of saline followed by a liter of 2% paraformaldehyde. The brains were removed from the calvarium and immersed in 10% buffered formalin for fixation and subsequent paraffin embedding, sectioning and staining with hematoxylin eosin.

3. RESULTS

Hypothermia significantly reduced tissue necrosis volume as quantitated from the T2-weighted images, which were significantly lower in the hypothermia treated monkeys over ten days (Fig. 1). Edema volume was significantly lower at 40 hrs after TBI and cooling, the effect apparently disappeared at days 7 and 10 after TBI (Fig. 2).

Interestingly, despite the reduction in necrosis volume, the histological changes observed around the necrotic core was worse in the hypothermia-treated brains than in normothermia as shown in the two examples, one hypothermic and the other normothermic (Fig.3).

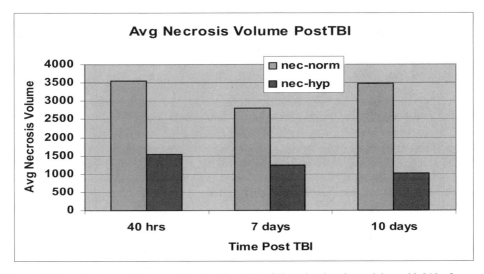

Figure. 1. Necrosis volume after traumatic brain injury (TBI) followed at three h post injury with 24 h of hypothermia (hyp)(cortical temp. = 15°C) or normothermia (norm)with cortical temp at 37°C and rewarming at 2.5°C/ h over ten hours. Necrosis volume is significantly lower in the hypothermia treated monkeys compare to the normothermia monkeys over the 10 days post TBI.

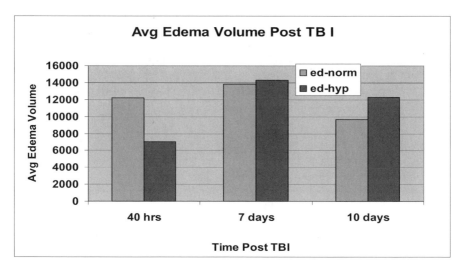

Figure. 2. Edema volumes after traumatic brain injury (TBI) in monkeys treated with hypothermia, cortical temperature of 15°C for 24 h beginning at three h after TBI or normothermia, cortical temperature at 37°C followed by rewarming at 2.5° C/hr over ten hours. Edema volumes were not significantly different (P>0.05) between the two groups over the 10 days of recovery.

Curiously, neuropathological analysis showed that the severity of histological changes were greater in the hypothermia treated monkeys compared to the normothermic despite the infarction volumes were approximately 50% smaller in the hypothermia treated monkeys. Whereas the hypothermia monkeys showed severe histological changes the normothermic monkeys showed mainly moderate changes.

4. DISCUSSION AND CONCLUSIONS

It is clear that the volume of necrosis was smaller by about 50% in the hypothermia-treated monkeys even with only two monkeys in each group, compared to the normothermia monkeys. However, we found an apparent discrepancy in that the histological findings were greater in the hypothermia-treated monkeys. How can we explain this discrepancy?

If we assume that the volume of injury sustained after controlled cortical impact (CCI) is the same in the hypothermia and normothermia treated monkeys and that it remained the same at three hours after injury when hypothermia was begun, hypothermia decreased the transition of the injured tissue into necrosis by 50% compared to normothermia. The result is that the tissue that was spared would appear to be worse than where the tissue was allowed to transition to necrosis as occurred in the normothermic monkeys as shown in Fig. 4. Following this line of reasoning, it is expected that histological analysis would reveal more severe histological changes in the hypothermia compared to the normothermia monkeys but yet, with smaller volumes of necrosis.

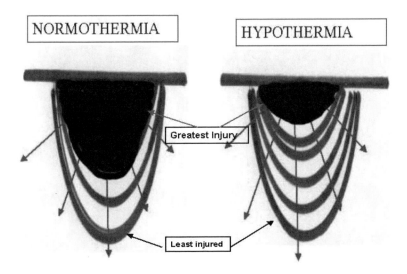

Figure. 4. Hypothetical injury pattern in monkeys treated with normothermia and hypothermia with a 50% reduction in necrosis volume in the hypothermia treated monkeys. The arrows indicate the direction of decreasing histologic damage. The black areas show the volume of tissue necrosis. The sparing of the tissue from transition to necrosis by hypothermia results in more severely injured tissue being represented in the hypothermia treated monkey and an apparently worse histological damage compared to normothermia.

5. REFERENCES

1. Clifton GL. Jiang JY. Lyeth BG. Jenkins LW. Hamm RJ. Hayes RL, 1991, Marked protection by moderate hypothermia after experimental traumatic brain injury. J Cereb Blood Flow & Metab **11,** 114.
2. Clifton GL. Miller ER. Choi SC. Levin HS. McCauley S. Smith KR Jr. Muizelaar JP. Wagner FC Jr. Marion DW. Luerssen TG. Chesnut RM. Schwartz M, 2001, Lack of effect of induction of hypothermia after acute brain injury. NEJM **344**(8), 556.
3. Clifton GL. Miller ER. Choi SC. Levin HS. McCauley S. Smith KR Jr. Muizelaar P. Marion DW. Luerssen TG., 2002, Hypothermia on admission in patients with severe brain injury. [Clinical Trial. Journal Article. Multicenter Study. Randomized Controlled Trial] J Neurotrauma. **19**(3), 293.

50.

SAFETY OF DIRECT LOCAL COOLING (15° C) OF THE CEREBRAL CORTEX WITH THE CHILLERSTRIP™ DURING FOCAL CEREBRAL ISCHEMIA IN MONKEYS

Edwin M. Nemoto, Charles Jungreis, Tudor Jovin, Gutti Rao, Timothy Robinson, Todd Sanders, Kate Casey, and John Kirkman[*]

1. INTRODUCTION

Systemic hypothermia to 32-33°C has been adopted for cardiopulmonary resuscitation for cardiac arrest from ventricular fibrillation by the American Heart Association[1]. Aside from its use in protecting the brain during ischemia on cardiopulmonary bypass, the use of hypothermia has not been proven in stroke or head injury. A recent NIH supported clinical trial failed to show improved outcome with hypothermia induced between eight and twelve hours after traumatic brain injury (TBI) [2]. However, patients who were less than 45 years of age and arrived in the hospital hypothermic (~35C) had better outcome[3]. One of the problems in the induction of systemic hypothermia is that cooling the entire body mass is difficult and likely to delay cooling. In addition, the induction of systemic hypothermia especially in conscious patients induces a tremendous stress response whereby the body elicits every available mechanism to maintain normothermia which may, itself, detrimentally affect ischemic or traumatic brain injury.

The induction of systemic hypothermia has several limitations. First, cooling the entire body is slow and difficult. Second, cooling of a conscious patient with systemic hypothermia is intolerable and elicits a tremendous stress response causing inappropriate vasoconstricton and metabolic activation.

[*] Edwin M. Nemoto, Charles Jungreis, Tudor Jovin, Gutti R. Rao, University of Pittsburgh, Pittsburgh, PA 15213. Timothy Robinson, Todd Sanders, Kate Casey, John Kirkman, Seacoast Technologies. Inc. 222 International Drive, Suite 145, Portsmouth, NH , 03801

Seacoast Technologies, Inc., (Portsmouth, NH) has developed a cooling device (ChillerPad) that can be applied directly onto the dura and the cerebral cortex to cool the brain to 15°C. The aim of this study was to show that the application of hypothermia directly onto the cortical surface using the ChillerStrip™ does not adversely affect the severity of ischemic injury in a monkey model of mild focal cerebral ischemia (FCI) of about 20 min duration followed by spontaneous rewarming beginning five min after reperfusion of the brain.

2. METHODS AND RESULTS

2.1 Study Design

Six male monkeys (M. Mulatta) weighing 6.3-9.6 kg body weight (8.1 ± 1.5, mean ± SD) were studied. (Table 1). Two monkeys served as sham controls that did not have the ChillerStrip placed on the cortical surface whereas the others did. The overall study paradigm was:

1)   Surgical placement of the ChillerStrip™;
2)   Systemic and cortical cooling five min before FCI;
3)   Endovascular occlusion of the bifurcation of the internal carotid artery;
4)   Stable xenon/CT measurement of cerebral blood flow (CBF) during FCI;
5)   Deflation of the occlusion and reperfusion;
6)   Stable Xe/CT measurements at 10 and 20 min after reperfusion; and
7)   Removal of the ChillerStrip,™ closing of the bone flap, and recovery.

The duration of FCI varied from the projected 20 min because in some cases, there were problems in obtaining the CBF measurements within the 20 min time limit. The duration of FCI ranged from 20 to 27.8 min.

Table 1: Study design. Two monkeys served as sham controls. Cortical and rectal temperatures were achieved 5 min prior to induction of FCI. Durations of ischemia varied because some studies had difficulties in initiating the first Xe/CT CBF measurement during ischemia and before the 20 min time limit. Rectal temperature was reduced to 32°C by passive cooling.

Group (N)	ChillerStrip	Cortical Temp (°C)	Rectal Temp (°C)	FCI (min)	Recovery
Contr (2)	No	37	37	20.0/20.0	7 days
Exp 1 (2)	Yes	15	37	27.8/24.0	7 days
Exp 2 (2)	Yes	15	32	22.0/20.0	7 days

2.2 Surgical Procedures

The monkeys were intubated and mechanically ventilated on isoflurane/70% nitrous oxide/29% oxygen for surgical anesthesia. The monkeys were placed prone and their heads fixed in a stereotactic device with monitoring of arterial blood pressure to ensure that they were adequately anesthetized in the device. The head was aseptically prepped with betadine. A U-shaped incision on the scalp was made with a monopolar cautery with the bottom of the U facing anteriorly and the skin flap is retracted posteriorly. The temporal muscles on the right hemisphere were retracted using a Wheatlander self-retaining retractor and the periosteum was removed with a periosteal elevator. The skull was marked with a 1.5 mm drill bit to demarcate the boundaries of a rectangular craniotomy 35 X 25 mm over the right hemisphere. The craniotomy was made 5 mm lateral to the midline with the anterior border approximately one cm from the frontal pole.

After completion of the craniotomy, the dura was carefully separated from the skull using a dural separator and the bone flap removed. The dura was incised and carefully reflected to expose the cortical surface with an area of 25 X 20 mm. Three thermocouples (Seacoast Technologies Contact Probe T-308A) were placed on three gyri on the frontal-parietal cortex in the anterior – posterior direction in the middle of the dural opening. The procedures for each of the study groups differed (see Table 1). Control monkeys (n=2) did not have ChillerStrips placed on the brain but instead, had saline-soaked sponges placed on the cortex and the galeoperiosteal flap and skin loosely sutured over the sponges. The other four monkeys studied in Expt groups one and two had the ChillerStrip™ placed on the cortex and the galeoperiosteal flap and skin were sutured to secure it in place. After completion of the surgery, in all studies, anesthesia was converted to fentanyl at a dose of about 100 ug/kg/ hr

2.3 ChillerStrip™ Cooling Performance

The ChillerStrip™ cooled the cortical surface to 15°C in 11:42 ± 4:22 min:sec (mean ± SD). Spontaneous rewarming to 37°C occurred within 35:18 ± 5:11 min:sec. Temperatures at depths of 9, 5, 10, 15 and 20 mm from the cortical surface approximated 15,20,27,33 and 35°C.

2.4 Endovascular Occlusion of the Bifurcation of the Internal Carotid Artery

In the angio suite, the right femoral artery was exposed and a sheath inserted. A balloon (Hyperform #104-4470, maximum inflated diam, 4 mm X 7mm, Microtherapeutics, Inc. Irvine, CA) was inserted into the internal carotid artery (ICA) at its bifurcation, which occluded the ICA and the MCA. Although the diameter of the uninflated balloon appeared to be occlusive, contrast injection was used to verify lack of flow around the balloon (Fig.1A &1B).

A   B

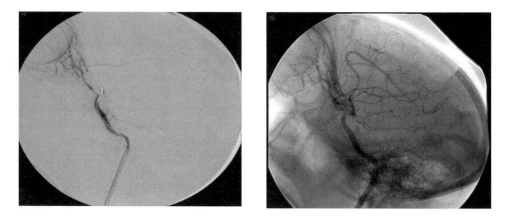

Fig.ure 1. Contrast arteriogram of the right hemisphere before (A) and after (B) balloon occlusion at the bifurcation of the internal carotid artery in a monkey.

2.5 Stable Xenon/ CT Cerebral Blood Flow Measurements.

After verification of the occlusion of the internal carotid artery at its bifurcation by angiography, the monkeys were quickly transported to the CT scanner for measurement of cerebral blood flow (CBF) using stable xenon/CT as previously described (Yonas et al, 1990). The monkeys were ventilated on 28% medical-grade xenon gas in 40% oxygen (XeScan stable xenon in oxygen USP, Praxair Pharmaceutical Gases, Praxair, Inc., Danbury, CT) via facemask, for 4.3 minutes. Two scans were obtained at each of four CT levels before and six scans during Xe inhalation (General Electric Systems, Milwaukee, WI). CBF measurements were made during ischemia and at 10 and 20 min after reperfusion (Fig. 1).

A CBF map from one of the studies shows that occlusion of the bifurcation of the ICA produces FCI over the frontal-parietal cortex with some involvement of the centrum semiovale. At 10 min after reperfusion, a relatively dramatic increase in CBF is seen in the ischemic hemisphere indicating reperfusion hyperemia which later subsides to some extent after 20 min.

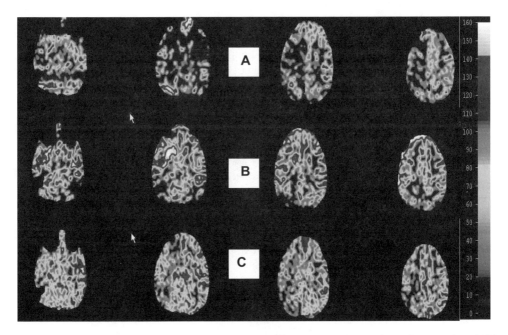

Figure. 2. Stable xenon CT cerebral blood flow (CBF) maps of M206-02 during ischemia (A) and at 10 (B) and 20 (C) min after FCI by occlusion of the right internal carotid artery at the bifurcation. The right hemisphere is on the left and the left hemisphere is on the right. The CBF gray scale illustrates flow values from 0 to 160 ml/100g/min. Note the reduction in CBF in the right frontal area during ischemia.

The changes in CBF in the three study groups are shown in Table 2 along with the neuropathology scores. CBF during ischemia in the normothermic control group was about a third higher in the cooled groups. The degree of postischemic hyperemia was significantly reduced in the systemically cooled monkeys at 10 and 20 min postischemia but not in the monkeys whose rectal temperatures were kept normothermic. The neuropathology scores indicate that the least damage was observed in the 15/32 group.

Table 2. Cerebral Blood Flow (CB F) (mean ± SD) in monkeys subjected to occlusion of the right internal carotid artery bifurcation for 20 min followed by reperfusion and CBF measurements at 10 and 20 min postischemia (PI).

Study Group	Ischemia Min	Ischemia R	L	10 Min PI R	L	20 Min PI R	L
37C/37C	20/20	$31 \pm 3*$	48 ± 5	60 ± 19	51 ± 19	53 ± 20	51 ± 15
15C/37C	28/24	$19 \pm 9*$	36 ± 9	36 ± 16	28 ± 2	45 ± 26	50 ± 18
15C/32C	22/20	$19 \pm 5+$	31 ± 4	$25 \pm 11**$	$27 \pm 5**$	$24 \pm 4**$	$22 \pm 4**$

*= $P<0.05$ compared to the Left (L) Hemisphere
**= P,0.03 Compared to 37C/37C

3. CONCLUSIONS

Direct cooling of the cerebral cortex with the ChillerStrip™ to 15°C followed by spontaneous rewarming to 37°C is safe. Direct cooling of the brain reduces the severity of the ischemic insult as judged by the reduction in the hyperemia after reperfusion which appeared to be directly related to the temperature of the brain.

4. REFERENCES

1. Nolan, J.P., Morley, P.T., Vanden Hoek, T.L., Hickey, R.W. 2003, Therapeutic Hypothermia After Cardiac Arrest. *Circulation* **108:**118
2. Clifton GL. Miller ER. Choi SC. Levin HS. 2001,McCauley S. Smith KR Jr. Muizelaar JP. Wagner FC Jr. Marion DW. Luerssen TG. Chesnut RM. Schwartz M. Lack of effect of induction of hypothermia after acute brain injury. New England Journal of Medicine. **344**(8):556.
3. Clifton GL. Miller ER. Choi SC. Levin HS. McCauley S. Smith KR Jr. Muizelaar JP. Marion DW. Luerssen TG. 2002, Hypothermia on admission in patients with severe brain injury. [Clinical Trial. Journal Article. Multicenter Study. Randomized Controlled Trial] Journal of Neurotrauma. **19**(3):293.

51.

BRAIN INJURY FOLLOWING REPETITIVE APNEA IN NEWBORN PIGLETS

Gregory Schears*, Jennifer Creed, Diego Antoni, Tatiana Zaitseva, William Greeley**, David F. Wilson and Anna Pastuszko[1]

1. INTRODUCTION

In spite of the vast literature on the pathophysiology of hypoxia, the mechanism(s) by which repetitive apnea affects the metabolic pathways that sustain the normal processes of cell survival and cell death in the brain. More importantly, there is a need to elucidate the identity of the parameters that, alone or in combination, determine the pathogenic outcome of the apneic insults. Our early studies show that, in newborn piglets, repetitive apnea causes a progressive decrease in cortical oxygenation[1]. In the present studies, the changes in brain oxygenation were correlated with changes in extracellular dopamine, hydroxyl radicals and with neuronal injury. The studies were done on striatum, "selectively vulnerable" dopaminergic region of the brain, which is shown to be uniquely susceptible to recurrent hypoxic events.

Dopamine is widely believed to play a critical role in the physiopathology of brain function. Dopaminergic system of the striatum of a newborn piglet's brain has been shown to be very sensitive to the local oxygen pressure[2-5]. It has been suggested that release of dopamine during ischemic/hypoxic conditions may play a major role in mediating neuronal damage, particularly in the striatum. Massive increase in extracellular dopamine in the striatum has been reported to occur in different models of cerebral ischemia/hypoxia. There is evidence that this increase in extracellular dopamine may promote or accentuate brain damage during reoxygenation by increase of level of free radicals. Both spin trapping[6,7] and direct[8] electron paramagnetic resonance measurements have indicated that reactive oxygen metabolites are formed in the first minutes of reperfusion after cerebral ischemia. Free radicals are probably the major cause of endothelial damage and brain edema after asphyxia. The newborn infant, particularly the preterm infant, is thought to be prone to tissue damage from oxidative stress because of reduced total antioxidant capacity (see rev.[9]). The brain is very susceptible to damage from free radicals because of its high concentration of polyunsaturated fatty acids (see revs. [10-12]), low activity of catalase, superoxide

[1] Department of Biochemistry & Biophysics, School of Medicine, University of Pennsylvania, Children's Hospital of Philadelphia, Department of Anesthesiology & Critical Care**, Philadelphia, PA 19104, U.S.A. and Department of Anesthesiology & Critical Care*, Mayo Clinic, Rochester, NY.

dismutases and glutathione peroxidase, high activity of xanthine oxidase[10, 13], high levels of iron[14] and oxidizable catecholamines[10], and low content of transferrin[15] etc.

In the present studies the level of ortho-tyrosine in striatum was used as a measure of *in vivo* hydroxyl radical production. Fluoro-Jade (FJ) was use to determine neuronal injury following repetitive apnea. FJ sensitively and reliably stains the cell bodies, dendrites, axons and axon terminals of degenerating neurons but does not stain healthy neurons, myelin or vascular elements of neuropil[16]. FJ was shown to mark apoptotic cells in mice[17] and neurons injured by methamphetamine[18], excitotoxic (kainic acid) and neurotoxic (trimethyltin) compounds[19].

2. METHODS

2.1. Animal Model

Newborn piglets, age 3-5 days, were used for all studies. Anesthesia was induced with halothane (Halocarbon Laboratories, Augusta, SC; 4% mixed with 96% oxygen), and 1.5% lidocaine-HCl (Abbott Laboratories, North Chicago, IL) was used as a local anesthetic. Halothane was withdrawn entirely after the tracheotomy, pancuronium was used to induce respiratory paralysis (Gensia Pharmaceuticals, Irvine, CA; 1.5 mg/kg). Fentanyl-citrate (Elkins-Linn, Inc., Cherry Hill, NJ) was intravenously injected at 30 μg/kg, and the animals were mechanically ventilated with a mixture of oxygen and 0.5% isoflurane (Baxter Healthcare Corp., Dearfield, IL). The femoral artery and femoral vein were then cannulated and the piglet was maintained on a D_5LR infusion with 10 mcg/kg/hr of Fentanyl-citrate throughout the experiments. The head was placed in a Kopf stereotaxic holder and an incision was made along the circumference of the scalp. The scalp was removed to expose the skull and a hole approximately 5 mm in diameter was made in the skull over one parietal hemisphere for measuring the oxygen pressure. In all experiments, the blood pressure, body temperature and respiratory rate were monitored. The blood pH, $PaCO_2$ and PaO_2 were measured using a Chiron/Diagnostics Rapidlab 800 blood gas machine.

2.2 Repetitive Apnea Model

Each animal underwent 10 episodes of apnea and recovery. Apnea was initiated by disconnecting the animal from the ventilator and completed by reconnecting it to the ventilator. The apneic episodes were terminated 30 sec after the heart rate reached the bradycardic threshold of 60 beats per minute. This simulates the delayed response time to resuscitation. The normal heart rate range for a piglet is between 140-190, similar to that of human infants. The piglets had a recovery period of 10 minutes on mechanical ventilation between periods of apnea.

2.3 Measurement of Extracellular Dopamine

The extracellular level of dopamine in striatum was measured as described in earlier publications[2,3,5].. The microdialysis samples were collected at the end of every apneic episode. Identification and quantitation of dopamine was by comparison with chromatograms of standard solutions. The values for the level of dopamine in the dialysate are presented after correction for relative recovery by the microdialysis probe.

2.4. Determination of Striatal Ortho-tyrosine

Striatal tissue (approx. 1mg protein/ml) was hydrolyzed with 6N HCl under nitrogen atmosphere for 12 hrs at 100°C. The hydrolysates were then dried, resuspended in mobile phase and analyzed for o-tyrosine using a BAS Liquid Chromatography System with electrochemical detection. The mobile phase consisted of 0.1M chloroacetic acid, 0.1M KH_2PO_4, 1mM EDTA, 1% methanol, 1mg/ml sodium octyl sulfonate, 10 mM NaCl and was adjusted to a final pH of 3.00.

2.5 Staining with Fluoro-Jade

Fluoro-Jade (Histo-Chem Inc.) is an anionic tribasic fluorescein derivative with a molecular wt 445 daltons. It has an emission peak at 550 nm and excitation peaks at 362 and 390 nm, respectively. The brain sections were deparaffinized and rehydrated, then were transferred to a solution of 0.06% potassium permanganate for 15 min and were gently shaken on a rotating platform. The slides were rinsed for 1 min in distilled water and then transferred to the 0.001% Fluoro-Jade solution (in 0.1% acetic acid). After staining, the sections were rinsed with distilled water and the slides were rapidly air dried on a slide warmer or with a hot air gun. When dry, the slides were immersed in xylene and then covered with DPX (Sigma) mounting media. Sections were examined with an epifluorescence microscope using a filter system suitable for fluorescein fluorescence.

2.6 Statistical Evaluation

All values are expressed as means for n experiments ± SD or SEM. Statistical significance was determined using one way analysis of variance with repeated measures by Wilcoxon signed-rank test. $p < 0.05$ was considered statistically significant.

3. RESULTS

3.1 Effect of Repetitive Apnea on Extracellular Level of Dopamine in Striatum of Newborn Piglets

In vivo microdialysis was used to measure the extracellular level of dopamine in striatum of newborn piglets. It has been shown that repetitive apnea cause a progressive increase in extracellular dopamine (Figure 1).

The significant increase in extracellular dopamine, up to 2314±2069% of control (p<0.05), was observed following the fifth episode of apnea. The extracellular dopamine increased progressively with an increase in the number of apnea and, at the end of ten apneic episodes, it had increased up to 58200±35943% from baseline levels. The dopamine level decreased to a level not significantly different from control values after 50 min of postapneic recovery.

3.2 Effect of Repetitive Apnea on Level of Striatal Ortho-tyrosine

The level of ortho-tyrosine (o-tyr) was used to determine the level of hydroxyl radicals in the brain of newborn piglets. The level of o-tyrosine in the striatum of sham operated animals was 0.53± 0.16 nmoles/g tissue (Table 1). After 10 episodes of apnea and two hours of recovery the level of o-tyrosine increased significantly to 1.67 ± 0.63 nmoles/g tissue (p<0.05 for difference from pre-apnea conditions).

3.3 Striatal Injury Following Repetitive Apnea in Newborn Piglets

Fluoro-Jade, a fluorescent indicator of neuronal damage, was used to identify neurons injured 12 hours after repetitive apnea (Figure 2). Fluoro-Jade sensitively and reliably stains the cell bodies, dendrites, axons and axon terminals of degenerating neurons but does not stain healthy neurons, myelin or vascular elements of neuropils.

Figure 2 and Table 2 show that the striatum of apneic piglets has many more Fluoro-Jade positive neurons than do control animals (56±29 versus 9±10, n=7; p<0.05).

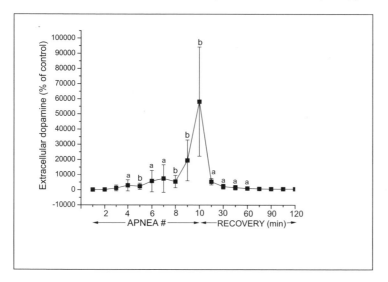

Figure 1. The effect of repetitive apnea on the extracellular concentrations of dopamine in striatum. Microdialysis probes were implanted in the striatum of newborn piglet and perfused with Ringer solution at 1ul/min. Collection of the microdialysis samples began 2 h after the probe was inserted. Samples were collected at the end of every apnea. Three measurements during the preapnea period were averaged for the baseline (100%). The results are means for 11 experiments ± SD. a- p< 0.05, b-p<0.005 for significant difference from control values as determined by one-way analysis of variance, followed by the Wilcoxon signed-rank test.

Table 1. Effect of repetitive apnea on the level of striatal o-tyrosine in newborn piglets: The results are expressed as the means ± SD (n=7). *p < 0.05 for significant difference from control values as determined by one way analysis of variance, followed by the Wilcoxon signed-rank test.

Experimental Conditions	Ortho-tyrosine nmol/g striatal tissue
Control, Sham operated	0.53 ± 0.16
Apnea (2-hrs Recovery)	1.67 ± 0.63*

Table 2. Effect of repetitive apnea on the numbers of injured cells in the striatum of newborn piglets: The results are means ± SD (n=7). *p < 0.05 for significant difference from control values as determined by one way analysis of variance, followed by the Wilcoxon signed-rank test.

Experimental Conditions	Number of damaged striatal neurons per 1 mm^2
Control, Sham operated	9 ± 10
Apnea (12-hrs Recovery)	56 ± 29*

Figure 2. Photomicrograph of Fluoro-Jade positive neurons in control (left) and apneic (right) striatum of newborn piglets. The piglets were subjected to 10 episodes of apnea and 12 hours of post-apneic recovery. Image dimensions: 1.05mm x 0.7mm (10x magnification).

4. DISCUSSION

Neonatal apnea may trigger cellular and molecular changes that can lead to metabolic disturbance and/or cell death. Our early studies show that both the pre- and post-apneic cortical oxygen pressures decreased with increasing numbers of apnea[1]. During each repetitive apnea, the cortical oxygen pressure decreased to the same minimum value. The maximum values that occur early in post-apneic recovery decreased significantly with increasing numbers of apnea, falling from about 76 mm Hg to 50 mm Hg by apnea 6 and then remaining nearly constant. The first and second post-apneic recoveries had an "overshoot" in the oxygen levels indicating a reactive hyperemia. This overshoot was markedly attenuated with increasing numbers of apnea, consistent with the lack of full recovery between apnea. These early results are consisted with data presented in this study which shown progressive increase in striatal extracellular dopamine with numbers of apnea. As was described in the Introduction this increase in extracellular dopamine can be one of the major contributors to neuronal injury. Oxidation of the excess dopamine released during each apnea by molecular oxygen, which may occur during reoxygenation, can result in the substantial formation of oxygen radicals. The enzymatic oxidation of dopamine by monoamine oxidase can also result in the formation of hydrogen peroxide, a hydroxyl radical "precursor". It has been proposed that pargyline, a monoamine oxidase inhibitor, protects against CNS oxygen toxicity in the rat by decreasing intracellular H_2O_2 generation from the oxidation of catecholamines.

Our studies have shown that repetitive apnea increases the level of o-tyr within the striatum of newborn piglets, indicating increased production of hydroxyl radicals within the tissue. It can be assumed that this increase of hydroxyl radicals in the striatum is directly related to the increase in extracellular dopamine. We have previously shown that, in newborn piglets, inhibition of dopamine synthesis abolished the post ischemia increase in striatal hydroxyl radicals[20].

Following 12 hrs of post apneic recovery, the numbers of injured neurons were significantly higher than in control animals. The observed increase in extracellular dopamine and increase in hydroxyl radical in striatum of newborn piglets is likely to be responsible for this increase in number of injured cells.

5. CONCLUSION

Repetitive apnea is associated with a significant increase in extracellular dopamine, generation of free radicals as determined by o-tyrosine formation and increase in Fluoro-Jade staining of degenerating neurons. This increase in extracellular dopamine and of hydroxyl radicals in striatum of newborn brain is likely to be at least partly responsible for the neuronal injury and neurological side effects of repetitive apnea.

Support by Grants NS-31465 and HD041484

6. REFERENCES

1. Schears, G., Creed, J., Zaitseva, T., Schultz, S., Wilson, D.F., and Pastuszko, A. (2004) Cerebral Oxygenation during repetitive apnea in newborn piglets. Adv. Exptl. Med. Biol., in press.
2. Pastuszko A, Lajevardi SN, Chen J, Tammela O, Wilson DF, Delivoria-Papadopoulos M. Effects of graded levels of tissue oxygen pressure on dopamine metabolism in striatum of newborn piglets. J Neurochem 1993;60:161-6.
3. Pastuszko A. Metabolic responses of the dopaminergic system during hypoxia in newborn brain. Biochem Med Metab Biol 1994;51:1-15.
4. Huang Ch-Ch, Lajevardi NS, Tammela O, Pastuszko A, Delivoria-Papadopoulos M, Wilson DF. Relationship of extracellular dopamine in striatum of newborn piglets to cortical oxygen pressure. Neurochem Res 1994;19:649-55.
5. Yonetani M, Huang Ch-Ch, Lajevardi N, Pastuszko A, Delivoria-Papadopoulos M, Wilson DF. Effect of hemorrhagic hypotension on extracellular level of dopamine, cortical oxygen pressure and blood flow in brain of newborn piglets. Neurosci Lett 1994;180:247-52.
6. Zini, I., Tomasi, A., Grimaldi, R., Vannini, V. and Agnati, L.F. (1992) Detection of free radicals during brain ischemia and reperfusion by spin trapping and microdialysis. Neurosci. Lett. 138, 279-282
7. Phillis, J.W. and Sen, S. (1993) Oxypurinol attenuates hydroxyl radicals production during ischemia/reperfusion injury of the rat cerebral cortex: an ESR study. Brain Res. 628, 309-312.
8. Hasegawa, K., Yoshioka, H., Sawada, T. and Nishikawa, H. (1993) Direct measurement of free radicals in the neonatal mouse brain subjected to hypoxia: an electron spin resonance spectroscopic study. Brain Res. 607, 161-166.
9. Taylor, D. L., Edwards, D. and Mehmet, H. (1999) Oxidative metabolism, apoptosis and perinatal brain injury. Brain Pathology. 9, 93-117.
10. Traystman RJ, Kirsch JR, Koehler RC. Oxygen radical mechanisms of brain injury following ischemia and reperfusion. J Appl Physiol 1991;71:1185-1195
11. Floyd RA. Role of oxygen free radicals in carcinogenesis and brain ischemia. FASEB J 1990;4:2587-2597.
12. Halliwell B. Reactive oxygen species and the central nervous system. J Neurochem 1992;59:1609-1623.
13. Cohen G. (1988) Oxygen radicals and Parkinson's disease. in: Halliwell B., ed. Oxygen radicals and tissue injury. Bethesda : FASEB, 130-135.
14. Youdim M.B.H. (1988) Iron in the brain: implications for Parkinson's and Alzheimer's disease. Mt. Sinai J. Med. 55, 97-101.

15. Gruener N., Gonzlan O., Goldstein T., Davis J., Besner I., and Iancu T.C. (1991) Iron, transferrin and ferritin in cerebrospinal fluid of children. *Clin. Chem.* 37, 263-265.
16. Schmued, L.C., Albertson, C. and Slikker Jr., W. (1997) Fluoro-Jade: a novel fluorochrome for the sensitive and reliable histochemical localization of neuronal degeneration. Brain Res., 751, 37-46.
17. Hornfelt, M., Edstrom, A. and Ekstrom, P.A. (1999) Upregulation of cytosolic phospholipase A2 correlates with apoptosis in mouse superior cervical and dorsal root ganglia neurons. Neurosci. Letters. 265, 87-90.
18. Eisch, A.J. and Marshall, J.F. (1998) Methamphetamine neurotoxicity: dissociation of striatal dopamine terminal damage from parietal cortical cell body injury. Synapse. 30, 433-445.
19. Noraberg, J., Kristensen, B.W. and Zimmer, J. (1999) Markers for neuronal degeneration in organotypic slice cultures. Brain Res. 3, 278-90.
20. Olano, M., Song, D., Murphy, S., Wilson, D.F. and Pastuszko, A. (1995) Relationships of dopamine, cortical oxygen pressure, and hydroxyl radicals in brain of newborn piglets during hypoxia and posthypoxic recovery. J. Neurochemistry, 65, 1205-1212.

52.

CEREBRAL OXYGENATION AND PUSH-PULL EFFECT

Cong C. Tran, Xavier Etienne, André Serra, Muriel Berthelot, Gérard Ossard, Jean-C. Jouanin, and Charles Y. Guézennec [*]

1. INTRODUCTION

Pilots of high performance aircraft are exposed to $+G_z$ accelerations (gravity in the foot-to-head direction expressed as Earth gravity equal to 1 G), which may induce loss of consciousness, caused by a failure of cerebral blood flow. Indeed, $+G_z$ accelerations enhance the cardiovascular responses observed during moving abruptly from supine to standing. Over the heart level, the arterial blood pressure decreases inducing an increase in heart rate and in vascular resistance. Arterial blood pressure is maintained with a sympathetically mediated increase in vascular resistance.[1] These hemodynamic changes are responsible for the decrease in cerebral perfusion pressure and in oxygenation. Such a decrease is kept limited via cerebral autoregulatory mechanisms inducing a cerebral vasodilatation.[1] Under $-G_z$ acceleration, the arterial blood pressure is increased over the heart level, and a cerebral vasoconstriction occurs in response to prevent brain overperfusion.

The push-pull effect occurring in flight is known to be involved in many aircraft accidents, when the levels of $+G_z$ accelerations are less than those involved in loss of consciousness. The push-pull effect is defined as a decreased tolerance to $+G_z$ acceleration subsequent to a less than $+1$ G_z exposure.[2] Objective physiological tolerance to $+G_z$ acceleration is usually evaluated by change in blood pressure.[2] As a decreased blood pressure may lead to a decreased cerebral perfusion pressure, we wondered if the push-pull effect could induce a greater decrease in cerebral oxygenation. Furthermore, the delay in sympathetic drive during subsequent $+G_z$ exposure may be one of the mechanisms of the push-pull effect.[3] However, by using a vertical rotating-table to simulate the push-pull manoeuvre,[4] it has been demonstrated that cerebral vasoconstriction occurred to prevent brain overperfusion during head-down tilt. During subsequent head-up tilt, the elevated resistance of the cerebral vessel remained at the higher level for about 20 s, and may have worsened the cerebral perfusion from exposure to $+G_z$ acceleration.

[*] Cong C. Tran, Xavier Etienne, André Serra, Muriel Berthelot, Jean-C. Jouanin, Charles Y. Guézennec, Institut de médecine aérospatiale du service de santé des armées, BP 73, 91223 Brétigny/Orge, France. Gérard Ossard, Laboratoire de médecine aérospatiale, CEV, 91228 Brétigny/Orge, France.

One goal of the present work was to use a human centrifuge to evaluate the implication of the cerebral vasoconstriction occurring under $-G_z$ acceleration in the mechanisms of the push-pull effect. Transcranial Doppler (TCD) velocimetry was used for monitoring cerebral blood flow velocity (CBFV) and cerebral vascular resistance. Near infrared spectroscopy (NIRS) method was used for monitoring cerebral oxygenation and cerebral blood volume changes. Another goal was then to evaluate the accuracy of the NIRS method in the assessment of changes in cerebral vasomotion, by comparison between TCD index and NIRS determined-cerebral blood volume changes. We hypothesized that brief exposure to $-G_z$ acceleration may reduce cerebral oxygenation during subsequent $+G_z$ acceleration.

2. METHODS

2.1. Subjects

Four healthy male non-pilot volunteers participated in this preliminary study. The subjects averaged 35 yr in age (range = 32-42 yr), 79.5 ± 10.1 kg in weight, and 180 ± 3 cm in height. They were informed of the purpose of the experiment, and the study was carried out with the approval of the national Ethics Committee.

2.2. Acceleration Setup

Acceleration profiles were generated using the 8-m-radius human centrifuge of the Laboratoire de médecine aérospatiale (LAMAS), Brétigny-sur-Orge, France. Each subject was seated and secured inside the centrifuge gondola in a 15° reclined seat.

2.3. Physiological Measurements

2.3.1. Cerebral oxygenation

Cerebral oxygenation was measured with NIRS, which allows to measure continuously and non-invasively changes in brain oxygenation concentration and in cerebral hemodynamics.[5,6] This method was also validated under $+Gz$ acceleration,[7,8,9,10] particularly with the NIRO-300G,[8,9] a modified devised instrument (Hamamatsu Photonics, Hamamatsu, Japan). Our study was performed using the NIRO-300 monitor which produces four wavelengths of near-infrared light (775, 810, 850, and 910 nm) allowing monitoring changes in oxygenated (oxy-Hb) and deoxygenated (deoxy-Hb) hemoglobin concentrations, and cerebral blood volume as changes in total hemoglobin (total-Hb = oxy-Hb + deoxy-Hb) with a modified Beer-Lambert equation.[8] Furthermore, the NIRO-300 measures TOI (tissue oxygenation index = oxy-Hb/total-Hb) by using the spatially resolved technique.[8] By using a self-adhesive pad, the optodes (light source and light detector) were placed 4.0 cm apart on the right forehead, avoiding the temporal muscle regions. The differential path length factor was set at the level recommended for the adult head[11] to quantify the NIRS data in μM, and the sample frequency was set at 2 Hz.

2.3.2. Blood flow velocity

CBFV was monitored in the middle cerebral artery by a pulsed TCD technique. For the present study, we used a 2-MHz TCD prototype developed by Medical Biophysics Laboratory, F. Rabelais University, Tours, France. This TCD prototype was previously validated[12] and capable of working in the high G environment. The ultrasonic transducer was incorporated into an adjustable head-mounted support system to allow precise positioning of the probe before it was locked in place when optimal TCD signal was obtained. Several tests were performed to verify that there was no interference between NIRS and TCD signals. The standard algorithm implemented by the analyser allowed the calculation and plotting of instantaneous and mean CBFV (CBFV$_{mean}$) in the vessel.

2.3.3. Arterial blood pressure

Beat-to-beat arterial blood pressure was measured by a plethysmographic method using a Portapres monitor (TNO TPD Biomedical Instrumentation, Amsterdam, The Netherlands). The height correction transducer was positioned at eye level, allowing direct measurements of blood pressure at eye level whatever the G$_z$ level was.

2.3.4. Recordings

The analogue NIRS signals, CBFV, arterial blood pressure, and also electro-cardiogram, thigh and abdomen muscles electro-myogram from surface electrodes, were sampled simultaneously at 500 Hz per channel and stored by using a digital recorder (MP 100, Biopac, CA, USA). Integrated Acknowledge software allowed off-line data analysis after re-sampling at 50 Hz.

2.4. Protocol

Each subject was submitted to two consecutive runs including a plateau of +2G$_z$ for 30 s, separated by 15 min of rest. The first exposure was as control run, and the following was as experimental run with the +2 G$_z$ plateau preceded by –2 G$_z$ for 10 s. The G-onset rate was 1 G$_z$/s. The subjects remained relaxed during the two runs, and were not wearing an anti-G protection.

2.5. Data Analysis

The 5 s period before the G-onset of each run was regarded as the baseline values, and a 5 s-period in the middle of each +2 G$_z$ plateau was considered for the analysis. The comparison between control and experimental run was analysed by comparing the respective changes at +2 G$_z$ plateau.

All the values were averaged over each considered 5 s-period. As TOI signals were lost during centrifugations, we used Hbdiff index (oxy-Hb – deoxy-Hb changes), knowing that a change in hematocrit occurs only after 5 min at +4 G$_z$.[7] Arterial blood pressure and CBFV were analysed on a beat-to-beat basis. Mean arterial pressure (MAP) was calculated as one third of systolic plus two third of diastolic pressure. Cerebral vascular resistance was evaluated via the resistance index (RI), calculated as MAP at eye level (MAP$_{eye}$) divided by CBFV$_{mean}$.[13,14]

Data were expressed as means ± SD. A paired *t*-test was used to compare the change during each +2 G$_z$ plateau and between the two runs.

3. RESULTS

Each subject experienced the two runs without any muscular contractions (confirmed by the electro-myograms).

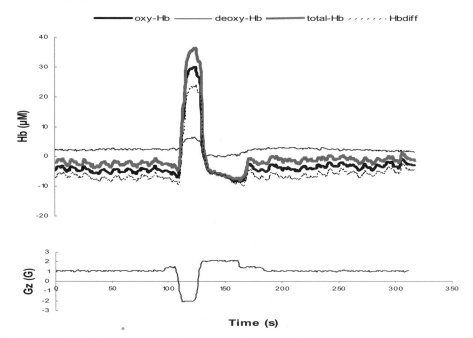

Figure 1. Typical changes in cerebral oxyhemoglobin (oxy-Hb), deoxyhemoglobin (deoxy-Hb), total hemoglobin (total-Hb) concentrations and in Hbdiff (oxy-Hb – deoxy-Hb) during the experimental run.

Table 1. Changes in cerebral oxygenation, mean cerebral blood flow velocity, resistance index and mean arterial blood pressure at eye level, at +2 G_z plateau during control and experimental run[a]

Variable	Control run	Experimental run
oxy-Hb	-9 ± 4*	-7 ± 3*
total-Hb	-8 ± 6	-8 ± 5*
Deoxy-Hb	0.2 ± 2	-0.9 ± 3
Hbdiff	-9 ± 3*	-6 ± 3*
$CBFV_{mean}$	-0.8 ± 0.8	-2.0 ± 0.4**§
RI	-1.0 ± 0.3**	-1.2 ± 0.5**
MAP_{eye}	-27 ± 10**	-40 ± 11**§

[a] Values are means ± SD. oxy-Hb, total-Hb, deoxy-Hb (oxyhemoglobin, total hemoglobin, and deoxyhemoglobin concentrations, respectively) and Hbdiff (oxy-Hb – deoxy-Hb) in μM; $CBFV_{mean}$ (mean cerebral blood flow velocity) in cm/s; RI (resistance index) in mmHg/cm/s; MAP_{eye} (mean arterial pressure at eye level) in mmHg. Significantly different from baseline: *P<0.05; **P<0.01. Significantly different from control: §P<0.01

Figure 1 shows typical changes of the NIRS parameters during the experimental run. When compared to the baseline values, such changes at +2 Gz appeared not to be different during the control run (not shown).

The mean data obtained during the two runs at $+2G_z$ plateau are given in Table 1. During the control run, there was a significant decrease in oxy-Hb (P<0.05), RI (P<0.01) and MAP_{eye} (P<0.01). There was no change in total-Hb and in $CBFV_{mean}$.

During the experimental run, the decrease in total-Hb (P<0.05) and in $CBFV_{mean}$ (P<0.01) was significant.

The comparison between the control and the experimental runs showed that the decrease in MAP_{eye} (P<0.01) and in $CBFV_{mean}$ (P<0.01) was more important during the experimental run. However, the change in oxy-Hb, total-Hb and in RI was not different.

4. DISCUSSION AND CONCLUSIONS

The main finding of this study was that the decrease in cerebral vascular resistance assessed by transcranial Doppler index, and cerebral blood volume assessed as cerebral total hemoglobin concentration changes, were not modified when the $+2 G_z$ plateau was preceded by a $-2 G_z$ exposure. Furthermore a greater decrease in cerebral blood flow velocity occurred with unchanged cerebral oxygenation.

The experimental run used in this study has induced a push-pull effect. Indeed, the decrease in blood pressure was greater when the $+2 G_z$ plateau was preceded by a $-2 G_z$ exposure. Furthermore, in our experimental conditions, brief exposure to $-2 G_z$ reduced cerebral blood flow velocity during subsequent $+2 G_z$ acceleration. However, the decrease in blood flow velocity was small when compared to the decrease in mean arterial pressure at eye level, with no difference in the change in cerebral oxygenation. These results suggest that the efficiency of cerebral oxygenation control was maintained despite the occurrence of a push-pull effect.

No difference in the change in total hemoglobin and in resistance index was observed under the push-pull effect. This result suggests that cerebral vasodilatation under $+2 G_z$ acceleration was not modified by the cerebral vasoconstriction occurring at $-2 G_z$ acceleration. Furthermore, the similar changes observed between total hemoglobin and in cerebral vascular resistance demonstrated the accuracy of the NIRS method in the evaluation of changes in cerebral vasomotion under $+G_z$ acceleration. Finally, cerebral vasoconstriction occurring under $-2 G_z$ exposure did not seem to be a major contributor in the mechanism of the push-pull effect, in our experimental conditions.

By using a vertical rotating-table to simulate on the ground the push-pull manoeuvre,[4] it has been demonstrated that cerebral vasoconstriction, assessed by transcranial Doppler indices, occurred to prevent brain overperfusion during head-down tilt. During subsequent head-up tilt, the elevated resistance of the cerebral vessel remained at the higher level for about 20 s, and may have worsened the cerebral perfusion from exposure to $+1 G_z$. Although the 20-s delay was consistent with our experimental conditions (i.e. analysis in the middle of the $+2 G_z$ plateau), we did not find such a result, since there was no difference in the change in total hemoglobin concentration and in cerebral vascular resistance. However, when various methods to simulate on the ground the effects of $+G_z$ accelerations on cardiovascular responses are very helpful, they cannot recreate the hypergravity conditions (over than $+1$ G_z). For example, by using the NIRS method coupled to electro-corticogram and laser-Doppler flowmetry measurements, we have demonstrated that loss of consciousness induced by $+G_z$ centrifugation was correlated to the decrease in cerebral blood volume, and not to the decrease in cerebral oxygenation, even when a decrease in cerebral oxygenation has preceded the decrease in cerebral blood volume.[15] Under head-up tilt conditions, a decrease in cerebral blood flow and in cerebral oxygenation has also been described, but the occurrence of symptoms of syncope was associated with a further sudden and marked drop of cerebral oxygenation.[16] We speculate that specific physiological adaptation mechanisms might be involved under $+G_z$ accelerations allowing the efficiency of cerebral oxygenation control, despite the occurrence of a push-pull effect.

5. REFERENCES

1. R. R. Burton, S. D. Leverett, and E. D. Michaelson, Man at high sustained $+G_z$ acceleration: a review, *Aerosp. Med.* **45**, 1115-1136 (1974).
2. R. D. Banks, J. D. Grissett, G. T. Turnipseed, P. L. Saunders, and A. H. Rupert, The "push-pull effect", *Aviat. Space Environ. Med.* **65**, 699-704 (1994).
3. L. S. Goodman, R. D. Banks, J. D. Grisset, and P. L. Saunders, Heart rate and blood pressure responses to $+G_z$ following varied-duration – G_z, *Aviat. Space Environ. Med.* **71**, 137- 141 (2000).
4. W. X. Zhang, C. L. Zhan, X. C. Geng, X. Lu, G. D. Yan, and X. Chu, Cerebral blood flow velocity by trancranial Doppler during a vertical-rotating table simulation of the push-pull effect, *Aviat. Space Environ. Med.* **71**, 485-488 (2000).
5. M. Ferrari, L. Mottola, and V. Quaresima, Principles, techniques, and limitations of near infrared spectroscopy, *Can. J. Appl. Physiol.* **29**(4): 463-487 (2004).
6. P.L. Madsen, and N.H. Secher, Near-infrared oximetry of the brain. Prog. Neurobiol. 58, 541-560 (1999).
7. D.H., Glaister, and Jöbsis-Vander Vliet, F.F., A near-infrared spectrophotometric method for studying brain O_2 sufficiency in man during $+G_z$ acceleration, *Aviat. Space Environ. Med.* **5**:199-207 (1988).
8. A. Kobayashi, and Y. Miyamoto, In-flight cerebral oxygen status: continuous monitoring by near-infrared spectroscopy, *Aviat. Space Environ. Med.* **71**, 177-183 (2000).
9. A. Kobayashi, A. Kikukawa, and A. Onozawa, Effect of muscle tensing on cerebral oxygen status during sustained high $+G_z$, *Aviat. Space Environ. Med.* **73**, 597-600 (2002).
10. L.D. Tripp, T. Chelette, S. Savul, and R.A. Widman, Female exposure to high G: Effects of simulated sorties on cerebral and arterial O_2 saturation, *Aviat. Space Environ. Med.* **69**, 869-874 (1998).
11. P. van der Zee, M. Cope, S.R. Arridge, M. Essenpries, L.A. Potter, A.D. Edwards, J.S. Wyatt, D.C. McCormick, S.C. Roth, E.O.R. Reynolds, and D.T. Delpy, Experimentally measured optical pathlengths for the adult head, calf and forearm and the head of the newborn infant as a function of interoptodes spacing. *Adv. Exp. Med. Biol.* **316**, 143-153 (1992).
12. G. Ossard, J. M. Clère, M. Kerguelen, F. Melchior, and J. Seylaz, Response of human cerebral blood flow to $+G_z$ accelerations, *J. Appl. Physiol.* **76**, 2114-2118 (1994).
13. R. L. Bondar, P. T. Dunphy, P. Moradshahi, M. S. Kassam, A. P. Blader, F. Stein, and R. Freeman, Cerebrovascular and cardiovascular responses to graded tilt in patients with autonomic failure, *Stroke* **28**, 1677-1685 (1997).
14. D. Dan, J .B. Hoag, K. A. Ellenbogen, M. A. Wood, D. L. Eckberg, and D. M. Gilligan, Cerebral blood flow velocity declines before arterial pressure in patients with orthostatic vasovagal presyncope, *JACC* **39**, 1039-1045 (2002).
15. C. C. Tran, G. Florence, E. Tinet, D. Lagarde, J. C. Bouy, P. Van Beers, A. Serra, S. Avrillier, and J. P. Ollivier, Cerebral hemodynamics and brain oxygen changes related to gravity-induced loss of consciousness in rhesus monkeys. *Neurosci. Lett.* **338**, 67-71 (2003).
16. K. Krakov, S. Ries, M. Daffertschofer, and M. Hennerici, Simultaneous assessment of brain tissue oxygenation and cerebral perfusion during orthostatic stress, *European Neurology* **43**, 39-46 (2000).

53

CANCER AND THE RESPIROME

D. Maguire*

1. INTRODUCTION

It is now over 80 years since Warburg first described the tumour respiratory phenotype. During that time great advances have taken place in terms of our understanding of the mechanisms and control of respiration and our understanding of the genetic processes that predispose some individuals to a higher risk of cancer. Warburg's original description of the cancer phenotype as being hypoxic or anoxic still applies to many advanced cancers but we now know that during tumourigenesis there is great variability in the oxygen metabolic status of individual tumours.

A fundamental contribution to our perception of the cancer process was the description of genes that were altered in individuals with familial cancers. That discovery emerged from basic science research into cancers that were virally transmissible between animals. From those studies emerged the concept of oncogenes and their translation products the oncoproteins. Oncogenes, being dominantly expressed, exert their phenotypic influence by gain of function. By contrast, the tumour suppressor genes, whose description followed soon after the discovery of oncogenes, are recessive genes and are expressed through loss of function. Although the majority of oncogenes and tumour suppressor genes were discovered before the end of the last millennium, it is only in the light of the publication of the results of the human genome projects that we can begin to analyse whether there might be any relationship between the genetic injury in a particular cancer and its respiratory phenotype. Potentially, this has important implications for the way in which treatment might be approached in particular patients or groups of patients.

The two human genome projects, one driven by government altruism and the other by commercial interest were commenced in the late 1980s and their results published in concurrent articles in December 2000. Even a cursory analysis of those results leaves the impression that there is much yet to be learnt about our inheritance. Nevertheless, the

* Genomics Research Centre, Griffith University, Nathan, Brisbane, Australia, 4111

resources provided by these two endeavours provide us with unprecedented opportunities to examine links between genotype and phenotype in cancer mutation sites and respiratory components. The competition to win the human genome race has also spawned a valuable coterie of new techniques and technology that make this task achievable on even modest research budgets.

One of the important outcomes of the discoveries to emerge from the human genome projects has been a revision of the central dogma of molecular biology, which was already shaking under the impact of the discoveries of reverse transcriptase viruses and of prions. The central dogma of molecular biology as initially enunciated, described the process by which the genetic information of any organism is first transcribed from DNA into messenger RNA (mRNA) and then translated into protein sequence.

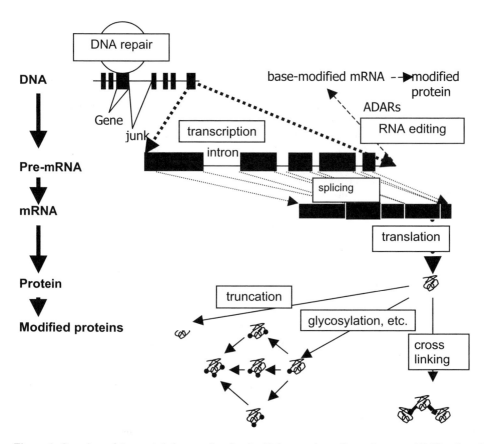

Figure 1. Overview of the central dogma of molecular biology as it applies to humans; ADAR, adenosine deaminase acting on RNA.

2. REVISED CENTRAL DOGMA OF MOLECULAR BIOLOGY

Advances in our understanding of how genes are decoded and expressed have forced a revision of each of the steps in this process. The process, as it operates in humans, is represented in Figure 1. In this revised version of the original central dogma of molecular biology, it is recognised that there are editing processes at each step in gene expression. Significant mechanisms exist in this process for repair of damaged DNA, editing of pre-mRNA, alternate splicing of pre-mRNA to create a coterie of mRNA molecules from one gene transcript, and a variety of protein-editing processes. Included among these latter editing processes are protein truncation, amino acid residue modification such as variable patterns of glycosylation (and many other modifications) and intra-protein or inter-protein cross-linking. The net result of these events is increased variability in terms of both control and range of expression as is required for tissues to perform their specific activities. It is this ability to extensively modify the basic architecture of gene expression that distinguishes highly evolved multi-cellular organisms from the bacteria rather than a predicted expansion in the total gene complement. There is now also interest in the role that 'junk' DNA might play in modulating expression in higher organisms.

3. ADVANCED TECHNIQUES OF MOLECULAR BIOLOGY

Among the new techniques that have emerged during the genomic era are a number of methods that make it possible to effectively 'paint' chromosomes in order to highlight the location of genes or segments of genes of interest. The simplest such painting approach is fluorescent in-situ hybridisation (FISH), now widely used as a routine diagnostic procedure in cytogenetics laboratories. This method allows precise localisation of specific genes and is applied routinely to the diagnosis of inborn errors of metabolism including cancers such as breast cancers in cancer-prone families.

A somewhat more challenging technique is comparative genomic hybridisation (CGH), a painting approach with the additional advantage of providing quantitative information about the chromosomal regions investigated. We have used this technique to identify aberrant chromosomal regions in non-melanotic skin cancers and in the leucocytes of a cohort of Australians exhibiting pre-cancerous, non-melanotic skin lesions [1,2]. The results of those studies are summarised in Figure 2.

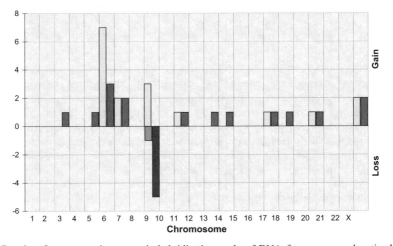

Figure 2. Results of a comparative genomic hybridisation study of DNA from a non-melanotic skin cancer cohort [1,2]. Frequency of all CNAs obtained in 15 BCC cases. *x axis*, chromosomes 1-22, X, *y axis*, number of cases. Ascending bars represent gains on the short p arm (*hatched bars*) or long q arm (*dark bars*), whereas descending red bars represent losses on the short p arm (*grey bar*) or long q arm (*hatched bar*) of the respective chromosomes.

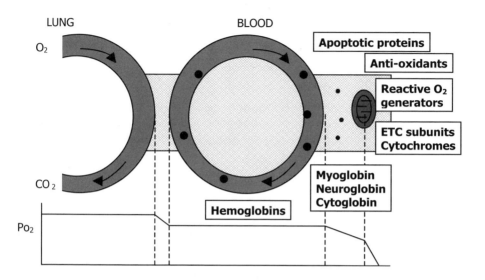

Figure 3. The human respirome, modified from the cascade model of Erdmann [5]. Tissues, cells, organelles and pathways associated with the transport, storage, consumption, synthesis and detoxification of oxygen metabolites are represented as modular components in this diagram.

Real-time PCR is another technique that has expanded the quantitative capacity of a tried and proven technique of the genomic era. The advantage over CGH is its relative simplicity and low cost for batched throughput investigations. We have used this technique to reveal aberrations in a non-melanotic skin cancer cohort [3,4].

4. THE HUMAN RESPIROME AND THE CASCADE MODEL OF OXYGEN FLOW FROM AIR TO CELL

The passage of oxygen from air to its ultimate consumption can be represented by a modified version of Erdmann's model (Figure 3). Individual components of this model include; (1) oxygen transport within blood, (2) oxygen exchange between blood and cells, (3) oxygen transport/storage within cells, (4) cellular respiration, (5) antioxidant processes, (6) free radical generation and (7) programmed cell death (apoptosis). Focus in this presentation will be on those areas in which advances have been made in cancer research since the post-genomic era began (January 2001) but for completeness, reference is made to each of these aspects of oxygen metabolism.

4.1 Oxygen Transport within Blood

Although additional information has become available about this, the oldest and best known module of the respirome, there appears to have been no discoveries relevant to cancer. A wealth of information is available about polymorphisms and mutations in the eleven identified hemoglobin genes (Table 1) that are encoded in the human genome. However there are no known causal relationships between these mutations or polymorphisms and any cancer. Conversely there are no known protective advantages to possession of the aberrant haemoglobins in any cancer. Has anyone looked for either such association, however? There has, in fact, been one report of such a study; that by Dawkins et al.[6], who investigated the incidence of cancer in patients with sickle cell disease (SCD). Those studies revealed a cancer incidence of 1.74 per 1,000 patient years compared to an incidence of 1.36 per 1,000 patient years in non-SCD patients. In a somewhat different vein, Miller, Schultz and Ware[7] expressed concern over hydroxyurea therapy in sickle cell patients with cancer.

Table 1. Human globin genes, disorders and polymorphisms

► Chromosome 16
  - α1-globin
  - α2-globin
  - ψα1-globin
  - ι1-globin
  - i2-globin
► Chromosome 11
  - β-globin
  - δ-globin
  - Aγ-globin
  - Gγ-globin
  - ψβ1-globin
  - ε-globin

4.2. Oxygen Transport and Storage within Cells

It is within this module of the human respirome that new important information has become available in the post-genomic era. Specifically, the discovery of two new oxygen-binding molecules has forced a re-evaluation of this step in the oxygen cascade. Myoglobin, the oxygen-binding molecule of muscle cells has been extensively studied since its discovery; its function in muscle has always been assumed to be as an oxygen sink within that tissue. The discovery, firstly of neuroglobin [8], predominantly localised in specific areas of the brain and later cytoglobin [9] has spurred a rash of interest in these new players in the respirome. One of the important facets of cytoglobin that has been discovered is that it may function as a sensor of hypoxia.[10,11,12] Cytoglobin appears to be uniformly distributed, albeit at varying levels, throughout all tissues.[14,15] Its localization exclusively within the nucleus suggests it may have important roles to play in transcriptional control. There have been no reports to date of its involvement in any aspect of tumour genesis or metastasis. The involvement of cytoglobin as a sensor of hypoxia suggests it may be important, at least in advanced cancers.

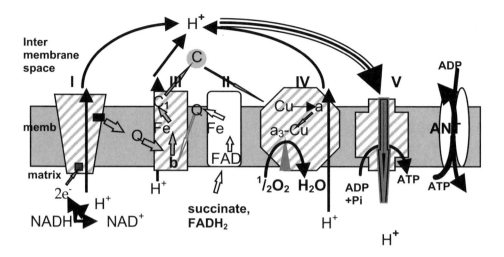

Figure 4. The electron transport chain (ETC) and oxidative phosphorylation components and activities. Complexes are shown embedded in the inner mitochondrial membrane. Complexes that are solely encoded by the nucleus are shown in white. Those with mitochondrial DNA (mtDNA) coding contribution are hatched. An emerging association between mutations in nuclear-encoded subunits, particularly of complex II and specific cancers, presents challenges to pre-conceived notions about the role of ETC subunits in carcinogenesis.

4.3. Cellular Respiration

This area of the oxygen cascade has also seen considerable advances in the post-genomic era, particularly in terms of the detailed description of a number of additional components of the electron transport chain (ETC) complexes. Despite these advances, it is likely that there are yet more discoveries to be made in this area. Detailed genetic and proteomic information about the 13 subunits encoded by the mitochondrial genome is well documented. However, less than one hundred or so nuclear-encoded components of the electron transport chain have been identified to date and the complete description of most of those components has not yet been achieved. Of particular relevance to cancer is the emerging account of the direct association of mitochondrial dysfunction and mutations in electron transport chain complex subunits with particular cancers. Mitochondrial DNA dysfunction has been identified in oncocytic hepatocytes.[16] Mitochondrial mutations have been described in p53-deficient human cancer cells. Taylor et al.[17] suggested that mitochondrial complex I is deficient in renal oncocytomas. In the same vein, Zhou et al.[18] demonstrated mitochondrial impairment in human colonic crypt stem cells. Up-regulation of CO I in colon cancer cell lines with p53 over-expression was shown by differential display[19]. Simmonet et al.[20] have demonstrated that low mitochondrial respiratory chain content correlates with tumour aggressiveness in renal cell carcinoma. Using reverse phase protein arrays, Herrmann et al.[21] demonstrated altered cytochrome C oxidase (complex IV) composition in prostate cancer. In their study, a shift in the ratio of mtDNA-encoded (CO I, II & III) sub-units compared to nuclear-encoded sub-units was observed. The CA125 and UQCRFS1 FISH studies of ovarian carcinoma undertaken by Kaneko et al.[22] also add evidence to the involvement of mitochondrial genes in cancers.

The most convincing association between mutations in an ETC complex with a particular cancer is that of complex II sub-unit succinate dehydrogenase D (SDHD) with Paraganglioma in head and neck. That connection was first was first reported by Burmester et al.[8] and the association between complex II mutations and cancer has been confirmed by Cascon et al., Dannenberg et al., Rustin et al., Lee et al. and Taschner et al.[23,24,25,26,27]. Cascon et al, and Taschner et al.[23,27] noted that there is an association between complex II mutations and another cancer, pheochromocytoma. Other subunits apart from suubnit D, namely B and C, have also been implicated in these associations. A review of this important field has been presented by Baysal et al .[28]

It is difficult to conceptualise any mechanism by which a defective electron transport chain component might exert a direct carcinogenic activity other than by the widespread generation of excess free radicals. Why such excess free radical production from one ETC subunit would be so specifically linked to a particular cancer or group of cancers is a mystery. Moreover, were this to be the mechanism, it would be expected that the growing cohort of carriers and expressers of the more than twenty mitopathies identified to date would have been identified as a particular cancer cohort. It is possible that the observed particular association of SDH with mutations with specific cancers reflects the fact that complex II contains no mtDNA-encoded components.

Our own studies on nuclear genomic and phenomic correlations in non-melanotic skin cancers[1,2,3,4,29] may be pertinent in this regard. Those investigations have demonstrated mutations and copy number aberrations in a non-melanotic skin cancer cohort and shown mutations in both cancer genes and nuclear-encoded ETC subunits. If the mutation in a particular nuclear ETC gene exerts some influence close to an area of chromosomal instability, then it may be that the incidence of chromosomal re-arrangement in individuals carrying that mutation may have a greater risk of carcinogenesis. In the case of SDHD, however, the closest cancer gene is MMP12 and that is about forty genes away from the SDHD locus.

Alternatively, the possession of a defective electron transport chain may offer some somatic advantage to a developing tumour. It is certainly true that mtDNA deletions in patients with particular mitopathies endow a competitive advantage to the cells with those deletions, such that this event alone explains the progressive nature of those disorders.

Finally, the recent discovery of copy number polymorphism in apparently normal populations[30] is germane to this discussion. None of the studies into cancer-associated complex II mutations have involved any estimation of copy number and it may be that there is concomitant copy number aberration in these cases, leading to the over-expression of the relevant complex II subunit.

4.4. Free Radical Generation

Free radicals can be formed as a result of the Fenton or Haber-Weiss reaction between O_2^- and H_2O_2 (Figure 6). There is now a host of information available regarding free radical generation within cells and the importance of free radicals in disease processes has also become appreciated in recent years. There is still much to discover about free radical generating enzymes. The roles of free radicals in the aging process[31,32,33] and in carcinogenesis are areas of current interest.

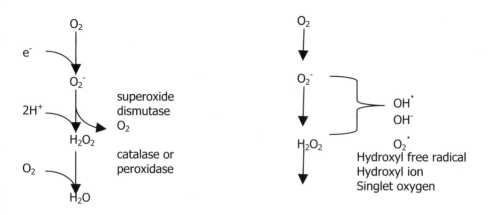

Figure 5. Free radical formation from reaction between and superoxide and hydrogen peroxide (after Erdmann, 1992)[5]

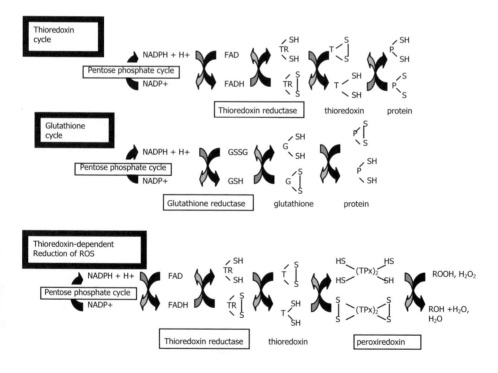

Figure 6. Antioxidant pathways operating in humans.

4.5. Antioxidant Processes

The role of vitamin C in general health has been recognised for almost two centuries, although its role as a general antioxidant is a relatively recent discovery. The recognition of the existence of other small organic antioxidant molecules of endogenous origin, such as ferredoxin, is much more recent. The processes and principle enzymes involved in the endogenous antioxidant processes are summarised in Figure 6. Interest in the role of the endogenous antioxidant pathways has developed in recent years. Using a functional proteomics approach, Young et al.[34] demonstrated that there is activation of antioxidant pathways in ras-mediated oncogenic transformation of human surface ovarian epithelial cells. It should not be surprising if elevated levels of endogenous antioxidant synthesis are a common event in a host response to tumourigenesis. However it should not be assumed that such events are causally associated with any cancer. Indeed quite the opposite; the diminution in these pathways might be expected to expose cells to the activity of increased reactive oxygen species.

4.6. Apoptosis

There has been an explosion in knowledge about the process of programmed cell death (apoptosis) since its discovery by Kerr, Wyllie and Currie in 1972 [35]. Several themes are central to the theories that have been advanced to describe the bewildering array of reaction pathways that interlink to lead to the organised and well controlled recovery of valuable molecules such as the components of DNA. A pivotal point in this process is the opening of the mitochondrial permeability transition pore (mptp) leading to the release of the water soluble electron carrier, cytochrome C from mitochondria.

A pre-genomic view of apoptosis is shown in Figure 7. The importance of mitochondrial contribution to this process is clear. For stark contrast, a recent post-genomic version of this same pathway is shown in Figure 8, in which the significant new discoveries in the post genomic era are highlighted. It can be seen from Figure 8 that there is a considerable body of new knowledge about apoptosis that has been revealed since the end of the last millennium. Not only are there now many more components known to participate in the process of apoptosis but there are complete modules in the process that were unknown until very recently. Of particular importance among the new discoveries in the apoptotic pathways are the roles played by the tumour suppressor protein p53 and the protein that has been shown to be mutated in Ataxia Telangectasia, namely ATM. The central role of the mitochondrion in this process has not been diminished by these revelations; rather the new discoveries have heightened our awareness of mitochondrial involvement in apoptosis. The important and by now well-established role of the mitochondrial permeability transition pore in releasing cytochrome C is a central feature of one pathway leading to apoptosis. It appears that this is not the only role for mitochondria, however. It has been shown that the tumour suppressor p53 exerts its role by promoting the production of reactive oxygen species within the mitochondria, a process that leads to the initiation of caspase independent pathways. This is an addition to the previously known involvement of p53, via BAX and Bcl2 in modulating the release of cytochrome C from mitochondria. As if these two roles are not of sufficient importance, it is also believed that p53 is a component in the sensing of hypoxia and that oncogenes exert their apoptotic influence by modulating the activity of p53. A number of recent reports have emphasised the importance of apoptotic processes in drug responses but an extensive discussion of this aspect of the pathway is considered outside the scope of the present review.

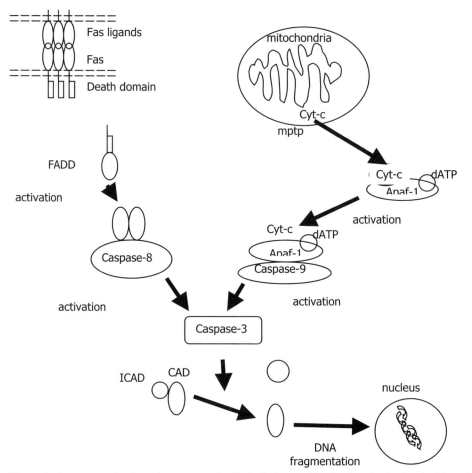

Figure 7. A pre-genomic view of programmed cell death (apoptosis), showing the basic principles of the process, including the crucial role of the mitochondrion.

5. CONCLUSIONS

It can be seen from this brief review that the short period of time since the publication of the human genome sequences has been an era of considerable discovery in the human respirome. This era has revealed new molecules involved in intra-cellular oxygen binding and as components of ETC complexes, while new pathways have emerged in the process of apoptosis. In terms of the involvement of the human respirome in cancer susceptibility and responses, important new discoveries have also been reported.

The impact of some of these discoveries on our understanding of the oxygen cascade will be significant, requiring major adjustments in the models that have been developed in the past three decades. The impact of the revelations regarding association of mutations in human respirome components with particular cancers is of even wider significance. It is perhaps appropriate that these associations will force a reconsideration of the role of cellular respiration in cancer initiation and proliferation.

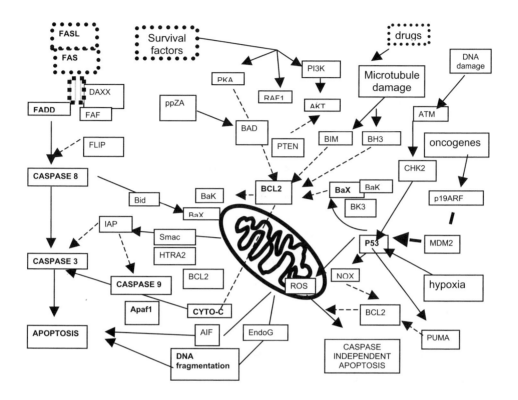

Figure 8. A recent overview of apoptosis. By comparison of this figure with the preceding one, it is obvious that a considerable body of new knowledge has arisen in the brief post-genomic era. Those emerging themes do not diminish the central role of the mitochondrion as a modulator of caspase-dependent and caspase-independent pathways within apoptosis. Legend; solid arrows – activation, dashed arrows – inhibition,

6. REFERENCES

1. Ashton KJ, Weinstein SR, Maguire D and Griffiths LR (2001) Molecular cytogenetic analysis of basal cell carcinoma DNA using comparative genomic hybridisation. J. Invest Derm, **117**, 683-686

2. Ashton KJ, Weinstein SR, Maguire D and Griffiths LR (2003) Chromosomal aberrations in squamous cell carcinoma and solar keratosis revealed by comparative genomic hybridisation. Arch Dermatol, **139**, 876-882

3. Maguire DJ, Lintell NA, McCabe M, Griffiths LR, Ashton K (2003) Genomic and phenomic correlations in the respiration of basal cell carcinomas. Adv Exp Med Biol. ;**540**:251-6.

4. Maguire DJ, Lintell NA, McCabe M, Griffiths LR, and Ashton KJ, (2005) The role of electron transport complex subunit genes in the somatic evolution of non-melanotic skin cancers, Adv Exp Med biol, in press

5. Erdmann W, (1992) The oxygen molecule and its course from air to cell. Adv Exp Med Biol. 317, 7-17

6. Dawkins FW, Kim KS, Squires RS, Chisholm R, Kark JA, Perlin E, Castro O. (1997) Cancer incidence rate and mortality rate in sickle cell disease patients at Howard University Hospital: 1986-1995. Am J Hematol. **55(4)**:188-92.

7. Miller MK, Zimmerman SA, Schultz WH, Ware RE. Hydroxyurea therapy for pediatric patients with haemoglobin SC disease. J Pediatr Hematol Oncol. 2001 Jun-Jul;**23**(5):306-8.

8. Burmester T, Weich B, Reinhardt S, and Hankeln T (2000) A vertebrate globin expressed in the brain. Nature, **407**, 520-523

9. Burmester T, Ebner B, Weich B, and Hankeln T (2003) Cytoglobin: a novel globin type ubiquitously expressed in vertebrate tissues. Mol Biol Evol. **19**, 416-421

10. Handane D, Kiger L, Dewilde S, Green BN, Pesce A, Uzan J, Burmester T, Hankeln T, Bolognesi M, Moens L, Marden MC, (2003) The redox state of the cell regulates the ligand binding affinity of human neuroglobin and cytoglobin. J Biol Chem, **278**, 51713-51721

11. Fordel E, Guens E, Dewilde S Rottiers P, Carmeliet P, Groten J, and Moens, L (2004) Cytoglobin expression is upregulated in all tissues upon hypoxia: an in vitro and in vivo study by quantitative real-time PCR. Biochem Biphys Res Commun, **319**, 342-348

12. Wystub S, Ebner B, Fuchs C, Weich B, Burmester T and Hankeln T, (2004) Interspecies comparison of neuroglobin, cytoglobin and nyoglobin: sequence evolution and candidate regulatory elements. Cytogenet Genome Res. **105**, 65-78

13. Schmidt M, Gerlach F, Avivi A, Laufs T, Wystub S, Simpson JC, Nevo E, Saler-Reinhardt S, Hankeln T and Burmester T, (2004) Cytoglobin is a rspiratory protein in connective tissue and neurons, which is upregulated in hypoxia. J Biol Chem, **279**, 8063-8069

14. Pesce A, Bolognesi M, Bodeci A, Ascenzi P, Dewilde S, Moens L, Hankeln T, Burmester T. (2002) Neuroglobin and cytoglobin. Fresh blood for the vertebrate globin family. EMBO report, **3**, 1146-1151

15. Geuens E, Brouns I, Flamez D, Dewilde S, Timmermans JP, Moens L. (2003) A globin in the nucleus! J Biol Chem. **278(33)**:30417-20.

16. Tanji K, Bhagat G, Vu TH, Monzon L, Bonilla E and Lefkowitch JH (2003) Mitochondrial DNA dysfunction in oncocytic hepatocytes. Liver Int, 23, 397-403

17. Taylor AC, Shu L, Danks MK, Poquette CA, Shetty S, Thayer MJ, Houghton PJ, Harris LC. P53 mutation and MDM2 amplification frequency in pediatric rhabdomyosarcoma tumors and cell lines. Med Pediatr Oncol. 2000 Aug;**35**(2):96-103

18. Zhou S, Kachhap S and Singh, KK, (2003) Mitochondrial impairment in p53-deficient human cancer cells. Mutagenesis, **18**, 287-292

19. Okamura S, Ng CC, Koyama K, Takei Y, Arakawa H, Monden M, Nakamura Y. (1999) Identification of seven genes regulated by wild-type p53 in a colon cancer cell line carrying a well-controlled wild-type p53 expression system. Oncol Res. **11**(6):281-5.

20. Simmonet H, Alazard N, Pfeiffer K, Gallou C, Beroud C, Demont,J, Bouvier R, Schagger H, and Godinot C, (2002) Low mitochondrial respiratory chain content correlates with tumour aggressiveness in renal cell carcinoma. Carcinogenesis **23**, 759-768

21. Herrmann PC, Gillespie JW, Charboneau L, Bichsel VE, Paweletz CP, Calvert VS, Kohn EC, Emmert-Buck MR, Liotta LA, Petricoin EF 3rd. Mitochondrial proteome: altered cytochrome c oxidase subunit levels in prostate cancer. Proteomics. 2003 Sep;**3**(9):1801-10.

22. Kaneko SJ, Gerasimova T, Smith ST, Lloyd KO, Suzumori K, and Young SR, (2003) CA125 and UQCRFS1 FISH studies of ovarian carcinoma. Gynecol Oncol, **90**, 29-36

23. Cascon A, Ruiz-Llorente S, Cebrian A, Telleria D, Rivero JC, Diez J, Lopez-Ibarra PJ, Jaunsolo MA, Benitez J and Robledo M (2002) Identification of novel SDHD mutations in patients with phaeochromocytoma and/or paraganglioma. Eur J Hum Genet. **10**, 457-461

24. Dannenberg H, Dinjens WN, Abbou M, Van Urk H, Pauw BK, Mouwen D, Mooi WJ and de Krijger RR, (2002) Frequent germ-line succinate dehydrogenase subunit D gene mutations in patients with apparently sporadic parasympathetic paraganglioma. Clin Cancer Res **8**, 2061-2066

25. Rustin, P, Munnich, A and Rotig, A, (2002) Succinate dehydrogenase and human diseases: new insights into a well-known enzyme. Eur L Hum Genet. **10**, 289-291

26. Lee SC, Chionh SB, Chong SM and Taschner PE, (2003) Hereditary paraganglioma due to the SDHD M1I mutation in a second Chinese family: a founder effect? Laryngoscope, **113**, 1055-1058

27. Taschner PE, Brocker-Vriends AH and van der Mey AG, (2002) From gene to disease; from SDHD, a defect in the respiratory chain, to paragangliomas and pheochromocytomas. Ned Tijdschr Geneeskd, 962, 48-60

28. Baysal BE, Willett-Brozick JE, Lawrence EC, Drovdlic CM, Savul SA, Mcleod DR, Yee HA, Brackman DE, Slattery WH 3rd, Myers EN, Ferrell RE and Rubenstein WS, (2002) Prevalence of SDHB, SDHC and SDHD germline mutations in clinic patients with head and neck paragangliomas. J Med Genet. **39**, 178-183

29. Lintell NA, Maguire DJ, Griffiths LR, and McCabe M, (2005) Focussing on genomic and phenomic aberrations in non-melanotic skin cancers, Adv Exp Med Biol, this volume

30. Sebat J, lakshmi,B, Troge J, Alexander J, Young J, Lundin P, Maner S, Massa H, Walker M, Chi M, Naviv N, Lucito R, Healy J, Hicks J, Ye K, Reiner A, Gilliam T, Trask B, Patterson N, Zetterberg A, and Wigler M (2004) Large scale copy number polymorphism in the human genome. Science, **305**, 525-528

31. Muller-Hocke J, Aust D, Rohrbach H, Napiwotsky J, Reith A, Link TA, Seibel PA, Holzel D, and Kadenbach B, (1997) Defects of the respiratory chain in the normal human liver and in cirrhosis during aging. Hepatology, **26**, 709-719

32. Huang H, and Manton KG, (2004) The role of oxidative damage in mitochondria during ageing: a review. Front Biosci **9**, 1000-1117

33. Merlo Pich M, Raule N, Catani L, Fafioli ME, Faenza I, and Cocco L, (2004) Increased transcription of mitochondrial genes for complex I in human platelets during aging. FEBS lett **558**, 19-22

34. Young TW, Mei FC, Yang G, Thomson-Lanza JA and Liu J.Cheng (2004) Activation of antioxidant pathways in ras-mediated oncogenic transformation of human surface ovarian epithelial cells revealed by functional proteomics. Cancer Res, **64**, 4577-4584

35. Kerr JF, Wyllie AH, Currie AR. (1972) Apoptosis: a basic biological phenomenon with wide-ranging implications in tissue kinetics. Br J Cancer. **26(4)**:239-57. Review.

54

LIGHTGUIDE SPECTROPHOTOMETRY TO MONITOR FREE TRAM FLAPS

Jenny Caddick[*], Cameron Raine, David Harrison, Matt Erdmann

1. INTRODUCTION

With improvement in microvascular techniques within the field of plastic surgery, free flaps are increasingly used in the reconstruction of many anatomical defects. Breast reconstruction following mastectomy is perhaps one of the most important areas in terms of the size of the patient group affected and the psychological impact of the oncological surgery and subsequent reconstruction.[1, 2] Here, the free transverse rectus abdominus myocutaneous (TRAM) flap has achieved popularity because of the superior cosmetic result, particularly in reconstructing larger breasts.[3]

While the importance of a meticulous surgical technique cannot be over-emphasised, equally important in ensuring flap survival is the post-operative maintenance of tissue perfusion. Free flaps fail due to a thrombosis within an artery or a vein. This initiates a cascade of events within the microvasculature allowing microthrombi and sludge to accumulate.[4] Early recognition of these events is vital in determining flap salvage should a problem arise. [5, 6, 7]

2. FREE FLAP MONITORING

The importance of early recognition of flap compromise before it becomes irreversible is undisputed. Nevertheless, human observation, which is inherently subjective, remains the most practical monitoring method among many surgeons.

Various alternative monitoring devices have been explored both experimentally and clinically over the past 6 decades.[4, 8] Despite the invasiveness and technical expertise required to use many of these devices, a recent study[9] suggests that 90% of microsurgeons in the USA use some kind of monitoring device routinely.

[*] Jenny Caddick, Cameron Raine, Matt Erdmann, Department of Plastic Surgery, University Hospital of North Durham DH1 5TW, UK. David Harrison, Regional Medical Physics Department, Durham Unit, University Hospital of North Durham.

In a plastic surgical context, visible wavelength range lightguide spectrophotometry (LGS) was first described by Jones et al.[10] using a rat epigastric flap model. The authors found that at specific wavelengths they were able to identify failing flaps and to differentiate between arterial and venous thrombosis. However, the heavy, cumbersome equipment precluded continuous monitoring and the careful calibration necessary for each measurement made the instrument impractical for clinical use.

More recently Wolff et al.[11] used the Erlangen Microlightguide Spectrophotometer II (EMPHO II) to measure intracapillary haemoglobin oxygenation from the spectrum of backscattered light. The device was shown to be useful in both animal models and human studies but could not differentiate between arterial and venous occlusion. It was also large and cumbersome limiting its practicality to laboratory work.

3. METHODS

3.1 Lightguide Spectrophotometry

A Whitland Research RM200 SO_2 monitor was used for this study. The features of the instrument are described elsewhere in this volume.[12] Values of tissue oxygenation (SO_2) and relative haemoglobin concentration (HbI)[13] were recorded every 10 seconds. The data were subsequently then smoothed to create a 5 minute moving average.

3.2 Patients

Consecutive women undergoing TRAM flap breast reconstruction at the University Hospital of North Durham were recruited between November 2001 and June 2004. A total of 14 women were studied and mean demographic data are shown in table 1. All gave informed consent to the study.

Patients were monitored on return from theatre for up to 72 hours post operatively using LGS and the two most clinically applicable alternatives: clinical observation (by a standardised patient care pathway) and laser Doppler flowmetry (Moor DRT4). Probes were attached to the flap as shown in figure 1. Some modification to probe design was made during the study. In the later part of the study, probe placement was standardised to zone 1 of the flap (the area of the flap with the most reliable blood supply determined by the anatomical distribution of the vessels supplying it).

Table 1. Basic demographic data for patients included in the study.

Total patients	14
Mean Age in Years (Range)	44 (37-64)
Mean BMI (Range)	27 (23-37)
Smokers	4
Previous Radiotherapy	6
Previous Chemotherapy	6

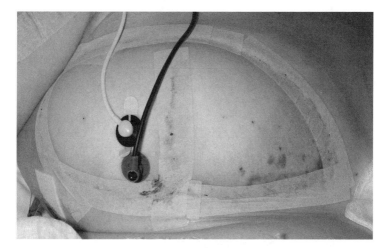

Figure 1. Positioning of probes on a TRAM flap breast reconstruction (LGS black; LD white).

4. RESULTS

For each patient a trace for laser Doppler (LD) and SO_2 measurement was obtained for the monitoring period. This ranged from 4-72 hours with a mean time of 53 hours. Traces were compared with documented clinical events occurring within the monitoring period. In all cases the monitoring equipment was well tolerated by patients. Some technical problems occurred in the early part of the study: probe detachment (3 cases), computer crash (1 case), LED failure (1 case). One flap failed.

Overall the SO_2 traces appear to correlate well with clinical events. Figure 2 shows the combined traces for SO_2, LD and HbI in a patient who was transfused two hours post operatively. It is interesting to note in this trace that there is a clear rise in SO_2 as the blood transfusion increases oxygen transport to the tissue, yet the flow remains unchanged.

Figure 3 shows the traces for a patient who developed an erythematous rash (thought to be secondary to a drug eruption) which included the flap tissue 50 hours post operatively. At this point the HbI trace begins to rise as vasodilation increases blood flow and thus total haemoglobin (irrespective of state of oxygenation) within the flap. This trace also illustrates well the increased sensitivity of LD to movement artefact when compared to LGS.

A further important finding is the potential ability of the LGS not only to detect flap failure but also, with the addition of the HbI trace, to differentiate between venous and arterial occlusion. Figure 4 shows traces for a flap following re-exploration. In the first two hours after the salvage procedure the flap was clinically viable and acceptable SO_2 and flux traces were recorded. The patient then developed severe venous congestion. The traces show a decrease in SO_2 and flux associated with a dramatic rise in haemoglobin concentration at 2 hours post operatively, suggesting the cause in the falling SO_2 to be venous congestion (in which haemoglobin accumulates) rather than arterial obstruction.

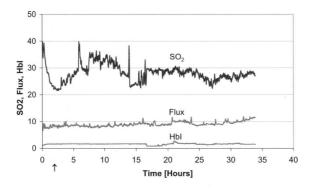

Figure 2. SO_2, LD flux and HbI measurements for a 40-year-old woman undergoing uncomplicated TRAM flap breast reconstruction. The patient was given a blood transfusion 2 hours post-operatively (arrow) causing a rise in SO_2 measurement as the transfusion increases oxygen transport to the flap tissue. Ordinate Scale: SO_2 %; Flux and HbI Arbitrary Units.

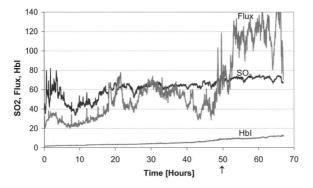

Figure 3. SO_2, LD flux and HbI measurements for a 46-year-old woman undergoing TRAM flap breast reconstruction. At 50 hours (arrow) the patient developed an erythematous rash. Ordinate Scale: SO_2 %; Flux and HbI Arbitrary Units.

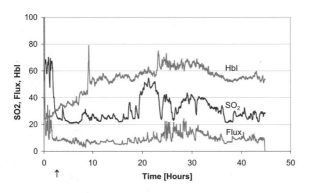

Figure 4. SO_2, LD flux and HbI measurements following a salvage procedure to restore circulation to a TRAM flap. From 2 hours (arrow) the flap showed increasing clinical signs of venous congestion. The rise in HbI seen here associated with a fall in SO_2 reflects this accurately. SO_2 %; Flux and HbI Arbitrary Units.

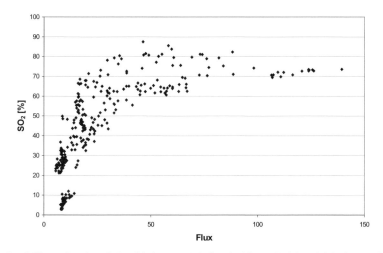

Figure 5. Graph illustrating the relationship between LD flux (Arbitrary Units) and SO_2 for 7 patients in whom measurements were all taken from zone 1 of the flap.

Although the numbers in this study are small, there is also evidence to suggest that the sensitivity of SO_2 monitoring is superior to that of LD flux at low flow rates where determination of viability is particularly crucial. This is illustrated in figure 5 which shows the relationship between LD flux and SO_2 for the combined results of seven patients. These seven were selected because in each case the probes were documented to have been sited in the same flap zone (zone 1). The importance of recording this was not appreciated in the early part of the study hence inclusion of that data could be misleading.

The rapid, early rise in SO_2 relative to LD seen in the graph suggests an increased sensitivity of the former to detect changes in saturation at low flow rates. Thus for a small, possibly undetectable change in the LD flux trace there will be a large variation in the SO_2 trace across what may well be a critical level.

5. CONCLUSIONS

This pilot study has shown the LGS to be a useful technique in flap monitoring, comparing favourably with both laser Doppler flowmetry and clinical observation. The device provides an absolute value of oxygenation rather than just a trend as is given with an LD trace. There is some reduction in the movement artefact seen which it is hoped can be improved upon further with modifications to the SO_2 probe. The LGS appears to be able to differentiate between arterial and venous occlusion by the incorporation of the HbI measurement. Importantly the device is maximally sensitive at low flow rates. In addition a poor signal can be differentiated from a low reading reducing the potential error rate of the equipment.

From a clinical perspective the portable device was easy to use and well tolerated by all patients. It therefore approaches more closely than previously described tools the criteria for an ideal monitoring device.[14]

Further refinements to this study now need to be made in order both to standardise our results and to establish a saturation level likely to predict flap failure. Authors using similar devices have suggested a critical saturation level of between 10 and 15%.[15] Within our own unit the same tool has been used successfully to assess tissue viability prior to lower limb amputation. In this application a level of 15% saturation has been shown to be clinically sensitive and specific in determining amputation level.[16] A similar figure now needs to be clarified in the context of free flap monitoring.

6. ACKNOWLEDGEMENTS

Thanks are due to the Education Centre, University Hospital of North Durham and Spectrum Medical, Cheltenham, UK for financial support towards attending ISOTT 2004.

7. REFERENCES

1. Office for National Statistics, Cancer Trends for England and Wales, 2000-2003; http://www.statistics.gov.uk.
2. L. A. Stevens, M. H. McGrath, R. G. Druss, S. J. Kister, F. E. Gump, K. A. Forde, The psychological impact of immediate breast reconstruction for women with early breast cancer, *Plast. Reconstr. Surg.* **73**, 619-625 (1984).
3. M. E. Beasley, The pedicled TRAM as preference for immediate autogenous tissue breast reconstruction, *Clin. Plast. Surg.* **21**(2), 191-205 (1994).
4. H. Furnas, J. M. Rosen, Monitoring in microvascular surgery, *Ann. Plast. Surg.* **26**, 265-272 (1991).
5. C. L. Kerrigan, R. G. Zelt, R. K. Daniel, Secondary critical ischaemia time of experimental skin flaps, *Plast. Reconstr. Surg.* 74, 522 (1984).
6. J. W. May, L. A. Chait, B. M. O'Brien, J. V. Hurley, The no-reflow phenomenon in experimental free flaps, *Plast. Reconstr. Surg.* **61,** 256 (1978).
7. W. A. Goodstein, H. J. Buncke Jr., Patterns of vascular anastomoses versus success of free groin flap transfers, *Plast. Reconstr. Surg.* **64,** 37-40 (1979**)**.
8. B. M. Jones, Monitors for the cutaneous microcirculation, *Plast. Reconstr. Surg.* **73**, 843-849 (1984).
9. M. B. Hirigoyen, M. L. Urken, Free flap monitoring: a review of current practice, *Microsurgery* **16**, 723-726 (1995).
10. B. M. Jones, R. Sanders, R. M. Greenhalgh, Monitoring skin flaps by colour measurement, *Brit. J. Plast. Surg.* **36**, 88-94 (1983).
11. K. D. Wolff, C. Marks, B. Uekermann, M. Specht, K. H. Frank, Monitoring of flaps by measurement of intracapillary haemoglobin oxygenation with EMPHO II: experimental and clinical study, *Brit. J. Oral Maxillofac. Surg.* **34**, 524-529 (1996).
12. C. L. Ives, D. K. Harrison, G. Stansby, Relationships between muscle SO2, skin SO2 and physiological variables, *Adv. Exp. Med. Biol.*, This volume (2005).
13. D. K. Harrison, S. D. Evans, N. C. Abbot, J. Swanson Beck, P. T. McCollum, Spectrophotometric measurements of haemoglobin saturation and concentration in skin during the tuberculin reaction in normal human subjects, *Clin. Phys. & Physiol. Meas.* **13**, 349-363 (1992).
14. M. F. Stranc, M. G. Sowa, B. Abdulrauf, H. H. Mantsch, Assessment of tissue viability using near-infrared spectroscopy, *Brit. J. Plast. Surg* **51**, 210-217 (1998).
15. S. Schultze-Moshau, J. Wiltfang, F. Birklein, F. W. Neukam, Micro-lightguide spectrophotometry as an intraoral monitoring method in free vascular soft tissue flaps, J. Oral Maxillofac. Surg. 61, 292-297 (2003).
16. D. K. Harrison, I. E. Hawthorn, Amputation level viability in critical limb ischaemia: Setting new standards, Adv. Exp. Med. Biol. In press (2004).

55.

THEORETICAL SIMULATION OF TUMOUR OXYGENATION - PRACTICAL APPLICATIONS

Alexandru Daşu and Iuliana Toma-Daşu[*]

1. INTRODUCTION

Tissue oxygenation has been recognised many years ago as a major factor that influences the outcome of radiotherapy[1]. It has also been observed that tumours usually have a poor vascular network that results in an impaired supply of oxygen which in turn leads to tissue hypoxia. Furthermore, it has been shown that tumour hypoxia can be divided into two types according to the mechanisms that have led to its appearance. Thus, diffusion limited hypoxia is caused by the limited diffusion of oxygen into tissue due to cellular consumption. Its appearance in human tumours has been reported for the first time by Thomlinson and Gray[2]. Several years later it has been suggested by Brown[3] that transient regions of hypoxia might also appear in tumours due to perfusion-related events such as the temporary closure of blood vessels. The existence of this type of hypoxia, termed perfusion limited hypoxia, has been demonstrated experimentally through mismatch techniques[4-6]. In contrast to the diffusion limited hypoxia that changes very slowly in time, the pattern of perfusion limited hypoxia has a short lifetime ranging from minutes to hours.

Due to the importance of tumour oxygenation, many attempts have been made to characterise it both as practical measurements and as theoretical modelling studies for the quantitative modelling of treatment outcome. However, most experimental methods are invasive and usually do not offer dynamical information. Non-invasive methods on the other hand provide a geometrical resolution that does not allow a detailed quantification of the tumour microenvironment. Theoretical simulation may thus be an alternative method that can provide quantitative data for a whole range of applications. Thus, the results of the simulations may be used to study the influence of the parameters characterising the tissue on the oxygenation pattern. As the results of the theoretical modelling are least affected by experimental artefacts, they can also be used to investigate the accuracy of various experimental techniques for measuring the tissue oxygenation. Furthermore, they can be used as input parameters for other simulations, such as the biological modelling of tumour response for treatment planning.

---

[*] Department of Radiation Sciences, Umeå University, 901 87 Umeå, Sweden

2. MATERIALS AND METHODS

We have developed a theoretical model that simulates the tissue oxygenation starting from the distribution of intervessel distances and taking into account the diffusion and cellular consumption of oxygen[7]. A Monte Carlo method was used to generate tissues with parallel blood vessels with distributions of intervascular distances in agreement with the experimental study by Konerding and co-workers[8]. The pO$_2$ value in the blood vessels was set to a low value (40 mmHg) in agreement with the statement of Vaupel and co-workers[9] that most tumour blood vessels come from the venous side of the vasculature. The tissue oxygenation was calculated for the simulated tissue through a numerical method, starting from the differential equation describing the movement of oxygen given by diffusion and consumption. The result thus described the oxygenation resulting from limitations in the oxygen diffusion and therefore reflected only the diffusion limited hypoxia. The appearance of perfusion limited hypoxia was modelled by randomly closing a fraction of the blood vessels in the tissue and recalculating the oxygenation map. This approach also allowed the simulation of different patterns of hypoxia as may appear at different time points during the treatment. The results of the simulations were expressed as histogram distributions of individual oxygen tensions (pO$_2$) and they were used to calculate the predicted response of the tissue to a fractionated radiation treatment of 30 fractions of 2 Gy each. For this it was assumed that the same vascular structure exists throughout the whole treatment. This is a reasonable assumption, since on one hand the increase in cell number due to proliferation is accompanied by the creation of new blood vessels through angiogenesis and on the other hand tumour shrinkage is accompanied by vascular damage.

The linear quadratic (LQ) model[10-12] was used to calculate the predicted cell survival. The parameters of the LQ model have been chosen to give an oxic cell survival fraction of 0.5 at 2 Gy while assuming $\alpha/\beta=10$ Gy. It was also assumed that the formula proposed by Alper and Howard-Flanders[13] describes the variation of the radiosensitivity with the oxygenation level. Tissue responses were compared in terms of tumour control probabilities (TCP) calculated with a Poisson function from the total cellular survival and the number of clonogenic cells in the tumour.

3. RESULTS AND DISCUSSION

Figure 1 shows the results of a typical simulation of tissue oxygenation. The shape of the log-normal distribution function of the inter-vessel distances used to simulate the tissue is shown in the left-hand panel. The parameters of the distribution (mean intervascular distance of 80 μm and a relative standard deviation of 0.05) are within the range of values that have been encountered in experimental studies of tumour vasculature[8]. It has to be mentioned that all the parameters of the distribution of inter-vessel distances are necessary to describe accurately the tissue oxygenation. The middle panel of Figure 1 shows the resulting tissue oxygenation under the assumption that all the tumour blood vessels are open while the right hand panel shows the results of the simulation under the assumption that some blood vessels have collapsed.

The calculated oxygenation maps, such as those in figure 1, were used to study various aspects of the influence of tissue hypoxia on the predictions of fractionated treatment outcome. In particular, the influence of the temporal variation of the pattern of perfusion limited hypoxia was taken into consideration. The amount of tissue hypoxia can now be measured both at the beginning of the treatment as well as at different time points during the treatment. This information can be quantified either as individual oxygenations existing at each fraction or as an average oxygenation throughout the whole treatment. Indeed, the use of an average oxygen distribution might seem a convenient way to avoid the sometimes large inter-fraction variations. However, it is very important how one calculates this average value and how the averaging process influences the predictions of outcome. In order to investigate these aspects, we have calculated the corresponding tumour control probability (TCP) values that were predicted from the use of individual histograms distributions of pO_2 values at each treatment fraction and have compared them to the values calculated from the histogram of the average oxygenation of the tissue. The predicted TCP values for a population of 10^7 clonogens are given in Table 1 for ten different oxygen distributions resulting from the random closure of the same fraction of blood vessels.

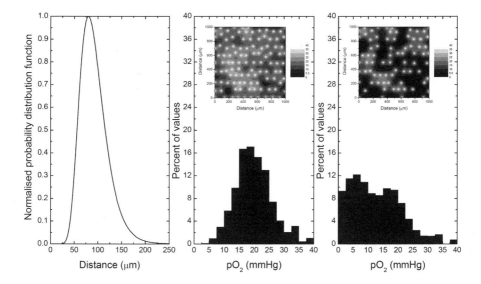

Figure 1. Simulation of tumour oxygenation. Left hand panel: probability distribution function for intervascular distances. Middle panel: oxygen distribution when all the tumour blood vessels are open. Right hand panel: oxygen distribution when some tumour blood vessels are closed.

The results show that the use of single, individual tissue oxygenations to estimate the treatment outcome results in large variations of the predictions. In the particular case considered, the predicted TCP from individual measurements of tissue oxygenation ranges from 57 to 82% illustrating the large variability in the amount of tissue hypoxia that may appear from the random closure of blood vessels non-uniformly distributed throughout the tissue. Indeed, we have shown previously[7] that the fraction of hypoxia that appears in tumours due to the temporary closure of some blood vessels is not constant and it does not correlate with the fraction of blood vessels that are closed because it also depends on the spatial distribution of the blood vessels. The histogram distributions in figure 1 support this finding as it can be seen that the fraction of closed blood vessels (28% in this case) is quite different from the fraction of hypoxia that appears in the histogram distribution (9%).

The predicted TCP at the end of the treatment can also be calculated in a cumulative fashion, when the effect of each individual fraction is added to the cumulated effect of all the previous fractions. Since this mode of calculation simulates most closely the actual treatment process, it is expected to yield the most accurate results. Thus the variability of the effect produced by each individual fraction is quickly dampened down by the increased weight of all the cumulated previous fractions. It is interesting to note that the prediction of the cumulative calculation (a TCP of 71%) is quite close to the mean of the TCP values obtained from the individual distributions (70%).

Table 1. Predicted tumour control probabilities (TCP) obtained with different methods of calculation.

Fraction no.	Individual histograms TCP (%)	Cumulative calculation TCP (%)	Average tissue oxygenation TCP (%)
1	71.5	71.5	71.5
2	61.4	66.7	78.8
3	77.1	70.5	83.5
4	82.1	73.9	85.2
5	78.4	74.8	86.1
6	62.4	73.0	85.6
7	69.4	72.5	85.8
8	56.6	70.8	85.4
9	64.9	70.2	85.4
10	73.8	70.6	85.5

In contrast to these two ways of calculation, one could calculate the TCP from the histogram distributions corresponding to the average tissue oxygenation over periods of time extending over many fractions (i.e., the average oxygenation over several periods of time is first calculated and then used to estimate the predicted TCP value). This mode of calculation simulates the use of the results of experimental measurements collected over relatively long periods of time in which the pattern of perfusion limited hypoxia has changed. The results in table 1 suggest that this way of calculation leads to an overestimation of the treatment outcome. The explanation for this effect is quite simple and resides in the nonlinear modification of the radioprotection conferred by the various levels of hypoxia. Indeed, the response of the tumour with a small number of cells protected very much by the local variations of pO_2 at each fraction is quite different from that estimated from the average tissue oxygenation where the individual variations generated by perfusion limited hypoxia are levelled out and most of the cells have therefore only a limited protection. This latter approach results in an apparent reduced radioprotection to tumour cells and hence in an increased probability of tumour control that does not have practical significance. Thus, this method of calculating the predictions for treatment outcome or the use of results from measurements protocols that pool together data collected over relatively long periods of time may be quite dangerous if the modelling is used for such quantitative approaches as the analysis and ranking of plans. Furthermore, it stresses the importance of interpreting correctly the conditions of measurement and how they might influence the results.

All these results suggest that many aspects have to be taken into consideration when hypoxia is incorporated into the theoretical modelling of tissue response.

4. CONCLUSIONS

Theoretical simulation of tissue oxygenation is a robust method that can be used to quantify the tissue oxygenation for a variety of applications. However, it is necessary that the relevant input parameters are used for the model describing the tumour microenvironment. The results of the simulations presented in this article suggest that the accuracy of the simulations depends very much on the method of calculation of the effects of the temporal change of the hypoxic pattern due to the opening and the closure of blood vessels. Thus, the use of average oxygenations might lead to dangerous overestimations of the treatment response. This indicates that care should be taken when incorporating hypoxia information into the biological modelling of tumour response.

5. ACKNOWLEDGEMENTS

This project was supported by grants from the Cancer Research Foundation in Northern Sweden.

6. REFERENCES

1. L. H. Gray, A. D. Conger, M. Ebert, S. Hornsey, and O. C. A. Scott, The concentration of oxygen dissolved in tissues at the time of irradiation as a factor in radiotherapy, *Br. J. Radiol.* **26**, 638-648 (1953).
2. R. H. Thomlinson and L. H. Gray, The histological structure of some human lung cancers and the possible implications for radiotherapy, *Br. J. Cancer* **9**, 539-549 (1955).
3. J. M. Brown, Evidence for acutely hypoxic cells in mouse tumours, and a possible mechanism of reoxygenation, *Br. J. Radiol.* **52**, 650-656 (1979).
4. D. J. Chaplin, P. L. Olive, and R. E. Durand, Intermittent blood flow in a murine tumor: radiobiological effects, *Cancer Res.* **47**, 597-601 (1987).
5. J. Bussink, J. H. A. M. Kaanders, P. F. J. W. Rijken, J. P. W. Peters, R. J. Hodgkiss, H. A. M. Marres, and A. J. van der Kogel, Vascular architecture and microenvironmental parameters in human squamous cell carcinoma xenografts: effects of carbogen and nicotinamide, *Radiother. Oncol.* **50**, 173-184 (1999).
6. P. F. J. W. Rijken, H. J. J. A. Bernsen, J. P. W. Peters, R. J. Hodgkiss, J. A. Raleigh, and A. van der Kogel, Spatial relationship between hypoxia and the (perfused) vascular network in a human glioma xenograft: a quantitative multi-parameter analysis, *Int. J. Radiat. Oncol. Biol. Phys.* **48**, 571-582 (2000).
7. A. Daşu, I. Toma-Daşu, and M. Karlsson, Theoretical simulation of tumour oxygenation and results from acute and chronic hypoxia, *Phys. Med. Biol.* **48**, 2829-2842 (2003).
8. M. A. Konerding, W. Malkusch, B. Klapthor, C. van Ackern, E. Fait, S. A. Hill, C. Parkins, D. J. Chaplin, M. Presta, and J. Denekamp, Evidence for characteristic vascular patterns in solid tumours: quantitative studies using corrosion casts, *Br. J. Cancer* **80**, 724-732 (1999).
9. P. Vaupel, F. Kallinowski, and P. Okunieff, Blood flow, oxygen and nutrient supply, and metabolic microenvironment of human tumors: a review, *Cancer Res.* **49**, 6449-6465 (1989).
10. K. H. Chadwick and H. P. Leenhouts, A molecular theory of cell survival, *Phys. Med. Biol.* **18**, 78-87 (1973).

11. B. G. Douglas and J. F. Fowler, Fractionation schedules and a quadratic dose-effect relationship, *Br. J. Radiol.* **48**, 502-504 (1975).

12. G. W. Barendsen, Dose fractionation, dose rate and iso-effect relationships for normal tissue responses, *Int. J. Radiat. Oncol. Biol. Phys.* **8**, 1981-1997 (1982).

13. T. Alper and P. Howard-Flanders, Role of oxygen in modifying the radiosensitivity of *E. Coli* B., *Nature* **178**, 978-979 (1956).

56

CHANGES IN PERFUSION PATTERN OF EXPERIMENTAL TUMORS DUE TO REDUCTION IN ARTERIAL OXYGEN PARTIAL PRESSURE

Oliver Thews[*], Debra K. Kelleher[**], and Peter Vaupel[**]

1. INTRODUCTION

In many solid tumors, oxygen deficiency (hypoxia) can be found which results from a discrepancy between tissue O_2 supply and O_2 consumption by the cells. The presence of low O_2 partial pressures (pO_2) in tumors can influence the efficacy of O_2-dependent treatment modalities and also the biological properties of tumor cells, e.g., by modulating expression of hypoxia-dependent proteins (for a review see Höckel & Vaupel, 2001)[1]. Major parameters determining the O_2-supply are the convective transport with the blood as well as the O_2-diffusion from blood to the cells which is mainly a function of the diffusion distance. Convective O_2-transport in tumors is fundamentally different from that found in normal tissues, since tumor vessels show many structural abnormalities such as blind vessel endings, irregular branching patterns or loss of vascular hierarchy [2] but also functional differences compared to normal blood vessels, e.g., intermittent flow stop or pronounced arterio-venous shunt perfusion.

Since hypoxia can limit the efficacy of non-surgical treatment modalities (e.g., irradiation or O_2-dependent chemotherapy) [3, 4] new treatment alternatives have been developed which preferentially target hypoxic tumor cells [5, 6]. In order to improve the efficacy of these treatments, it might be beneficial to further increase tumor hypoxia temporarily, e.g., by breathing a gas mixture with a reduced O_2 fraction. However, since it is known that hypoxia affects tumor biology and thus may have an impact on the malignant progression of the tumor [1], the question arises of whether a reduction of the arterial pO_2 during ventilation of a hypoxic gas mixture may change the metabolic microenvironment of the tumor. Since hypoxia at least partially affects the tumor by influencing gene expression, both acute and long-term effects of a lower tumor pO_2 have to be considered. This study focuses on changes in tumor perfusion during acute (20 min) and chronic (several days) inspiratory hypoxia.

[*]   Institute of Physiology, University of Würzburg, 97070 Würzburg, Germany
[**] Institute of Physiology and Pathophysiology, University of Mainz, 55099 Mainz, Germany

2. MATERIAL AND METHODS

2.1. Animals and Tumors

All studies were performed using the DS-sarcoma of the rat. Following subcutaneous injection of DS-ascites cells (0.4 ml; approx. 10^4 cells/µl) into the dorsum of the hind foot of male Sprague Dawley rats (Charles River Deutschland, Sulzfeld, Germany; body weight 190-240 g) experimental tumors grew as flat, spherical segments and replaced the subcutis and corium completely. The volume was determined by measurement of the three orthogonal diameters of the tumor and using an ellipsoid approximation with the formula: $V = d_1 \times d_2 \times d_3 \times \pi/6$. Perfusion distribution experiments were performed when tumors reached a target volume of 0.5 - 3.0 ml approx. 6 to 14 days after inoculation of tumor cells. Studies had previously been approved by the regional ethics committee and were conducted according to UKCCCR guidelines [7] and the German Law for Animal Protection.

2.2. Inspiratory Hypoxia

Animals were housed either under normoxic ambient conditions (room air; 21% O_2) or in a hypoxic atmosphere containing 8% O_2 (chronic hypoxia) for the whole period of tumor growth. For experiments analyzing the impact of acute hypoxia, animals were housed under normoxic conditions during tumor growth but breathed the hypoxic gas mixture for 20 min prior to and during measurements (e.g., oxygenation measurements).

2.3. pO$_2$-Measurement

The distribution of tumor O_2 partial pressures (pO$_2$) was measured polarographically using steel-shafted microelectrodes (outer diameter: 300 µm) and the pO$_2$ histography system (Eppendorf, Hamburg, Germany)[8]. A small midline incision was made in the skin covering the lower abdomen and the Ag/AgCl reference electrode placed between the skin and underlying musculature. For tumor pO$_2$ measurement, a small incision was made into the skin covering the tumor and the electrode was then moved automatically through the tissue in pre-set steps with an effective step length of 0.7 mm. Approximately 100 pO$_2$ values were obtained in less than 20 min from each tumor in up to 8 parallel electrode tracks. The O_2 status of each tumor was described by the median pO$_2$ and the fraction of pO$_2$ values $\leq$ 5 mmHg. Additionally, arterial blood gas analysis was performed immediately before and after tumor tissue pO$_2$ measurements using a pH/blood gas analyzer (type ABL 5, Radiometer, Copenhagen, Denmark) to ensure that values for arterial blood gases were within the physiological range during the measurement period.

2.4. Analysis of Perfusion Distribution

In order to assess parameters related to the perfusion distribution within a tumor, the fluorescent dye Hoechst 33342 (Sigma, Deisenhofen, Germany) was injected i.v. (25 mg/kg body weight). For this, the dye was dissolved in isotonic saline (10 mg/ml) and approx. 0.5 ml of this stock solution rapidly injected into a tail vein. Sixty seconds after injection, animals were sacrificed by an overdose of anesthetic and the tumors rapidly frozen in liquid nitrogen. Following subsequent tumor removal, cryosections were prepared. Using fluorescence microscopy (filter set #02, Zeiss, Jena, Germany, excitation wavelength: 365 nm, emission filter: low-pass 420 nm) at low resolution (range of vision

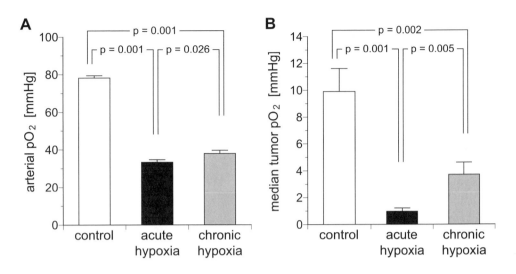

Figure 1. (A) Arterial oxygen partial pressure (pO$_2$) and (B) median tumor pO$_2$ during acute reduction of the inspiratory O$_2$-fraction for 20 min (acute hypoxia), chronic inspiratory hypoxia (chronic hypoxia), and under normoxic control conditions. Bars represent means ± SEM of at least 17 experiments.

3 x 2.5 mm), images of perfused vessels were taken from 2 to 3 different regions of each cryosection. Approx. 3 to 4 sections from each tumor were analyzed resulting in 9 to 10 images per tumor. In these images, each vessel was marked and the mean density of perfused vessels, the mean distance between neighboring vessels and the area of vascular domains were calculated using image analysis software (Optimas Ver. 6.2, Media Cybernetics, Silver Spring, MD, USA).

2.5. Statistical Analysis

Results are expressed as means ± standard error of the mean (SEM). Differences between the groups were assessed using the two-tailed Wilcoxon test for unpaired samples. The significance level was set at α=5%.

3. RESULTS

Exposing rats acutely (20 min) to an atmosphere containing a reduced O$_2$ fraction (8%) led to a significant reduction in arterial pO$_2$ (Fig. 1A) followed by a stimulation of ventilation as indicated by a lower arterial pCO$_2$ (48±1 *vs.* 32±1 mmHg) and a higher pH (7.41±0.03 *vs.* 7.50±0.01). If inspiratory hypoxia was maintained over the whole period of tumor growth ("chronic hypoxia"), the arterial pO$_2$ was slightly higher compared to acute hypoxia (38±1 *vs.* 33±2 mmHg), possibly indicating an adaptation of ventilation to a chronically reduced inspiratory O$_2$-fraction.

Reduction of arterial pO$_2$ strongly affected the tumor oxygenation status. In animals acutely exposed to an atmosphere containing 8% O$_2$, the median tumor pO$_2$ was reduced from 10 mmHg (control) to approx. 1 mmHg (Fig. 1B). In parallel, the fraction of hypoxic pO$_2$-values $\leq$ 5 mmHg significantly increased from 33±5% to 94±1% (p=0.001). However, during chronic hypoxia (lasting the whole period of tumor growth), the O$_2$-

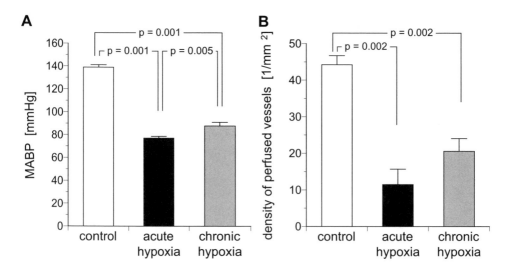

Figure 2. (A) Mean arterial blood pressure (MABP) and (B) mean density of perfused vessels in DS-sarcomas during acute reduction of the inspiratory O_2-fraction for 20 min (acute hypoxia), chronic inspiratory hypoxia (chronic hypoxia), and under normoxic control conditions. Bars represent means ± SEM of 10 experiments.

status was significantly improved (compared to acute hypoxia) with a median pO_2 of 4±1 mmHg (Fig. 1B) and a fraction of hypoxic pO_2-values ≤ 5 mmHg of 66±8% (in tumors of comparable volumes), indicating an adaptation of tumor oxygenation to chronic hypoxemia.

A possible explanation for the increased tumor pO_2 might be an improved convective O_2 transport via the blood or a reduction in O_2 diffusion distances through formation of a denser vascular network. Analysis of morphological properties of the vascular network (e.g., vascular density) by CD31 staining revealed no differences in either of the analyzed parameters (data not shown). While the number of blood vessels did not differ, analysis of the spatial distribution of tumor perfusion (assessed by Hoechst 33342 staining) showed pronounced differences upon breathing of hypoxic gas mixtures. In animals breathing 8% O_2 for 20 min ("acute hypoxia"), the density of perfused vessels was less than half that of control animals (Fig. 2B). In parallel, the tissue area supplied by a single blood vessel ("area of vascular domain") increased dramatically (Fig. 3A) resulting in intervascular distances much larger than the *in vivo* O_2 diffusion distances [9]. Assuming an O_2 diffusion distance of 90 μm, vessels > 180 μm apart from the next perfused tumor vessel are not able to supply the tissue sufficiently with oxygen, resulting in severe tumor hypoxia or even anoxia. In control tumors, the fraction of vessels > 180 μm apart was only 6% (Fig. 3B). However, when animals breathed the hypoxic gas mixture acutely, this fraction increased dramatically to >70% indicating pronounced heterogeneity of perfusion distribution. This worsening of tumor perfusion may result from a reduction in perfusion pressure (as the driving force of blood flow) as indicated by a significant reduction in mean arterial blood pressure (MABP) by 50-60 mmHg (Fig. 2A).

In animals kept under hypoxic atmospheric conditions for the whole period of tumor growth, the density of perfused vessels was higher than during acute inspiratory hypoxia though still significantly lower than in controls (Fig. 2B). The most striking difference in perfusion as compared to acute hypoxia was that the perfusion distribution became much less heterogeneous as indicated by a pronounced reduction of the area of vascular domain and a significant reduction of the fraction of vessels more than 180 μm apart from each other (Fig. 3), even though the arterial blood pressure was comparably low (Fig. 2A).

4. DISCUSSION

Reduction of the arterial pO_2 by breathing a hypoxic gas mixture has a strong impact on tumor oxygenation status. An acute reduction in the inspiratory O_2 fraction to 8% (resulting in an arterial pO_2 of approx. 33 mmHg) reduces the median pO_2 dramatically down to 1 mmHg within 20 min (Fig. 1) and increases the fraction of hypoxic pO_2 values significantly. Since the arterial pO_2 is a major parameter influencing O_2 diffusion into tissue, the reduction of tumor pO_2 is presumably the result of a reduced O_2 diffusional flux from blood to the tumor cell. However, tumor perfusion analysis under these conditions clearly demonstrates that acute reduction of arterial pO_2 leads to a significant reduction of perfused tumor vessels (Fig. 2A) whereby the remaining blood flow is very heterogeneously distributed with large tissue areas showing almost no detectable perfusion (Fig. 3). Since the MABP is a paramount factor influencing tissue perfusion, one reason for a reduced tumor perfusion is a significant hypoxia-induced decrease in MABP by 30-40% (Fig. 2A). Another possible reason for the hypoxia-induced decrease in tumor perfusion may be a redistribution of blood flow in favor of the surrounding normal tissue. In normal tissue (e.g., skeletal muscle or skin), the vascular tone of small arteries is mainly influenced by local NO concentration. A decrease in arterial pO_2 (e.g., due to inspiratory hypoxia) leads to an increase in the NO level, resulting in vasodilation. Newly formed tumor vessels often lack smooth muscle cells [2] so that these vessels cannot react to vasoactive agents [10]. For this reason, arterial hypoxemia probably leads to vasodilation in normal tissue but not in tumor whereby the blood flow may be redistributed to the disadvantage of the tumor resulting in a reduction of perfusion disproportionate to changes in perfusion pressure ("steal phenomenon") [11].

When arterial hypoxia was maintained over the whole tumor growth period ("chronic hypoxia"), the worsening of the O_2-status was less pronounced than under acute hypoxia although the arterial pO_2 was comparable to that of the acute situation. These results reveal some kind of tumor adaptation to chronic hypoxia. Since the morphological properties of the tumor (e.g., vascular density) were not affected by a chronic reduction of arterial pO_2, the question remains as to why, under these conditions, the tumor oxygenation was significantly improved as compared to the acute situation.

The most striking difference under acute *vs.* chronic hypoxia was the pattern of perfused tumor vessels. During **short-term** hypoxia, the number of perfused vessels decreased dramatically (Fig. 2B) and was very heterogeneously distributed (Fig. 3) with large intra-tumoral variability. Tumor regions with almost normal perfusion distribution were found adjacent to large areas of vital tumor tissue with almost no perfusion. In **chronic** hypoxia however, tissue perfusion was more homogeneously distributed as seen in a significant reduction of the fraction of blood vessels more than 180 μm away from the next perfused vessel (Fig. 3B). Even so, the causes for the reduced perfusion heterogeneity during chronic hypoxia remain unclear. Finally, the increased O_2-carrying capacity due to a higher hemoglobin level seen during chronic inspiratory hypoxia may also contribute to the improved tumor oxygenation seen in these animals.

In conclusion, pronounced long-term reduction of arterial pO_2 in an experimental tumor induces adaptive processes resulting in an improvement of the oxygenation status (as compared to acute hypoxia). However, the improved O_2-status does not result from hypoxia-induced angiogenesis but rather from a functional adaptation of the perfusion pattern. Acute hypoxia results in pronounced heterogeneity of the tumor blood flow distribution whereas chronic hypoxia leads to a more homogeneous pattern with significantly smaller distances between perfused vessels.

5. REFERENCES

1. M. Höckel and P. Vaupel, Tumor hypoxia: definitions and current clinical, biological, and molecular aspects, *J. Natl. Cancer Inst.* **93**, 266-276 (2001).
2. M.A. Konerding, W. Malkusch, B. Klapthor, C. van Ackern, E. Fait, S.A. Hill, C. Parkins, D.J. Chaplin, M. Presta, and J. Denekamp, Evidence for characteristic vascular patterns in solid tumours: quantitative studies using corrosion casts, *Br. J. Cancer* **80**, 724-732 (1999).
3. E.J. Hall, *Radiobiology for the Radiologist* (J.B. Lippincott, Philadelphia, 2000).
4. O. Thews, D.K. Kelleher, and P. Vaupel, Erythropoietin restores the anemia-induced reduction in cyclophosphamide cytotoxicity in rat tumors, *Cancer Res.* **61**, 1358-1361 (2001).
5. T. Shibata, A.J. Giaccia, and J.M. Brown, Hypoxia-inducible regulation of a prodrug-activating enzyme for tumor-specific gene therapy, *Neoplasia.* **4**, 40-48 (2002).
6. B.G. Siim, D.R. Menke, M.J. Dorie, and J.M. Brown, Tirapazamine-induced cytotoxicity and DNA damage in transplanted tumors: relationship to tumor hypoxia, *Cancer Res.* **57**, 2922-2928 (1997).
7. P. Workman, P. Twentyman, F. Balkwill, A. Balmain, D.J. Chaplin, J.A. Double, J. Embleton, D. Newell, R. Raymond, J. Stables, T. Stephens, and J. Wallace, United Kingdom Co-ordinating Committee on Cancer Research (UKCCCR) Guidelines for the Welfare of Animals in Experimental Neoplasia (2nd edit.), *Br. J. Cancer* **77**, 1-10 (1998).
8. P. Vaupel, K. Schlenger, C. Knoop, and M. Höckel, Oxygenation of human tumors: evaluation of tissue oxygen distribution in breast cancers by computerized O_2 tension measurements, *Cancer Res.* **51**, 3316-3322 (1991).
9. K. Groebe and P. Vaupel, Evaluation of oxygen diffusion distances in human breast cancer xenografts using tumor-specific in vivo data: role of various mechanisms in the development of tumor hypoxia, *Int. J. Radiat. Oncol. Biol. Phys.* **15**, 691-697 (1988).
10. O. Thews, D.K. Kelleher, and P. Vaupel, Disparate responses of tumour vessels to angiotensin II: tumour volume-dependent effects on perfusion and oxygenation, *Br. J. Cancer* **83**, 225-231 (2000).
11. D.G. Hirst. Tumor blood flow modification therapeutic benefit: is this approach ready for clinical application? in: *Gray Laboratory 1989 Annual Report*, edited by B.Michael and M.Hance (Cancer Research Campaign, London, 1989), pp. 14-17.

57

THEORETICAL SIMULATION OF TUMOUR HYPOXIA MEASUREMENTS

Iuliana Toma-Daşu, Alexandru Daşu and Mikael Karlsson[*]

1. INTRODUCTION

Tumour oxygenation is one of the most important factors that influence the treatment response and several techniques have been developed to characterise it. Among the experimental methods, the polarographic oxygen sensor has been widely used for *in vivo* measurements of oxygen and has been considered a sort of gold standard for measuring tumour hypoxia. Several studies have indeed shown that there is a general correlation between polarographic measurements and the treatment outcome[1-7]. Like many other experimental methods, however, the polarographic technique is quite invasive because it involves the physical insertion of the probe into the tissue and the consumption of oxygen thus perturbing the natural steady state. It would therefore be interesting to investigate the relationship between the electrode measurements and the real tissue oxygenation.

Our previous modelling on tissues with regular blood vessel networks[8-10] has shown that the electrode measurement is inevitably modified by the averaging of the pO_2 values in the measurement volume. In fact, given a certain distribution of oxygen values in the tissue, this averaging process determines the removal from the measured distribution of the extreme values especially if these values appear in regions smaller than the electrode measurement volume. Based on this observation we have concluded that generally it is not possible to use a polarographic electrode to measure hypoxia appearing in small rims, one or two cells wide, surrounding a blood vessel. In the light of these findings, the next step of the study was to expand the simulation for measurements in tissues with realistic vascular configurations in order to investigate the influence of the supposedly large regions of perfusion limited hypoxia caused by the temporary closure of some tumour blood vessels.

The aim of this article is to study the ability of the electrode to measure tumour hypoxia in various tissues having all or only a fraction of the blood vessels open and in particular the influence on the electrode measurements of the two types of hypoxia known to appear in tumours.

[*] Department of Radiation Sciences, Umeå University, 901 87 Umeå, Sweden

2. MATERIALS AND METHODS

The investigation was performed as a computer simulation of both the oxygen diffusion into the tissue and the individual electrode measurements. The oxygen distribution into tissue was calculated from the basic processes of diffusion from the blood capillaries and consumption at the cells. A Monte Carlo method was used to generate tissues with blood vessels placed according to distributions that were in agreement with the experimental measurements of Konerding and his co-workers[11]. A numerical method was then employed to calculate the oxygen distributions around the blood vessels. The pO_2 value in the blood vessels was set to 40 mmHg as most tumour vasculature comes from the venous side[12]. The oxygenation map thus obtained reflected only the distribution of diffusion limited hypoxia throughout the tissue. The effects of perfusion limited hypoxia were modelled by the random closure of some of the blood vessels in the simulated tissue and the recalculation of the tissue oxygenation. The details are described elsewhere[13]. We were thus able to control the amount and pattern of hypoxia in the tissue and therefore to investigate the influence of the two types of hypoxia on the electrode measured distributions of pO_2 values.

Electrode measurements were simulated according to the method described previously[8-9], where the probe was placed in a random point in the tissue and each individual measurement was calculated as the weighted average of oxygen tensions in the measuring volume according to the response function of the electrode. As it has been shown elsewhere[14], this approach closely mimics the clinical measurements. Simulations were performed for a large number of tumours covering a whole range of situations that may be encountered in practice. The results of the simulations (both real and measured oxygenations) were presented as discrete distributions of pO_2 values grouped in 2.5 mmHg intervals.

The theoretical approach used for this study avoided the artefacts that characterise most of the experimental measuring methods, thus allowing the determination of the tumour oxygenation in the least invasive manner. It was therefore possible to compare the results of the simulated measurements with the actual oxygenation of the tissue. This direct comparison also minimised the importance of the input parameters, and hence the use of higher or lower pO_2 values in the blood vessels would not have changed the concepts illustrated by the results in this paper. Furthermore, the controlled distribution of tumour hypoxia between diffusion limited and perfusion limited compartments allowed the investigation of the effects of each of them on the measurements.

3. RESULTS AND DISCUSSION

In practice, the electrode is placed and the oxygenation is measured in a limited number of more or less random points, chosen according to the tumour size. Thus, one important question that may arise regarding the use of a polarographic electrode for measurements of tissue oxygenation is the relevance of the measured sample for the whole tissue. In order to investigate how the process of sampling influences the result of the measurement, we have calculated the oxygen distribution in a tissue and then we have extracted the pO_2 values in a limited number of random points (50-1000), thus simulating the random placing of the electrode in the tissue but without including the averaging effect of the electrode measurement.

Figure 1 shows the distributions of pO_2 values in the whole tissue (grey bars) and the distributions of values in samples containing a limited number of points (black bars). Our simulations seem to indicate that a reasonably low number of samples may be relevant for the whole tissue under investigation. For smaller samples (below 100) the extracted distribution might deviate seriously from that of the whole tissue due to statistical uncertainties, thus raising the question of the relevance of measurements with a small number of readings.

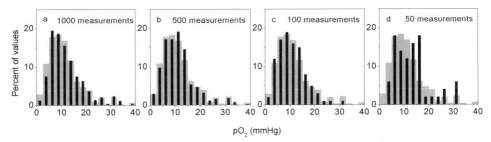

Figure 1. The influence of the sample size on the relevance of the measurement in a tumour tissue with heterogeneous vasculature. The number of measurements per sample is indicated in each panel. Grey bars: the existing oxygen distribution in tissue. Black bars: the extracted oxygen distribution.

In contrast, simulations performed for very hypoxic tumours (Figure 2) showed that the number of samples is less important than for well oxygenated tissues. The results in figure 2 are easily explained by the fact that poorly oxygenated tumours have very narrow distributions of pO_2 values, for which the importance of the number of samples is minimised.

These results led us to use samples with 1000 points for the subsequent electrode simulations in order to avoid the sampling artefacts and to investigate only the effects of the averaging implied by the polarographic method.

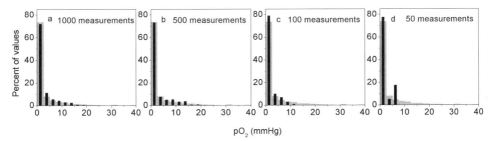

Figure 2. The influence of the sample size on the relevance of the measurement in a poorly oxygenated tissue. The number of measurements per sample is indicated in each panel. Grey bars: the existing oxygen distribution in tissue. Black bars: the extracted oxygen distribution.

Figure 3 shows the simulation of electrode measurements in a well oxygenated tissue that is characterised by a relatively narrow distribution of intervascular distances around a mean value of 50 μm. The grey bars represent the real distribution of pO_2 values in the tissue and the black bars represent the electrode measurements. One individual distribution represents the result of measuring the pO_2 in 1000 random points in the tissue.

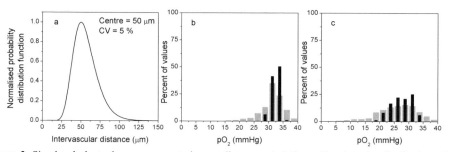

Figure 3. Simulated electrode measurements in a well oxygenated tissue. Panel a: the distribution of the intervascular distances. Panel b: the actual and measured pO_2 distributions when all the blood vessels are open. Panel c: the actual and measured pO_2 distributions when 25% of the blood vessels collapse. Grey bars: the actual oxygen distribution in the tissue. Black bars: the measured oxygen distribution.

Due to the small intervascular distances in figure 3, the oxygen supply to the tissue is very good, no cells being in the hypoxic compartment even when some blood vessels are closed. Thus, the random closure of some blood vessels worsens the oxygenation of the tissue (grey bars in panel 3c) compared to the situation where all the blood vessels are open (grey bars in panel 3b), but no hypoxic compartment appears. The measurement simulations show that the electrode could reflect the general change in the tissue oxygenation, but the measurements do not reflect the real situation in either case. Indeed, the extreme values of the real pO_2 distributions cannot be found in the measured distributions.

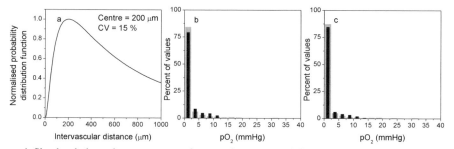

Figure 4. Simulated electrode measurements in a poorly oxygenated tissue. Panel a: the distribution of the intervascular distances. Panel b: the actual and measured pO_2 distributions when all the blood vessels are open. Panel c: the actual and measured pO_2 distributions when 25% of the blood vessels collapse. Grey bars: the actual oxygen distribution in the tissue. Black bars: the measured oxygen distribution.

In contrast, figure 4 shows the simulations for a tumour characterised by a broad distribution of values centred on a large mean intervascular distance (200 μm). The broad distribution of intervascular distances means that an important fraction of the blood vessels are in remote and isolated locations and that many cells are dependent on the limited oxygen supply from a single blood vessel. This results in very poor supply of oxygen to distant cells and therefore to a significant compartment of diffusion limited hypoxic cells (grey bars in panel 4b) that is measured quite accurately by the electrode (black bars in panel 4b). The closure of some blood vessels produces only minimal changes to the hypoxic compartment (grey bars in panel 4c), as most cells were anyhow hypoxic even when all the blood vessels were open. The measured distribution seems again very close to the real one, but this was expected for a tissue with this vascular configuration which is characterised by large hypoxic regions where the averaging process of the electrode plays only a negligible role. The situations in between these two extreme tumour tissues have shown that if some acute hypoxia is present in the tumour, the efficiency of the polarographic electrode is improved. This illustrates that the electrode can follow the general trend of tissue oxygenation. It also suggests that it is not the type of hypoxia that determines the efficiency of the polarographic electrode, but rather the geometrical disposition of hypoxia in the tumour.

It would have been very appealing to use the electrode as an indicator of acute hypoxia in the tumours, as there have been indications that the two types of hypoxia may have different radiobiological responses[15-16]. Unfortunately, it seems that this distinction may not be performed with the polarographic electrode and indeed the mixing of the two types of hypoxia may be one of the causes why the electrode yielded only some general qualitative correlations between measurements and treatment output.

4. CONCLUSIONS

Our simulations suggested that measurements performed in a limited number of points in the tumour can be representative for the situation in the whole tumour. It has further been shown that the polarographic electrode cannot be used to measure small regions of hypoxia. In fact it has been suggested that the most important factor that determines the efficiency of the polarographic electrode is the spatial distribution of the hypoxic cells and not their type, and therefore the polarographic electrode cannot be used to make the distinction between acute and chronic hypoxia. The simulations have also shown that it is reasonable to assume that the electrode measurement can be correlated to the situation in the whole tissue, even though the correlation is only qualitative. And because the electrode measurements are greatly influenced by the averaging process, the quantitative use of the electrode measurements may lead to erroneous results, especially for modelling the treatment response.

5. ACKNOWLEDGEMENTS

This investigation was supported by grants from the Cancer Research Foundation in Northen Sweden and the Medical Faculty, both at Umeå University and the Swedish Cancer Society. We would also want to thank to professors Jack Fowler and Bo Littbrand for help and encouragement.

6. REFERENCES

1. M. Höckel, C. Knoop, K. Schlenger, B. Vorndran, E. Baussmann, M. Mitze, P. G. Knapstein, and P. Vaupel, Intratumoral pO_2 predicts survival in advanced cancer of the uterine cervix, *Radiother. Oncol.* **26**, 45-50 (1993).
2. M. Nordsmark, M. Overgaard, and J. Overgaard, Pretreatment oxygenation predicts radiation response in advanced squamous cell carcinoma of the head and neck, *Radiother. Oncol.* **41**, 31-39 (1996).

3. D. M. Brizel, G. L. Rosner, J. Harrelson, L. R. Prosnitz, and M. W. Dewhirst, Pretreatment oxygenation profiles of human soft tissue sarcomas, *Int. J. Radiat. Oncol. Biol. Phys.* **30**, 635-642 (1994).
4. A. W. Fyles, M. Milosevic, R. Wong, M. C. Kavanagh, M. Pintilie, A. Sun, W. Chapman, W. Levin, L. Manchul, T. J. Keane, and R. P. Hill, Oxygenation predicts radiation response and survival in patients with cervix cancer, *Radiother. Oncol.* **48**, 149-156 (1998).
5. B. Movsas, J. D. Chapman, A. L. Hanlon, E. M. Horwitz, R. E. Greenberg, C. Stobbe, G. E. Hanks, and A. Pollack, Hypoxic prostate/muscle pO_2 ratio predicts for biochemical failure in patients with prostate cancer: preliminary findings, *Urology* **60**, 634-639 (2002).
6. C. M. Doll, M. Milosevic, M. Pintilie, R. P. Hill, and A. W. Fyles, Estimating hypoxic status in human tumors: a simulation using Eppendorf oxygen probe data in cervical cancer patients, *Int. J. Radiat. Oncol. Biol. Phys.* **55**, 1239-1246 (2003).
7. C. Parker, M. Milosevic, A. Toi, J. Sweet, T. Panzarella, R. Bristow, C. Catton, P. Catton, J. Crook, M. Gospodarowicz, M. McLean, P. Warde, and R. P. Hill, Polarographic electrode study of tumor oxygenation in clinically localized prostate cancer, *Int. J. Radiat. Oncol. Biol. Phys.* **58**, 750-757 (2004).
8. I. Toma-Daşu, A. Waites, A. Daşu, and J. Denekamp, Theoretical simulation of oxygen tension measurement in tissues using a microelectrode: I. The response function of the electrode, *Physiol Meas.* **22**, 713-725 (2001).
9. I. Toma-Daşu, A. Daşu, A. Waites, J. Denekamp, and J. Fowler, Theoretical simulation of oxygen tension measurement in the tissue using a microelectrode: II. Simulated measurements in tissues, *Radiother. Oncol.* **64**, 109-118 (2002).
10. I. Toma-Daşu, A. Daşu, A. Waites, and J. Denekamp, Computer simulation of oxygen microelectrode measurements in tissues, *Adv. Exp. Med. Biol.* **510**, 157-161 (2003).
11. M. A. Konerding, W. Malkusch, B. Klapthor, C. van Ackern, E. Fait, S. A. Hill, C. Parkins, D. J. Chaplin, M. Presta, and J. Denekamp, Evidence for characteristic vascular patterns in solid tumours: quantitative studies using corrosion casts, *Br. J. Cancer* **80**, 724-732 (1999).
12. P. Vaupel, F. Kallinowski, and P. Okunieff, Blood flow, oxygen and nutrient supply, and metabolic microenvironment of human tumors: a review, *Cancer Res.* **49**, 6449-6465 (1989).
13. A. Daşu, I. Toma-Daşu, and M. Karlsson, Theoretical simulation of tumour oxygenation and results from acute and chronic hypoxia, *Phys. Med. Biol.* **48**, 2829-2842 (2003).
14. I. Toma-Daşu, A. Daşu, and M. Karlsson, The relationship between temporal variation of hypoxia, polarographic measurements and predictions of tumour response to radiation, *Phys. Med. Biol.* **49**, 4463-4475 (2004).
15. J. Denekamp and A. Daşu, Inducible repair and the two forms of tumour hypoxia--time for a paradigm shift, *Acta Oncol.* **38**, 903-918 (1999).
16. A. Daşu and J. Denekamp, The impact of tissue microenvironment on treatment simulation, *Adv. Exp. Med. Biol.* **510**, 63-67 (2003).

58.

MONITORING OXYGENATION DURING THE GROWTH OF A TRANSPLANTED TUMOR

Anna Bratasz, Ramasamy P. Pandian, Govindasamy Ilangovan, and Periannan Kuppusamy[*]

1. INTRODUCTION

Several studies with human tumors have shown a strong correlation between oxygenation status and treatment outcome.[1,2] Tumors, in general, are known to be poorly oxygenated (hypoxic) and oxygen deficiency is associated with increased resistance of the tumor cells to treatments such as chemotherapy or radiation.[3,4] Another confounding factor that may potentially interfere with the prediction of the treatment outcome is the non-uniform distribution of oxygenation within the same tumor.[5-7] Many tumors are characterized with regions of hypoxia, even in well-oxygenated tissue. Thus, information on the distribution of tumor oxygenation and its alterations as a function of tumor size (growth), response to therapeutic stress (e.g., radiation), and preconditioning (e.g., hyperoxic treatment) would be valuable to help understand and develop effective treatment strategies for targeted cancer-cell killing. This information would require the availability of tools and procedures capable of repetitive noninvasive imaging of oxygen concentration in living tissues. Of the variety of methods that are available for tissue oximetry,[8] the magnetic resonance-based methods, such as MRI, electron paramagnetic resonance (EPR) spectroscopy and imaging (EPRI), have the advantage of noninvasive imaging of oxygen concentration in tissues.[9-12]

The EPR-based method, known as "EPR oximetry", uses spin probes whose EPR lines are broadened by molecular oxygen. The oxygen-induced broadening is usually linear with respect to pO_2, and hence the measured line-width can be converted to oxygen concentration using appropriate calibration curves. The measurements can be performed noninvasively and repeatedly over periods of months in the same site. This approach uses particulate such as lithium phthalocyanine (LiPc), and lithium butoxy-naphthalocyanine (LiNc-BuO) whose EPR line-widths are highly sensitive to local oxygen

* Center for Biomedical EPR Spectroscopy and Imaging, Davis Heart and Lung Research Institute, Department of Medicine, The Ohio State University, 420 West 12th Avenue, Room 114, Columbus, OH 43210. Tel: 614-292-8998; Fax: 614-292-8454; e-mail: kuppusamy.1@osu.edu

concentration.[13,14] These probes are stable in tissues, nontoxic, and biocompatible. They can be implanted at the desired site or, with a suitable coating, they can be infused into the vasculature for targeted delivery to tissues.[15] In addition, these probes can be internalized in cells enabling measurement of intracellular pO_2 with high accuracy.[13]

In this manuscript, we report the development of a novel procedure for *in situ* monitoring of oxygen concentration in growing tumors by EPR-based oximetry using embedded paramagnetic particulates. Unlike the existing methods of oxygen measurement wherein the oximetry probes (needle electrodes, optical probes, or EPR implants) are physically inserted during measurement, the new approach uses spin probes that are permanently embedded in the tumor. A particular advantage of this procedure is that it is noninvasive, both in terms of implantation of the probe, as well as when obtaining readouts of oxygen.

2.   MATERIALS AND METHODS

2.1 Reagents

Cell culture medium (RPMI 1640), fetal bovine serum, penicillin/streptomycin, sodium pyruvate, trypsin, and phosphate-buffered saline (PBS) were purchased from Gibco BRL (Grand Island, NY). In vitro MTT Toxicology Assay Kit was obtained from Sigma (St. Louis, MO). Lithium phthalocyanine (LiPc) and lithium octa-*n*-butoxynaphthalocyanine (LiNc-BuO) probes were synthesized as reported[13].

2.2 Mice

Female C3H mice were obtained from Frederick Cancer Research Center Animal Production (Frederick, MD). The animals were housed five per cage in a climate- and light-controlled room. Food and water were allowed *ad libitum*. The animals were 50-days old and weighed about 25 g at the time of the experiment. Animals were anesthetized with ketamine and xylazine (i.p.). The animals breathed either room air (21% O_2) or carbogen (mixture of 95% O_2 and 5% CO_2) delivered through a nose cone. During the measurements, the body temperature of the animal was maintained at $37\pm1°C$ by an infra-red lamp placed just above the animal. The body temperature was monitored using a rectal thermistor probe.

2.3 Tumor Growth and Implantation of the Oxygen Probe

Radiation-induced fibrosarcoma-1 (RIF-1) cells were used in the present study. The cells were grown in RPMI-1640 medium supplemented with 10% fetal bovine serum and 1% penicillin/streptomycin in an atmosphere of 95% air and 5% CO_2 at 37°C. Cells were trypsinized, centrifuged and suspended in PBS without calcium and magnesium ions. The oxygen-sensing probes were implanted in the tumor in three different ways. (i) *Surgical implantation:* One million RIF-1 cells in 0.04 ml PBS were injected into the hind limbs of C3H mice. The tumors were allowed to grow to about 8 mm diameter. About 10 μg of LiPc microcrystalline powder (size: 5 – 50 μm) was implanted in the tumor tissue at a depth of 2-3 mm. (ii) *Coinjection:* A suspension of $1x10^6$ RIF-1 cells mixed with 20 µg of LiPc in 0.06 ml of PBS was injected into the upper portion of the right hind limb of mice. (iii) *Intracellular internalization:* Cells were cultured in 75 cm^2 dishes for 4 passages, together with the particulates of LiNc-BuO (size < 1 µm). Then the cells were washed 4 times (to remove the extracellular particulates), trypsinized, centrifuged, and suspended in PBS. The uptake of the particulates into the cells was quantified by EPR spectroscopy. The cells ($1x10^6$), internalized with the particulates, were injected into the upper portion of the right hind limb of C3H mice and grown as solid tumor. The size of the tumor was measured by a Vernier caliper. The volume was determined from the orthogonal dimensions (d_1, d_2, d_3) by using the formula $(d_1*d_2*d_3)*\pi/6$.

2.4 EPR Oximetry and Imaging

Microcrystals of lithium phthalocyanine (LiPc) and lithium octa-*n*-butoxynaphthalo-cyanine (LiNc-BuO) were used as probes for EPR oximetry. The peak-to-peak line-width of the EPR spectra of the probes showed a linear relationship with pO$_2$. The value of pO$_2$ in tissue was obtained from a standard curve of the EPR line-width versus the oxygen concentration. The EPR measurements were performed by using an L-band (1.2 GHz) EPR imager with a surface coil resonator.

2.5 Cell Proliferation Assay

Cells were plated into 96-well dishes at a density of $5x10^4$ cells per well containing 100 µl RPMI-1640 (without phenol red), supplemented with 10% fetal bovine serum and 1% penicillin/streptomycin, incubated in an atmosphere of 95% air / 5% CO$_2$ at 37°C for 12 h, then assayed for cell proliferation using an MTT-based Toxicology Assay Kit. Measurements were taken using a Beckman Coulter AD 340 UV-VIS spectrophotometer set at a wavelength of 570 nm with a reference filter at 690 nm. The data were obtained as an average of 3-5 wells.

2.6 Cellular Membrane Integrity Assay

The toxic effect of the crystals on RIF-1 cells was studied by a cellular-membrane permeability assay. Cells were grown in 60-mm Petri dishes containing RPMI-1640 supplemented with 10% FBS and 1% penicillin/streptomycin. When the cells reached 60% confluence, 1 mg of LiPc was added to 4 ml of culture medium. The cells were incubated for 24 h, and then rinsed 5 times with PBS to remove the particulates. The cells were trypsinized, centrifuged, and stained with 1% Trypan Blue to assay the integrity of the cell membrane.

3. RESULTS AND DISCUSSION

3.1 Effect of Particulates on Cells and Tumor Growth

The effect of microcrystalline particulates of LiPc and LiNc-BuO on the RIF-1 cells was studied by a cell viability assay using Trypan blue. The RIF-1 cells, at 60% confluence in 60-mm culture dishes, were co-incubated with the particulates for 24 h and then assayed for cytotoxicity. The results showed that there was no significant effect on the membrane integrity of the treated cells when compared to that of the untreated cells (Figure 1). Since the sizes of these needle-shaped particulates were in the range of 60-80 μm, they were expected to remain primarily outside of the cells. In another set of experiments, the influence of nanocrystalline particulates of LiNc-BuO on cell proliferation was studied using an MTT assay. After 4 cycles of passages, the cells, both the untreated controls and those treated with LiNc-BuO particulates, were washed with PBS to remove the externally attached particulates and then trypsinized. The internalization of the particulates in the cells was confirmed by EPR. About 50,000 of these cells (per well) were then plated in 96-well dishes. After overnight incubation, the cells were subjected to the MTT assay. As seen in Figure 1, the growth rate of these cells containing the LiNc-BuO particulates was slightly lower.

The effect of the particulates on growth rate of the RIF-1 tumor was studied by tumor growth volume curve. The tumor volume was obtained from the 3 orthogonal diameters of the tumor. All three sets of tumors (untreated control, LiPc-co-injected, and LiNc-BuO-internalized) showed exponential growth rates. However, the co-injected and internalized groups showed significantly reduced growth volume curves compared to the control group (Figure 1).

3.1 Change of pO$_2$ as a Function of Tumor Growth

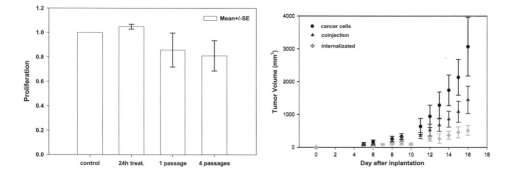

Figure 1. Effect of the oximetry/imaging particulates on the proliferation and growth of RIF-1 tumor cells. (Left) Cells were cultured in the presence of extracellular LiPc crystals (size: 20 – 26 μm) for 24 h or intracellular LiNc-BuO crystals (size: < 1 μm) assayed at the end of a single or 4 passages. (Right) Tumor growth volume curves are shown in the case of three different models of particle introduction into the tumor.

The intra- and extra-cellular pO_2 values in the RIF-1 cells within the growing tumor were measured daily for 2 weeks after implantation of the cells. The pO_2 values in the case of co-implantation (extracellular pO_2) were higher in the early stages of tumor growth and reached about 8 mmHg on day 9. On the other hand, the intracellular pO_2 levels remained almost the same (2-3 mmHg) during the whole period of tumor growth.

3.2 Mapping of Intracellular Oxygen Concentration

In order to determine the distribution of the co-injected or internalized particulates in the solid tumor, we performed additional experiments using EPR imaging. Spatial imaging was performed to obtain the distribution of the particulates in the tumor. It was observed that the particulates were confined to a small region (about 5x5x8 mm^3 in a tumor of about 10x12x12 mm^3 size), which is 15% of the tumor volume (data not shown). We further performed spectral-spatial EPR imaging on the same animals, at the same time, to map the distribution of the oxygen concentration in the tumor. The technique uses oxygen-dependent line-shape information from the collected projections to obtain pO_2 values at each voxel in the image of the particulates. Figure 2 shows the oxygen maps obtained at two different time points of tumor development, namely the 5th and 9th day, after inoculation. Figure 2 also shows the frequency plot (number of occurrences of a particular tumor pO_2 value in imaged region). In general, the intracellular pO_2 levels were very low from the beginning of tumor growth and stayed almost at the same level

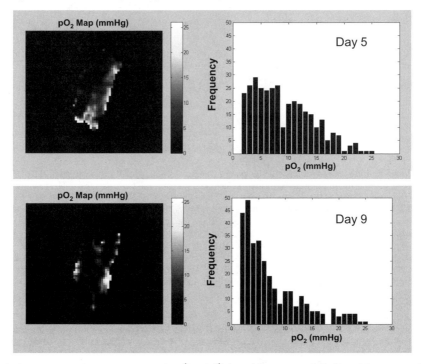

Figure 2. Oxygen map of RIF-1 tumor on the 5th and 9th day of tumor growth (left) and frequency of oxygen level in tumor tissue (right). Images were obtained by spectral-spatial EPR imaging using an L-band (1.32 GHz) EPR spectrometer.

during the entire period of tumor growth. The spread of the pO_2 values in the frequency plot is an indication of the heterogeneity of the pO_2 in the tumor. As observed in the present study, there is considerable heterogeneity in the distribution of oxygen in the RIF-1 tumor. It is also noteworthy that the spread in the frequency plot on day 9 is narrowed down with a shift to lower pO_2 values as compared to the data on day 5.

4. ACKNOWLEDGEMENTS

We thank Nancy Trigg for critical reading of the manuscript. This work was supported by NIH grants CA078886 and EB004031.

5. REFERENCES

1. P. Okunieff, M. Hoeckel, E. P. Dunphy, K. Schlenger, C. Knoop, P. Vaupel, Oxygen tension distributions are sufficient to explain the local response of human breast tumors treated with radiation alone, Int J Radiat Oncol Biol Phys 26, 631-636 (1993).
2. M. Hockel, C. Knoop, K. Schlenger, B. Vorndran, E. Baussmann, M. Mitze, P. G. Knapstein, P. Vaupel, Intratumoral pO2 predicts survival in advanced cancer of the uterine cervix, Radiother Oncol 26, 45-50 (1993).
3. P. Vaupel, D. K. Kelleher, M. Hockel, Oxygen status of malignant tumors: pathogenesis of hypoxia and significance for tumor therapy, Semin Oncol 28, 29-35 (2001).
4. M. Hockel, P. Vaupel, Tumor hypoxia: definitions and current clinical, biologic, and molecular aspects, J Natl Cancer Inst 93, 266-276 (2001).
5. K. G. Brurberg, B. A. Graff, E. K. Rofstad, Temporal heterogeneity in oxygen tension in human melanoma xenografts, Br J Cancer 89, 350-356 (2003).
6. S. M. Evans, S. M. Hahn, D. P. Magarelli, C. J. Koch, Hypoxic heterogeneity in human tumors: EF5 binding, vasculature, necrosis, and proliferation, Am J Clin Oncol 24, 467-472 (2001).
7. S. Hunjan, D. Zhao, A. Constantinescu, E. W. Hahn, P. P. Antich, R. P. Mason, Tumor oximetry: demonstration of an enhanced dynamic mapping procedure using fluorine-19 echo planar magnetic resonance imaging in the Dunning prostate R3327-AT1 rat tumor, Int J Radiat Oncol Biol Phys 49, 1097-1108 (2001).
8. H. B. Stone, J. M. Brown, T. L. Phillips, R. M. Sutherland, Oxygen in human tumors: correlations between methods of measurement and response to therapy. Summary of a workshop held November 19-20, 1992, at the National Cancer Institute, Bethesda, Maryland, Radiat Res 136, 422-434 (1993).
9. X. Fan, J. N. River, M. Zamora, H. A. Al-Hallaq, G. S. Karczmar, Effect of carbogen on tumor oxygenation: combined fluorine-19 and proton MRI measurements, Int J Radiat Oncol Biol Phys 54, 1202-1209 (2002).
10. J. F. Dunn, J. A. O'Hara, Y. Zaim-Wadghiri, H. Lei, M. E. Meyerand, O. Y. Grinberg, H. Hou, P. J. Hoopes, E. Demidenko, H. M. Swartz, Changes in oxygenation of intracranial tumors with carbogen: a BOLD MRI and EPR oximetry study, J Magn Reson Imaging 16, 511-521 (2002).
11. R. P. Mason, S. Hunjan, D. Le, A. Constantinescu, B. R. Barker, P. S. Wong, P. Peschke, E. W. Hahn, P. P. Antich, Regional tumor oxygen tension: fluorine echo planar imaging of hexafluorobenzene reveals heterogeneity of dynamics, Int J Radiat Oncol Biol Phys 42, 747-750 (1998).
12. M. C. Krishna, S. English, K. Yamada, J. Yoo, R. Murugesan, N. Devasahayam, J. A. Cook, K. Golman, J. H. Ardenkjaer-Larsen, S. Subramanian, J. B. Mitchell, Overhauser enhanced magnetic resonance imaging for tumor oximetry: coregistration of tumor anatomy and tissue oxygen concentration, Proc Natl Acad Sci U S A 99, 2216-2221 (2002).
13. R. P. Pandian, N. L. Parinandi, G. Ilangovan, J. L. Zweier, P. Kuppusamy, Novel particulate spin probe for targeted determination of oxygen in cells and tissues, Free Radic Biol Med 35, 1138-1148 (2003).
14. G. Ilangovan, H. Li, J. L. Zweier, M. C. Krishna, J. B. Mitchell, P. Kuppusamy, In vivo measurement of regional oxygenation and imaging of redox status in RIF-1 murine tumor: effect of carbogen-breathing, Magn Reson Med 48, 723-730 (2002).
15. G. Ilangovan, A. Bratasz, H. Li, P. Schmalbrock, J. L. Zweier, P. Kuppusamy, In vivo measurement and imaging of tumor oxygenation using coembedded paramagnetic particulates, Magn Reson Med 52, 650-657 (2004).

59.

FOCUSING ON GENOMIC AND PHENOMIC ABERRATIONS IN NON-MELANOTIC SKIN CANCERS

N.A. Lintell*, D.J. Maguire*, L.R. Griffiths[#], and M. McCabe*[1]

1. INTRODUCTION

Solar Keratosis is a precancerous lesion that is directly linked to long-term exposure to ultraviolet light (both ultraviolet B and A have been linked to DNA damage) [1]. These lesions have been categorised as Keratinocytic Intraepithelial Neoplasias or KIN. Solar Keratosis (SK) has three dermatologic stages, (I, II, and III), which the lesion progresses through before it is classified as a Squamous Cell Carcinoma (SCC) in situ [2]. The genes associated with SCC development that were analysed in this study are the MMP12, EMS1 and RASA1 genes[3,4]. Those associated with the progression of Basal Cell Carcinoma (BCC) are the PTCH and SMOH genes [3].

Mitochondria are the only extra-nuclear cellular organelle to contain their own DNA, and some of the genes it possesses encode several subunits of the ETC (13 out of the known 91)[5,6,7]. Mitochondria are also responsible for the initiation and regulation of apoptosis, which is a cellular process that removes diseased and damaged cells [5,6,7].

It is the malfunction of this pathway that has been fundamentally linked to the initiation of carcinogenesis [8]. The five genes that are involved in the ETC that were investigated are all encoded by the nucleus. They include NDUFA8, NDUFA5, and NDUFV1 from complex I, SDHD from complex II, and COXVIIc from complex IV[9].

This study analysed samples from a solar keratosis cohort via Real-Time PCR (polymerase chain reaction). This technique can measure the amount of target DNA (i.e. the gene that is being analysed) produced in real time as the reaction progresses[10]. It does this by detecting a particular compound that is bound to the DNA template (either a non-specific compound like SYBR Green I or a specific probe, such as a dual-labelled one), which fluoresces at a certain wavelength [11]. Each reaction cycle produces more DNA and the instrument gauges the total amount produced at the end of the PCR run [10]. Non-

---

[1] *School of Biomolecular and Biomedical Science and # School of Health Science, Griffith University Queensland 4111, Australia

specific compounds like SYBR Green I can also be used to calculate changes in DNA length by participating in melt-curve measurements, which utilises the fact that smaller DNA amplicons melt at a lower temperature than larger ones (if the G/C content is similar) [11].

2. METHODOLOGY

The first component of this study was to identify candidate genes. This process was aided by the submission of a PhD in 2002 by Kevin Ashton [12] who had analysed skin samples taken from BCC, SCC, and SK patients. He investigated these samples utilising the Comparative Genomic Hybridisation (CGH) technique, to analyse large areas on chromosomes for gains and losses. This study identified several areas of interest for the project to pursue, with the majority of these chromosomal areas associated with the location of NMSC and ETC genes. Indicated in Table 1 are the primer sequences devised for analysing specific exons of the genes in question. The study chose to analyse these gene sets via Real-Time PCR and SYBR Green I. The primers designed for each of the specific gene exons are depicted in Table 1, along with their associated PCR conditions.

Table 1. Primer sequences, MgCl2 concentrations, and melt temperatures

SDHD	Forward	AAGTAGCTTACCTATGGTCA	2.5uM, 56°C
SDHD	Reverse	AAGCAGCAGCGATGGAGA	2.5uM, 56°C
MMP12	Forward	CCATAGGTCATCTATTCTAG	2.5uM, 56°C
MMP12	Reverse	ACGTTGGAGTAGGAAGTCA	2.5uM, 56°C
NDUFA5	Forward	GGTAATATTTTAACCTATGG	3.0uM, 52°C
NDUFA5	Reverse	TTGGTTAAATGTTACACAAG	3.0uM, 52°C
SMOH	Forward	CCTAAGGTCACAGAATGGCC	3.0uM, 62°C
SMOH	Reverse	CTGTACCTTCAGGTCTGGGT	3.0uM, 62°C
NDUFA8	Forward	AGCAAGGCTATGTATTTGAG	3.0uM, 56°C
NDUFA8	Reverse	TCTGTATTTACAGAGACCCT	3.0uM, 56°C
PTCH	Forward	TTCATGGTCTCGTCTCCTAA	3.0uM, 58°C
PTCH	Reverse	AAGTGAACGATGAATGGACA	3.0uM, 58°C
NDUFV1	Forward	AGATCATCAGGCCCTCTCTT	2.5uM, 62°C
NDUFV1	Reverse	CGCAGAAGGGTGGTGAATAC	2.5uM, 62°C
EMS1	Forward	ATTGCTGCCCTGTCTCTCCA	2.5uM, 62°C
EMS1	Reverse	AACGCGTGATTACAGACCGT	2.5uM, 62°C
COXVIIc	Forward	CATCTGTCCTCATTCTCTGC	2.0uM, 60°C
COXVIIc	Reverse	CCCTTACACACTAACCTTCC	2.0uM, 60°C
RASA1	Forward	TTGTGAATCTGGTTTTAGGT	2.0uM, 54°C
RASA1	Reverse	GTTTCTGTATCAACTTACAG	2.0uM, 54°C

It was found that a standard stock of 1x BioTaq reaction buffer, 0.8 uM dNTPs, 0.5 uM primer and 1.5 units of BioTaq polymerase was needed for each of the primers as the majority of them had been specifically selected to have similar compositions and melt temperatures.

An initial denaturation step of 95°C for three minutes was implemented for all the primer sets. The cycling component was comprised of a 95°C step for 25 seconds, followed by a binding step of 20 seconds and finished with a 72°C for 35 seconds. These cycles were repeated 45 times. This protocol was implemented with SYBR Green I and two different Real-Time PCR machines. The first machine to be used was the Bio-Rad iCylcer, with a SYBR Green I concentration of 4 parts per 10000. The genes analysed on this machine were NDUFA8, PTCH, NDUFA5, and SMOH. The machine used for the remaining six genes (SDHD, MMp12, NDUFV1, EMS1, COXVIIc, and RASA1) was a Corbett Rotor-Gene, with a corresponding SYBR Green I ratio of 4 parts per 10000. These changes were made due to the higher sensitivity of the Rotor-Gene set-up and also the additional software that the machine utilised, which enabled a greater range of analytical references.

3. RESULTS

The results of the real-time component of this study are summarised in Table 2. They depict the analysis of the gene sets via SYBR Green I. The samples had to show up in triplicate as being aberrant due to the fidelity of real-time PCR based on laser excitation of a non-specific fluorophore like SYBR Green I. The reasons for this are discussed later on in the paper. A separate analysis of the samples was undertaken in duplicate for the samples having putative aberrations for the genes at the same time (i.e. ETC gene and a gene associated with NMSC from the same chromosomal area). This is also summarised in Table 2. This table highlights that the incidence of putative mutants is higher in both sets of analysis in the affected population when compared to the control.

Table 2. Summary of ETC/ NMSC gene analysis via Melt Curve & SYBR Green I

ETC/NMSC Gene Association	Affected Aberrants Duplicate	Control Aberrants Duplicate	ETC Gene	Affected Aberrants Triplicate	Control Aberrants Triplicate	NMSC Gene	Affected Aberrants Triplicate	Control Aberrants Triplicate
NDUFA8/ PTCH	23% (31/135)	1.4% (2/144)	NDUFA8	13% (18/135)	0.0% (0/144)	PTCH	8.9% (12/135)	1.4% (2/144)
NDUFA5/ SMOH	26% (35/135)	4.9% (7/144)	NDUFA5	13% (18/135)	2.8% (4/144)	SDHD	9.6% (13/135)	2.1% (3/144)
SDHD/ MMP12	21% (28/135)	3.5% (5/144)	SDHD	10% (14/135)	2.1% (3/144)	MMP12	6.0% (8/135)	2.8% (4/144)
NDUFV1/ EMS1	17% (23/135)	1.4% (2/144)	NDUFV1	20% (27/135)	1.4% (2/144)	EMS1	24% (32/135)	4.9% (7/144)
COXVIIc/ RASA1	15% (20/135)	1.4% (2/144)	COXVIIc	30% (40/135)	0.7% (1/144)	RASA1	24% (32/135)	0.0% (0/144)

Another observation is that in the triplicate single gene section an inverse relationship exists between incidence of putative aberrants and the distance between the genes of interest. This is shown in the following column graph (Figure 1). The legend along the bottom of the figure indicates the gene pairs (1 = NDUFB8 & PTCH, 2 = NDUFA5 & SMOH, 3 = SDHD & MMP12, 4 = NDUFV1 & EMS1, 5 = COXVIIc & RASA1).

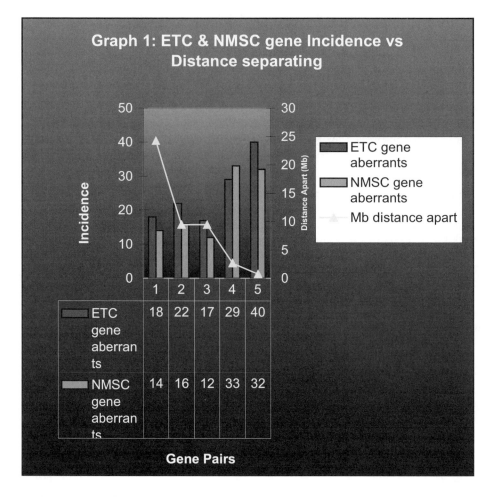

Figure 1. Graph showing ETC and NMSC gene incidence vs distance separating

4. DISCUSSION

Lesions of the skin originating from non-melanotic cells occur at a significantly higher rate than other neo-plastic malignancies within the Australian population, yet the knowledge of their genomic and proteomic manifestations falls below that of breast and colon cancer. These types of tumours, Non-melanotic skin cancers (NMSC), possess characteristics that make them ideal for clinical research, namely the abundance of medical samples available for study and their accessibility on the human body. One type of NMSC, Squamous Cell Carcinoma (SCC), has a dermatologically identifiable pre-cancerous state in which it can proliferate into a malignant lesion or regress into a benign tissue state. This stage of this disease is known as Solar Keratosis (SK), but as yet no clear genetic shifts have been identified in this stage of the disease for a progression into SCC, other than mutations in the gene encoding for the p53 cell cycle mechanism.

The main premise behind this project was to analyse chromosomal regions that contained both a nuclear-encoded electron transport chain gene and an oncogene/tumour suppressor gene associated with non-melanoma skin cancer for relatively large-scale mutations. In order to do this a real-time PCR approach was optimised, which contained both a quantitative component as well as a melt curve analysis [13, 14, 15].

This study analysed blood samples taken from an affected and control solar keratosis population. In all there 144 control samples and 135 affected samples analysed in at least duplicate (mostly triplicate but up to six times due to ambiguity) for aberrations in ten genes. These ten genes were subdivided into five pairs; one of the pair being a gene associated with the development of a NMSC, the other a gene encoding for a subunit of the ETC, each of these pairs exists in close proximity to one another on a particular chromosomal locale. As shown in Table 3 and on the column graph, the closer these two genes were to each other on the chromosome, the higher the incidence of putative aberrants (18 and 12 for NDUFA8 and PTCH, with a distance of 24Mb, compared to 40 and 32 for COXVIIc and RASA1, with a distance of 0.73 Mb). Another relationship that the data highlighted was that of a higher incidence occurring in the affected population when compared to the control. In fact out of the entire control run population, a total of 720, only 18 (2.5%) showed as being a putative aberrant, compared to 137 (20.3%) of the 675 affected population. This difference is also highlighted in the single gene triplicate run population, with the ETC gene component having 10 / 720 (1.37%) as being putative mutants, compared to 117 / 675 (17.3%) for the affected population. The NMSC gene component produced a 16 / 720 (2.22%) ratio for the control population, with the affected population having an incidence of 97 / 675 (14.4 %) for putative mutants.

The types of mutations that are readily identifiable via the SYBR Green I Real-Time PCR technique are mutations caused by exposure to ultraviolet radiation (UVR), which can give rise to genetic areas that, when amplified in a SYBR Green I Real-Time PCR run, melt in an a anomalous manner and present as a plateau like peak or an area-under the curve irregularity. These mutations are in the form of C to T and CC to TT transitions, known as UV `signature' mutations [3].

The cost of sequencing numerous genetic areas in a large sample cohort is usually prohibitive. Another precise method of analysing specific genes lies with screening the samples first with a non-specific fluorophore, with both the quantitation and melt-curve methods, then analysing the putative aberrant population from that study via single-labelled and dual-labelled probes. The former can accurately analyse a sample for amplicon melting anomalies, which can indicate the presence of a nucleotide aberrations down to the single-nucleotide level, whilst the latter is a precise way to detect the presence of gross copy-number aberrations in the genetic area of interest.

This project was the first, as far as a three-year literature review across numerous scientific databases could ascertain, to analyse precancerous lesions for abnormalities in genes that encode for subunits of the ETC.

Another point of interest is that the DNA utilised in this project did not come from an epithelial source, rather it was extracted from whole blood samples taken from a control and affected SK cohort, and thus the genetic malformations detected in this project have purportedly wider implications in terms of systematic affectations contributing to tumour initiation and promotion.

5. REFERENCES

1. Freedberg, I.M. Epithelial precancerous lesions. *Dermatol in General Med.* **1** (1999), pp.823- 826.

2. Bernhard, J.D. Actinic keratoses: scientific evaluation and public health implications. *J Amer Acad Dermatol.* **42 (**2000), pp: 1-30.
3. Prime SS, Thakker NS, Pring M, Guest PG, and Paterson IC. A review of inherited cancer syndromes and their relevance to oral squamous cell carcinoma. *Oral Oncology.* **37** (2001), pp.1-16.
4. Robert P. Takes. Staging of the neck in patients with head and neck squamous cell cancer: Imaging techniques and biomarkers. *Oral Oncology.* **40** (7) August (2004), pp. 656-667
5. Marcelino LA, and Thilly WG. Mitochondrial mutagenesis in human cells and tissues. *Mutation Research.* **434** (1999), pp.177-203.
6. Shoubridge EA. Nuclear genetic defects of oxidative phosphorylation. *Human Mol. Genetics.* **10**, No.20 (2001), pp.2277-2284.
7. Mandavilli BS, Santos JH, and Van Houten, B. Mitochondrial DNA repair and ageing. *Mutat. Res./ Fund. Mol. Mech. Mutagenesis.* **509** (2002), pp. 127-151.
8. Skulachev VP. Mitochondrial physiology and pathology; concepts of programmed death of organelles, cells, and organisms. *Mol. Aspects Med.* **20** (1999), pp.139-184.
9. NCBI. http://www.ncbi.nlm.nih.gov/htbin-post/Omim. **(October 2002, July 2004)**.
10. Bustin SA. Absolute quantification of mRNA using real-time reverse transcription polymerase chain reaction assays. *J. Mol. Endocrinology.* **25** (2000), pp.169-193.
11. http://dna-9.int-med.uiowa.edu/realtime.htm. Real-time or kinetic PCR. *University of Iowa DNA facility.* **(October 2002)**, pp.1-7.
12. Ashton KJ, Weinstein SR, Maguire DJ, and Griffiths LR. Molecular cytogenetic analysis of basal cell carcinoma DNA using comparative genomic hybridisation. *J Investigative Dermatol.* **117**, No. 3 (2001), pp.683-686.
13. Mikula, M., Dzwonek, A., Jagusztyn-Krynicka, K., and Ostrowski, J. Quantitative detection for low levels of *Helicobacter Pylori* infection in experimentally infected mice by real-time PCR. *J. Micro. Methods.* Nov (2003); **55** (2), pp351-9.
14. Moen, E.M., Sleboda, J., and Grinde, B. Real-time PCR methods for independent quantitation of TTV and TLMV. *J. Viro. Methods.* **104** (2002), pp. 59-67.
15. O'Mahony, J. and Hill, C. A real-time PCR assay for the detection and quantitation of *Mycobacterium avium* subsp. *Paratuberculosis* using SYBR Green and the Light Cycler. *J. Microbiol. Methods.* **51**(2002), pp.283-293.

AUTHOR INDEX

SUBJECT INDEX